For Thérèse

Animal and Human Aggression

PIERRE KARLI

Louis Pasteur University
Strasbourg

Translated by
S. M. Carmona and H. Whyte

OXFORD UNIVERSITY PRESS
1991

Oxford University Press, Walton Street, Oxford OX2 6DP

Oxford   New York   Toronto
Delhi   Bombay   Calcutta   Madras   Karachi
Petaling Jaya   Singapore   Hong Kong   Tokyo
Nairobi   Dar es Salaam   Cape Town
Melbourne   Auckland

and associated companies in
Berlin   Ibadan

Oxford is a trade mark of Oxford University Press

Published in the United States
by Oxford University Press, New York

A catalogue record for this book is available from the British Library

Library of Congress Cataloging in Publication Data
Karli, Pierre.
[L'homme agressif. English]
Animal and human aggression / Pierre Karli ; translated by S. M.
Carmona and H. Whyte.
p.   cm.
Translation of: L'homme agressif.
Includes bibliographical references and index.
1. Aggresiveness (Psychology)   2. Aggressive behavior in animals.
I. Title.
BF575.A3K3713   1991   155.2'32 — dc20   91–14351

ISBN 0–19–852186–3

Typeset by Footnote Graphics, Warminster, Wiltshire

Printed in Great Britain by
Dotesios Ltd., Trowbridge, Wilts.

Animal and Human Aggression

PREFACE

When considering the possibility of adding a preface to the English edition of my book, I soon realized that it would in fact be a postscript to the original French edition (1987). The time that elapsed between the two editions was all but empty and it therefore deserves to be looked upon in two closely related respects.

Intellectual honesty requires that the author be fully aware of the fact that the book he writes in the highly controversial field of aggression reflects not only the expertise he could acquire through his scientific activities, but also — to some extent — his personal beliefs and the kind of 'message' he more or less consciously intends to deliver. In the present case, such a message is basically, one of necessity and fate or, on the contrary, one of freedom, responsibility, and hope. When two multi-author volumes were published in 1988[1] and 1989[2] on the same topic, I was necessarily led to check not only the kind of facts put forward by the various authors, but also the general spirit in which they were discussed and the main conclusions that were eventually drawn from them. Suffice it to say that, in both cases, these conclusions are in perfect agreement with my own, namely that the notion of an aggression instinct which is claimed to be a biological fate is ill-founded, that human aggression is neither inevitable nor unalterable, and that the time has therefore come to get rid of the evil myth of the beast within. In my capacity as a neurobiologist, I may add briefly — before elaborating in the book — that no one has ever found within the brain any centre or neuronal system that could be considered to be the generator of an 'aggressiveness' supposed to be — and to act as — a natural entity, a causal reality, an endogenous driving force.

Not only does such a book reflect various aspects of one's own personality, but it may — and it actually did — affect in return the course of one's life. If this aspect of things is also referred to, it is because of its more general significance and interest, which go well beyond any strictly personal consideration. Shortly after the publication of my book, I retired prematurely from my posts as professor and head of a research department in behavioural neurobiology and launched, at my University's invitation, a new research centre devoted to studies in the history, philosophy, and sociology of science.[3] More than ever before it is of vital

---

[1] John Klama (1988). *Aggression. Conflict in animals and humans reconsidered*, Longman, Harlow. [John Klama is a pseudonym for the nine authors of the volume].

[2] Jo Groebel and Robert A. Hinde (ed.) (1989). *Aggression and war. Their biological and social bases*. Cambridge University Press.

[3] Centre de Recherches Transdisciplinaires sur les Sciences et les Techniques, Université Louis Pasteur, Strasbourg.

importance for the future of our human communities that the universities remain — or become — *the* scene of a more global and more objective
reflection on the growing development of science and technology with
all its social, cultural, and ethical implications. Since all major societal
problems are highly multidimensional, the universities must promote
inter-and trans-disciplinary studies if they are to contribute efficiently
to the understanding — and, one hopes, to the solution — of such
problems. A reductionist approach corresponds to a methodological
necessity whenever we analyse an isolated phenomenon, process, or
mechanism. But adopting a reductionist position on a complex problem
simply prevents us from fully understanding and possibly solving it.

Strasbourg   P.K.
June 1991

CONTENTS

1 Introduction   1

2 What is aggressiveness?   8

3 Interactions between the brain and behaviour:
   a common history   21

4 Behaviour and its motivations   52

5 Neurobiology of the processes of motivation and decision   75

6 How can behaviour be modified?   122

7 Factors which contribute towards determining the
   probability of aggression   139

8 The neurobiology of aggressive behaviour   181

9 By way of conclusion: what can be done?   219

   Glossary   253

   References   258

   Index   285

1   Introduction

Day after day, the media supply a regular dose of violence. If to some extent they indulge themselves while keeping us informed, going out of their way to revel in the bad news, the events that they report are all too real and tend to undermine that lucid and reasoned confidence which man ought to have in himself and in his future. For we have to choose between two opposing attitudes: either to accept the 'inevitable fate' of humanity, hoping ourselves to slip through the net while to some extent disregarding the fate of those trapped in it; or else to feel that we are involved in the future of the species of which our own adventure forms a part, and to wonder: must this destiny be considered inescapable? Would not a fuller awareness of our own responsibilities have the effect of changing a course of events that is truly worrying?

Philosophers, sociologists, and politicians have attempted to give their answers, but in partial and contradictory ways. Now a neurobiologist has chosen to intervene in the debate. Why? On what basis? Entrusted with what mission, clothed in what legitimacy? Primarily because he is often called on to adopt an attitude and give a response. In fact, the authorities are wont to turn to him for answers that are efficacious, quick, and simple. We need only recall Stanley Kubrick's *Clockwork orange*, or Milos Forman's *One flew over the cuckoo's nest*: in both these cases the authorities put an end to the aggressive behaviour of non-conforming individuals by means of a brain operation. The success of these films, which are in any case of high quality, also has to do with the disquiet they have aroused in people's minds: can a man's behaviour be changed so profoundly by a simple brain operation? And, if it can, should a 'caste' of specialists be given such far-reaching power? Could not other, better procedures replace this kind of operation, at least some of whose effects are irreversible?

No neurobiologist can remain insensitive to these worried questions, especially when he has to face his own questions. Some years ago I happened to take part in a meeting, chaired by a high magistrate from the public prosecutor's department, which set out to examine to what extent the modern investigative methods of neurobiology might help the courts decide on the criminal's degree of responsibility at the moment he 'takes action'. (On the basis of this concept, considered in article 64 of the French penal code, a choice may be made between sending him to prison and putting him into psychiatric care.) Apart from the fact that the researcher is encouraged in this way to exaggerate the status of his research results in his own eyes and in those of others, a temptation he does not always find easy to resist, he has real difficulty trying to appreciate all the ethical implications of this kind of approach.

Should he accept a role which, after all, may not be an appropriate one? Until the biological clues he is competent to provide are able to give real help to the courts in arriving at a better evaluation of a criminal's degree of responsibility, the role surely cannot be appropriate.

A neurobiologist who is questioned in this way may also feel that man's aggressiveness is not primarily a question for neurobiology but should first of all be considered in terms of *relations*: relations between man and his surroundings; relations between man and his fellows. This feeling, a very pertinent one, does not in any way remove human aggressiveness from the field of study of neurobiology, for it is indeed the brain which manages these relations, which positions them in a space and a history it has constructed, from the individual's birth to his death. A knowledge of the functions and mechanisms of the brain is therefore indispensable to anyone wishing to understand how these relations are established, how they are expressed, and how they evolve over time. To ignore the brain would be to deprive reality of one of its main dimensions. And we know from experience that a conception which distorts reality can too easily result in destructive acts.

For it is not only in fictional films but also in real life that we are constantly tempted to intervene directly in the brain, as part of our attempts to stem the rising flood of violence. If we do decide to resort to the methods of brain surgery and psychopharmacology, would it not be infinitely better to do so in full knowledge of the facts, using methods founded on solid scientific bases? This is to say that one should think very carefully about the way in which the brain is involved in the genesis of aggressive actions and of violent behaviour in general. Psychotherapy, moreover, a third and well tried approach to changing behaviour that is considered harmful, cannot be said to have 'immaterial' effects, any more than can the other two: it brings about a restructuring of certain internal representations which provide markers, reference points necessary to both the apprehension and the interpretation of a set of circumstances. Lastly, how can one imagine being able to operate some day on this or that aspect of the information carried by genes, using genetic engineering methods, without having a precise knowledge of the bases and the risks of such an approach? For, in the belief that the 'human instinct of aggression and self-destruction' constitutes a major peril for the future of humanity, some people are very explicitly asking to what extent it might not be appropriate to try a modification of 'human nature' through positive genetic engineering, with a view to eliminating its 'more harmful and dangerous side'.[1]

When, not long ago—in a confusion over the double Y chromosome —it was claimed that a 'chromosome for crime' existed, was not this just another (modern) chapter in a so-called *Natural history of evil* (Konrad Lorenz, as plainly stated below)? Is aggressive behaviour one of

---

[1] J. Glover 1984.

man's *natural* instincts? Let us beware of this so-called 'human nature', so frequently invoked: in the name of its 'laws', some people justify things that we may reasonably see as unjustifiable, as unacceptable forms of social and political violence. In this connection it cannot but be disturbing that, in the French edition of Lorenz's work, the title has undergone a highly significant shift of meaning (hardly without the author's knowledge). The original version is, in fact, entitled *Das sogenannte Böse. Zur Naturgeschichte der Aggression*, i.e. *That which is conventionally termed evil. A contribution to the natural history of aggression*. In its French version, the title becomes *L'Agression. Une histoire naturelle du Mal* (*Aggression. A natural history of evil*). A shift of this kind, from a natural history of aggression (a notion which has a scientific basis) to a *natural history of evil* is far from innocent, for it not only reflects the unjustifiable assimilation of aggression (observable behaviour) with evil (a type of moral judgement passed on *human intentions* that can be expressed by, among other things, an act of aggression), it also postulates that evil, as we experience it in our human communities, is the fruit of our 'natural history'. This postulate is both gratuitous and dangerous.

If, then (and I reaffirm this here for the last time), the neurobiologist is not simply authorized but also required to make pronouncements in the debate on aggressiveness, he is immediately struck by at least two kinds of difficulty: how, by arguments that are necessarily linear, can we account for an infinitely complex reality in which multiple interactions and their changing dynamic play a major role? And how can we expound in a lucid, objective, and unemotional way questions which, because of their implications, arouse strong reactions and, in everyday life, are most often approached with habits of thought and attitudes feeding on a number of prejudices and myths? It is important that the reader, as well as the author, should take full account of these difficulties, and it is therefore appropriate that we dwell on them a little longer.

A biologist's thinking is strongly characterized by those approaches that have allowed molecular biology to extend very remarkably our knowledge of the phenomena and elementary mechanisms of life. The most striking progress has been made in areas where a linear and unidirectional chain of processes was under consideration, the solution to the problem often necessitating the isolation of one of the parts of a system. Now, this way of looking at things ceases to be appropriate when one moves into a sphere where determinism is often circular, where any given process takes on its full meaning only when, after being analysed, it is reintegrated into the complex dynamic of its interactions, where evolution along a time axis — the historic dimension — plays a major role.

Furthermore, in order to understand our subject fully, we shall have to consider the human brain as an organ that generates meaning by virtue of the fact that it is the point of convergence, interaction, and

reciprocal structuring between biological, psychological, and sociological systems. Now the application of a *truly* interdisciplinary
approach, which would illuminate problems rather than obscure them,
runs into a three-fold obstacle here. On the one hand it might, for
instance, be tempting to combine the 'death instinct' of psychoanalysis,
the 'mimetic violence' spoken of by René Girard, and the data furnished
by neurobiology, in a fusion which—as an intellectual construct—
would be completely satisfactory. But this would have more to do with
literary creation than with the scientific approach. Similarly, our task is
to establish correlations—and, perhaps, concrete causal connections—
between phenomena and processes thrown up by different disciplines,
not between ideas which, rightly or wrongly, we feel obliged to abstract
from them. Lastly, one must accept an obligation to accord the same
status to different categories of phenomena and processes: whatever the
neurobiologist may think from time to time, the factuality of this or that
behavioural process is no less real than that of any molecular interaction.
If it does often happen that one accords unequal status to different
categories of fact, this is mainly because privileged commerce with one
of them, and the particular competence that is thought to flow from it,
give this category a denser, more real existence because it is more
immediately tangible. But this too results from an erroneous conception
of the relations between the brain and behaviour: the incorrect belief
(perhaps by analogy with the second law of thermodynamics) that these
relations are unidirectional and irreversible, the brain's functioning expressing itself, and exhaustively at that, in behaviour. Behaviour is put
'at the end of the chain', whereas it really forms part of a process that
starts out from the brain and returns there.

What about the objectivity and serenity without which there can be no
constructive debate? Even in discussions which are thought to be confined within the limits of scientific discourse, it seems that the protagonists have difficulty in debating aggressiveness in a non-aggressive way,
as has been regretted by Eibl-Eibesfeldt (1979). When, for example,
Ashley Montagu (1974) criticizes the work of Konrad Lorenz, stating
that it contains 'an amalgam of unsound assumptions, alleged statements of fact, erroneous implications and unwarranted inferences', we
cannot convict him of understatement. And when these themes are
taken over into political discourse, whether national or international, it
is not usually for the purposes of serene debate. When an American
colleague, David Adams, publishes in the Moscow *Psychological Journal*
(1984) an article in which he expounds the reasons leading him to
believe that no instinct exists driving man to war, the journal editor feels
obliged to add a short introductory note contrasting President Andro-
pov's peace policy with the militaristic one advocated by the Reagan
administration (what is, perhaps, 'fair in war' is not necessarily suited to
the promotion of peace!). Can we really say that the desire to find a
remedy takes precedence over all others when topics so serious as

'security' or 'racism' are taken over by party politicians and used to fuel semi-institutionalized polemic?

As for one's own personal attitudes, it should be stressed that our thinking and the positions we adopt in the matter under consideration are clearly influenced by value judgements that may be well or ill founded. The analysis that we make of an aggressive act and the attitude we adopt to it are clearly influenced, whether or not we are aware of it, by the *feelings* we undergo when we consider the aggressor, his victim, and the (imagined) motive for his action. Taking an extreme case to begin with, we can compare two men, each of whom has killed a fellow human being: one during a hold-up, the other while escaping from a prisoner-of-war camp. From a strictly biological point of view, i.e. setting aside all value judgements, behaviour is strictly the same in each case: the brain has set in motion the same means of action in order to obtain something (money, liberty) that it wants. But our viewpoint is going to be fundamentally different in each case: of the first we shall say that it was a 'cowardly attack' deserving exemplary punishment, while the second was a 'courageous deed' for which the perpetrator should be decorated. The question whether it is legitimate to kill one's fellow human being in one case or the other is not a biological but a moral one (either timeless or dependent on the situation). Even if we make the assumption that achieving liberty has supreme legitimacy, the prisoner might perhaps wonder, 'Is my liberty worth more than the other man's life?' Since, however, he will (with our help) have developed a firm contempt for and tenacious hatred of the 'enemy', the question will not even arise. In a much more general way, an aggressor makes every effort, consciously or otherwise, to denigrate his victim when trying to justify his action in his own eyes.

It is most often as a function of our convictions and our prejudices, which reflect a socio-cultural context and the place we occupy within it, that we analyse and appreciate an act of aggression or a case of violence. When an assassination takes place, one expects to know who was the intended victim and who 'claims' to have carried it out. And if there have been victims, the emotion evoked is far from the same in all cases (Bruno Frappat is right to wonder, in *Le Monde* of 10–11 February 1985: 'Why this parcelled-out emotion, *nuanced* so to speak?'). Where different forms of social and political violence are concerned, it goes without saying that they are not experienced and appreciated in the same way by those who undergo and suffer from them, and those who exploit and profit by them. And when a social group resorts to actions that may lead to acts of violence, on entering into the game it sometimes starts with an attempt at self-justification by declaring that it is 'angry'. Lastly, when violence is put at the service of a great cause (invariably considered a 'righteous' one), some go as far as to consider it a 'purifying force' capable of 'bringing back our self-respect'. This indicates to what extent, in this field, our way of thinking is directed by a whole complex of value judgements.

Being perfectly well aware of the difficulties outlined in the foregoing, I wondered at first whether I should not restrict myself to as 'objective' a scientific discourse as possible. But two closely complementary reasons led me to reject this approach. On the one hand, the subject here considered is one of those that are difficult to take out of their context without losing certain major dimensions. On the other, questions about aggressiveness are an integral part of a more general and more crucial debate altogether:[2] is the human brain essentially a source of constraints or an instrument of liberty? It is important for us to determine to what degree our brain provides us with the foundations—potentially, at the very least—of a true individual liberty or, in the opposite view, the constraints of a narrow and rigid determinism. And it is precisely the data furnished by a scientific examination of the relations existing between the brain and behaviour that have brought comfort to my 'personalist' vision of man, with his necessary social dimension. When the future of man is at stake, it is no longer admissible to maintain a total separation between that which belongs to knowledge (scientific certainty) and that which belongs to values (the choices to be made). My intention is certainly not to make an ethical work or add a personal contribution to some political philosophy. Rather, the point is to try to ensure that moral or political philosophy and concrete measures taken by decision makers, are not founded on false or, at the very least, questionable biological ideas.

Overall structure of the book

The structure of this book naturally reflects the preoccupations outlined above. To be specific: only by placing it within the most general context of the foundations of the dialogue carried on by the individual with his surroundings can one hope to respond to the anguished question, 'Is aggressiveness our fate or our responsibility?' For this reason, after an opening chapter whose aim will be to clarify the semantic content of the notion of aggressiveness, the work will comprise two parts which, while closely complementary, are none the less clearly distinct.

The main theme itself will be developed in the last three chapters (Chapters 7 to 9) which together make up the second part. First we shall list and analyse the many and various factors that may contribute to determining the probability of the appearance, in animals or in man, of aggressive behaviour (Chapter 7). Some of these factors are directly linked with individual modes of cerebral functioning, while others operate via the intermediary of the brain which 'processes' them, so conferring on them all of their significance and impact. In order to get a better grasp of the nature of the first kind of factor and understand the

[2] See P. Karli 1986.

processing done by the brain on the second, we then have to put together, so as to interpret them coherently, the correlations we have been able to establish between well-defined brain mechanisms and the probability of the initiation of aggressive behaviour when a given situation is confronted (Chapter 8). We shall then have at our disposal the information required for an attempt at answering the question 'What can we do?', i.e. to review and evaluate the various means available for operating on the genesis of individual and collective aggressive acts, in a way that is both useful and, from an ethical point of view, acceptable (Chapter 9).

But if we truly wish to understand the full significance of particular data about aggressive behaviour, its genesis and its possible control, we must constantly reposition them in the most general and global context of the dynamics of the relations between the brain and behaviour. Moreover, in the absence of those bases which alone allow the notions and facts expounded to be critically evaluated, such notions and facts run the risk of looking like a series of authoritative postulates and arguments. For this reason we shall try, in the first part, to establish a general framework that will serve as a structure to accommodate and refer to particular data that will be expounded and discussed later. To begin with we shall look at the relations between the brain and behaviour, stressing both the reciprocal nature of these relations and the leading role played by the dimension of time; for the fact that the brain and behaviour are structured by each other can be fully grasped only in the context of their common history (Chapter 3). We shall then see how motives originate and how the configuration of the 'causal field' underlying given behaviour develops (Chapter 4). The following chapter will be devoted to an analysis of the neurobiological mechanisms involved in the processes of motivation and decision making (Chapter 5). On these bases we shall finally be in a position to examine the means available for modifying behaviour, whether they be therapeutic approaches or 'manipulations', whose objectives are entirely different (Chapter 6). The information and concepts provided in this first part will greatly ease understanding of the matter dealt with in the second, and will contribute to giving it its full meaning.

2   What is aggressiveness?

We need only put some quotations side by side, chosen from many others, in order to illuminate the peremptory way in which some people invoke our biological heritage in order to 'explain' human violence, this inevitability allegedly linked to the persistence and resurgence of 'the beast in man'. We can quote, in particular, some statements extracted from two works, each of which has had great influence. In his well-known book *On aggression*, Konrad Lorenz (1967) stresses 'the destructive intensity of the aggression drive, still a hereditary evil of mankind.' And he sees 'man as he is today, in his hand the atom bomb, the product of his intelligence, in his heart the aggression drive inherited from his anthropoid ancestors, which this same intelligence cannot control'. For his part, Robert Ardrey (1961) states, in *African genesis* (in a chapter entitled *Cain's children*), that 'man is a predator whose natural instinct is to kill with a weapon', since 'the primate has . . . an attitude of perpetual hostility for the territorial neighbour.' He considers that our destiny's foundations 'must rest on the beds of our (animal) past for there is nowhere else to build'. And he does not hesitate to state that the delinquent 'lives in a perfect world created solely by himself, . . . (in a) world so consummately free', since 'this ingenious, normal adolescent human creature has created a way of life in perfect image of his animal needs.'

To affirm the existence of an 'instinct for aggression', of 'violent destructive impulses which we contain within us only with difficulty', leads one easily to conclude that it is 'probable that the denial or repression of our aggressive drives is liable to cause disharmony within ourselves, however desirable it may be that we should get rid of them'.[1] Storr, considering that these aggressive impulses have found in time of war 'an acceptable channel for discharge', but that this possibility is now available only 'in wars between nations who do not possess nuclear weapons', comes to the conclusion that 'our only hope is that we can continue war by other means than the primitive one of killing each other'.

Whether or not we are always aware of it, this way of seeing things pervades our minds to some degree. So much so that, the day after the tragic massacres of Sabra and Chatila in the Lebanon, the readers of the *Matin de Paris* (20 September 1982) were probably not at all surprised to read, from the pen of Roger Ikor, the following 'explanation' (or at least an attempt at a partial one): 'This animal called Man, in the process of emerging towards humanity, is plunging his roots ever more deeply into bestiality'. In a world where nothing is steadfast, where myths are born and disappear, we must wonder about the perennial nature of this

---

[1] A. Storr 1964.

myth of 'the beast within man'. The principal reason for it is surely that it serves as a convenient alibi or scapegoat, so allowing us to shirk our own responsibilities. Because it exists, the chorus of ancient tragedy can be altered to suit the taste of the day (for the 'wrath of the gods' has gone out of style): we bemoan this biological inevitability . . . and wash our hands of it. Furthermore, as a consequence of this notion of biological inevitability, we shall be tempted to think it enough to ask 'brain specialists' to operate on that organ (to prevent 'aggressiveness' either arising or being expressed in some kind of 'acting out'), so that we may avoid having to do something else.

But this myth has another major drawback. Durant (1981) rightly draws attention to the fact that particular notions of man and human society are projected into nature in order to be brought back later in a naturalistic form that is even stronger and more restricting. Now, when an appeal is made to nature in this way so as to confirm certain social values, the biological data invoked more often serve to 'set pessimistic limits to what is humanly possible' than to create 'a context of optimism and encouragement'. This is by no means trivial, since what men undertake and accomplish depends on what they believe to be possible.

Since we have just had to deal with one particular myth, it is not irrelevant to stress the role played — in a more general sense — by certain myths in the genesis of conflicts. A myth, then, is a motive for some, a pretext for others, and a justification for everyone. History is littered with wars fought for the sake of the 'True Faith', the 'Genuine Destiny of Man', or 'Civilization'. Above the door of the chapel of Notre-Dame-de-Buei (near Guillaumes, in the Alpes-Maritimes), we may read the inscription: '*Da nobis virtutem contra hostes tuos*' ('Give us strength against *Thy* enemies'). It suffices, then, to name those who, for whatever reason, are considered one's own enemies as the 'enemies of God', for war to become 'just', and for love of one's neighbour to give way, in all legitimacy, to hatred of the enemy. In order to illustrate some of the myths which have mobilized the French in the course of their history, a German writer, Sieburg, provides the following three quotations:[2] 'Those who make war on the Holy Kingdom of France make war on Jesus, the King' (Joan of Arc); 'The true strength of the French Republic must henceforth consist in preventing the emergence of any idea that does not belong there' (Bonaparte); 'God needed France' (the Bishop of Versailles, in 1917). Even in our own day, heads of state can be heard anathematizing the 'Great Satan' or the 'Evil Empire'. Since 'in politics, that which is believed becomes more important than that which is true' (according to Talleyrand, who was an expert on the subject), stereotyped language will be with us for a long time to come. And it does seem that this phenomenon is quite an ancient one, for as early a figure as

[2] Reported by E. Muraise 1982.

Confucius used to declare that, if he ever had to govern, he would start by 'restoring the meaning of words'.

For our purpose, it is the meaning of the word 'aggressiveness' that we must first of all examine and clarify.

The notion of 'aggressiveness'

This notion is at once vague and ambiguous, because more often than not it combines, whatever kind of discourse is being carried on, two meanings that ought to be carefully distinguished. Whether the word occurs in colloquial or in learned language, there is generally confusion of its descriptive virtue and the explanatory virtues ascribed to it.

The term 'aggressiveness' is a useful one, posing no problems as long as it is used purely descriptively. For the biologist, the concrete and primary reality involved is the existence, throughout the animal kingdom, of a set of behaviours referred to as 'aggressive' (they strike at the physical and/or psychic integrity of another living being, or at least threaten to do so). Because of the universality of similar behavioural phenomena, we are quite naturally led to abstract from them a generic character and to speak of 'displays of aggressiveness'. As long as the term is used to describe and evaluate a category of phenomena, of events, it is not necessary to give it a precise definition. Free rein can be given to any metaphorical usage (one may or may not say of a course of action marked by will, assurance, and dynamism that it is characterized by 'aggressiveness') and to everyone's free judgement (just as not everyone agrees on what ought to be called 'beautiful' or 'good', we may disagree in considering whether this attitude or that behaviour does or does not constitute a display of 'aggressiveness').

Quite other implications are involved when a semantic shift occurs from displays of aggressiveness to displays of 'Aggressiveness', and so a shift from the purely descriptive virtue of the notion towards explanatory, causal ones. In fact, the generic and abstract notion of aggressiveness in this way undergoes a process of reification ('thingification') and Aggressiveness becomes a natural entity, the primary reality from which ensue the various displays observed in the animal kingdom, including the human species. A shift takes place, in other words, towards an almost Platonic vision of Aggressiveness as a psychobiological 'idea', endowed with autonomous existence, i.e. independent of the mind that conceived it as an abstraction from a reality that is simultaneously diverse and universal. 'Aggressiveness' thus becomes the common source of all aggression.

The aggression instinct: myth or reality? The theory of Konrad Lorenz

Since, whenever we think about the causes of this or that form of violence, we refer more often than not to the scientific guarantee brought by Konrad Lorenz to the idea that there exists an aggression drive, it is important that we should adopt a position as to the value of that guarantee. Lorenz's conception is usually presented by saying that he postulates the existence of an aggression drive in the form of a specific, endogenous energy which, he supposes, accumulates progressively within an individual and is then necessarily externalized, 'discharged', in one way or another. Some allude to this postulate, rather irreverently but very tellingly, as the 'toilet-flush' theory, suggesting quite well the idea that something fairly small (a 'releaser') is enough to trigger the inevitable outpouring of the specific energy (the 'aggressiveness') previously accumulated. Eibl-Eibesfeldt, a pupil of Lorenz's who shares his views (but with important differences of emphasis), regrets, not without reason, that Lorenz's idea is often presented in a simplistic and therefore caricatural way.[3] But it really has to be said that, by his way of posing questions and answering them, Lorenz lays himself open to perfectly justifiable criticisms. With successive touches he constructs a system that, at first, may seduce the reader by its coherence and the universal character of its explanatory virtues. In reality, these virtues have largely to do with the fact that a new postulate is added to the preceding ones every time it appears that the latter are insufficient to account for this or that new observation. And the transition from hypothesis ('everything happens as if . . .') to assertion ('this is so') often seems rather a rapid one. Furthermore, since he is trying, in essence, to interpret real behaviour on the basis of events and processes which are supposed to have occurred during the course of evolutionary history, it is difficult to verify by means of appropriate experimental procedures how well-founded are the statements being made.

The debate about Lorenz's ideas will not be much advanced by a simple statement that one agrees or disagrees with him. But, if we do not share his ideas, there can be no question here of giving an exhaustive critical analysis of his work.[4] It will suffice to point out some ambiguities, contradictions, and lacunae which motivate our adopting a dissenting attitude. Given that the notion of the 'aggression instinct' occupies a central place in Lorenz's thinking, one would like to be able to grasp what precisely he understands by an 'instinct'. Now, we note that he makes no clear distinction between 'instinctive behaviour' that an observer can see from without, and the endogenous 'force' of which it is supposed to be the outward projection. When Lorenz writes, on the subject of relatively simple 'instinctive movements', that 'each of these hereditary coordinations has its own spontaneity', and when he considers the role played by 'aggression' (an observable form of behaviour) 'in the great parliament of instincts', one takes him to be using the notion of 'instinct' or 'drive' to allude to the genetically preprogrammed 'innate' character of a motor schema of some degree of complexity,

---

[3] See I. Eibl-Eibesfeldt 1979.   [4] K. Lorenz 1967.

together with some spontaneous activity of the nervous substrate that ensures its execution. But such is not the case, and 'instinct' must include a source of energy other than the mere activity of the generator of the movement, for he explains elsewhere that each 'hereditary coordination . . . forces the animal or human to get up and search actively for the special set of stimuli which elicit it and no other hereditary coordination'. Furthermore, if Lorenz says (in words quoted earlier) that man carries the aggression instinct 'in his heart', he certainly does not mean by this image simply that aggressive behaviour figures among the means of action of which the human being disposes in order to confront certain situations. Quite the contrary, the aggression drive springs 'spontaneously from the inner human being', and 'it is the spontaneity of this instinct that makes it so dangerous'. And Lorenz adds: 'If it were merely a reaction to certain external factors, as many sociologists and psychologists maintain, the state of mankind would not be as perilous as it really is, for, in that case, the reaction-eliciting factors could be eliminated with some hope of success.' But it is 'a completely erroneous view' to consider that 'animal and human behaviour is predominantly reactive and that, even if it contains any innate elements at all, it can be altered to an unlimited extent by learning'.

It thus appears that 'aggressiveness' or the 'aggression instinct' corresponds, for Lorenz, not to the mere existence of aggressive behaviour as such, but in fact to that of a specific endogenous energy that *must* be 'discharged' in the form of behaviour of this type. And yet an ambiguity reappears when we come to the role played in the evolution of behaviour by the *function* of aggression. In fact, Lorenz speaks indifferently of the function of aggression and the function of the aggressive drive. Now, if we accept that a process of natural selection can act on a certain form of aggression that serves—for each of the individuals displaying it—a certain purpose in particular circumstances, it is difficult to see how evolution could have given birth to an 'all-purpose' aggressive drive (since it is supposed to express itself in a variety of forms of aggression which serve a multitude of purposes under very varied circumstances, and since, when no fellow-creatures are within reach, some animals are reduced 'to discharge their anger on other objects'). An aggressive drive of this kind rather looks like a real biological nonsense, liable to lead the animal kingdom rapidly to its disappearance.

The notion of the 'ritualization' of aggression

Having thus postulated the existence of a highly 'formidable' instinct (since 'aggression' is defined by Lorenz as 'the fighting instinct in beast and man which is directed against members of the same species'), and aware of its effects, highly prejudicial to life, Lorenz is necessarily led (necessity makes the rules here!) to postulate the development of 'physiological mechanisms of behaviour whose function it is to prevent the injuring and killing of members of the same species'. He thence tells us of 'evolution's most ingenious expedient for guiding aggression into harmless channels', namely 'the redirection of the attack' by the process of 'ritualization'. Instinctive movements, driven by the aggression instinct, are thus supposed to change in both form and function. Indeed, 'just as the form of these movements, in the course of their progressive ritualisation has become different from that of the non-ritualised prototype, so also has their meaning'. Since the aggression instinct is bound, in the long run, to experience some discomfort in having to drive simultaneously movements that are supposedly aggress-ive and others, perfectly inoffensive ('ritualized') ones, Lorenz states a further postulate, namely that 'ritualized' movements will progressively be driven by an instinct of their own. Indeed, 'by the process of phylogenetic ritualisation a new and completely autonomous instinct may evolve' and 'in this parliament [of the instincts], it is particularly the drives that have arisen by ritualisation which are so often called upon to oppose aggression, to divert it into harmless channels, and to inhibit those of its actions that are injurious to the survival of the species'. It is not easy to see how a new and completely autonomous instinct can 'channel' aggression, which is the form of expression of another instinct. And it is difficult, above all, to see how this new instinct comes into being, starting from the aggression instinct: do we have to suppose, to use an image, that some river water goes back upstream to give rise to its own spring?

The pipe of peace

Speaking of the role played by rites in the human species, Lorenz imagines the scene in which 'for the first time, two enemy Indians became friends by smoking a pipe together'. He imagines that these two experienced old chiefs 'rather tired of war, have agreed to make an unusual experiment . . . settle the question of hunting rights . . . by peaceful talks instead of by war'. But it looks wrong to deduce from this that, in this case, too, the fortuitous creation of a rite (smoking the peace pipe) had the effect of modifying an instinct, or even creating another one. For something must have happened *before* and *in order that* this pipe should be smoked jointly for the first time. If the two protagonists are 'tired of war' and they have agreed to 'a mutual approach', is this not because they have thought things over and realize the absurdity of their usual procedure? No, Lorenz would say, for he states categorically that the aggressiveness man has inherited from his ancestors is an instinct 'which his intelligence cannot control'. Nevertheless, as we saw above, Lorenz considers the atom bomb an 'intelligent' expression of the 'aggression instinct'. Now, whatever potential for harm it may possess, this fruit of human intelligence has taken some singular liberties with the instinct that is supposed to have engendered it. Not only would the pressure of a finger on the fatal button have no connection with 'innate' behaviour: this press of a button would be motivated by ideas arising in a human mind and not by some natural 'releasing' stimulus (the target being perhaps thousands of kilometres away). Moreover, and above all, human intelligence has succeeded in preventing—for decades and, we must hope, forever—the 'aggression instinct' from putting this very special weapon into operation. Given all this, it is quite wrong to say that reason cannot control the said 'instinct'.

One last postulate of Lorenz's must be quoted, concerning a process described by the author as an 'ingenious feat' on the part of evolution (it may in fact be an authorial one!). For this is claimed to show how evolution was able, 'by the comparatively simple means of redirection and ritualisation', to transform behaviour motivated by the aggression instinct 'into a means of appeasement and further into a love ceremony which forms a strong tie between those that participate in it.' Here is how Lorenz envisages the origin of personal bonds: 'The aggression of a particular individual is diverted from a second, equally particular individual, while its discharge against all other, anonymous members of the species is not inhibited. Thus discrimination between friend and stranger arises, and for the first time in the world personal bonds between individuals come into being . . . As far as their origin and their original function are concerned, personal bonds belong to the aggression-inhibiting, appeasing behaviour mechanisms.' And Lorenz concludes from this, in a beautiful lyrical flight, that 'mutation and selection, the great "constructors" which make genealogical trees grow upwards, have chosen, of all unlikely things, the rough and spiny shoot of intra-specific aggression to bear the blossoms of personal friendship and love'. A complex construction, in other words, incorporating all the essential aspects of social intercourse, rests on—and, we might say, vanishes with—the initial postulate of the existence of an aggressive drive, a genetically preprogrammed natural entity.

Was Lorenz inspired by Freud?

Since for Freud too, aggression and auto-aggression are manifestations of the 'death instinct', and since Freud preceded Lorenz in time (and, as Lorenz explains, 'it was Freud who first pointed out the essential spontaneity of instincts, though he recognized that of aggression only rather later'), why do we only bring in the founder of psychoanalysis here, and at reduced length? Because Freud does not claim, as Lorenz does, the real—rather than hypothetical—nature of the processes described, or their universal explanatory value, any more than he does the existence of a direct and inescapable causal link between aggressive drive and aggression. In the interactions of the different 'agencies' of the 'psychical apparatus', Freud brings in exchanges of a drive 'energy', mirroring another form of energy which, according to his teacher E. Brücke, was thought to circulate in neural pathways. If he emphasizes the reduction of interior tension by discharges of energy, by processes of disorganization and dedifferentiation, this is because he is subject to the influence of the dominant scientific ideas of his day (the second principle of thermodynamics, with one form of energy degrading into another; biosynthesis and the degradation of molecules making up living matter). But, more and more, Freud will use these notions as metaphors, like the elements of a language full of imagery without whose aid he could not describe the processes he was interested in. And, while he considers that 'some of the analogies, connections and relations' he has established are 'worthy of consideration', Freud admits that in the space of some decades biology may perhaps give answers to the questions he has posed, 'such as to bring about the collapse of the artificial edifice of our hypotheses'.[5] After quoting these words, it is only right that the neurobiologist's response should be worthy of Freud's remarks, which are distinguished by their lucidity and humility. The neurobiologist is indeed bound to recognize that he is far from having provided all the answers to the questions Freud asked, and that the answers so far obtained lead one rather to say that mental life is not reducible to the data that can be brought to light by the investigative methods of neurobiology. For, as we shall see later, the relations between brain mechanisms and behavioural events cannot be considered simple, one-to-one causal relations.

The aggression instinct: an obsolete notion

As for aggressive behaviour, which alone concerns us here, it can be stated that at the present time the vast majority of those who actually study it in animals (ethologists and neurobiologists) or in man (specialists in social psychology, psychiatry, sociology, criminology, and war studies) consider that the notion of an instinct no longer has more than a historical interest. In particular, the notion of an instinct for aggression, of an aggressiveness seen as the internal source of a specific energy which allegedly discharges to the exterior in the form of the most diverse acts of aggression, has a heuristic value of zero: not only does it explain nothing, it also obscures the real problems by failing to lead us to ask truly pertinent questions. For this reason neurobiology has abandoned (as we shall see) the search for any 'centre' or system of neurons that might be the 'generator' of any such aggressiveness, considered as the initial cause of all aggression. In the context of an enquiry (*La pulsion, pour quoi faire?* (Why a Drive?) conducted by the French Psychoanalytical Association, Daniel Widlöcher (1984) wonders, for his part, whether it might not be appropriate to abandon 'a falsely explanatory model which

[5] See S. Freud 1981.

does not in reality lead to any progress in empirical knowledge', and substitute for the theory of drives a theory of the association of thought-acts. The psychic apparatus would then be seen as 'a set of potential acts, awaiting favourable circumstances in order to be realised', the appropriate thought at any time being 'that which grasps intersubjective reality and embraces it'.

Abandoning an exaggeratedly simplificatory 'explanation' does not mean (quite the contrary) that one should invoke another, equally simplificatory one and consider aggression as a reaction provoked by an 'aggressogenic' situation. From the latter viewpoint, aggression is no longer provoked by the accumulation of an excess of aggressive energy (for which 'Nature' is supposed to be responsible) but is aroused by aggressogenic external conditions (for which 'Society' is supposed to be responsible). This notion (an external evil driving an angel to act like a beast), like the preceding one (evil resides in us, the beast is in man), obscures essential aspects of the problem as it really exists. Let us take a concrete example, presenting it in intentionally black-and-white terms.

A little exemplary scene from married life

We, my wife and I, decided to watch a television programme. Channel 1 is showing a World Cup match, Channel 2 a great film. I would prefer to see the football match, but my wife would rather see the film. My behaviour in this situation might take either of two extreme forms: (1) my dearest wish being to give pleasure to my wife, I decided to watch the film with her; (2) I take up my .22 carbine and do away with my wife so as to be able to watch my cup match in peace and quiet. To explain the first behaviour by saying that I was able to convert my aggressiveness into love, via a 'sublimation' process, is to leave something lacking in the 'explanation'. To attempt to explain the second behaviour by emphasizing the aggressogenic nature of the situation offers no explanation of why the same situation should be aggressogenic in this case but not in the first one. In reality, in the one case as in the other, the behaviour that I exhibit is a *means* of expression and of action, and the situation functions as an *indicator*. In fact my behaviour in this situation reveals at one and the same time certain traits of my personality and the nature of my relations with certain things (football, television, etc.) and with my fellow humans—particularly my wife.

One must therefore start with the concrete event: the use of a certain behaviour as a *means* of expression and action, and ask oneself about the origin of the *motive* underlying the event. For the neurobiologist, any consideration of the motive that leads an individual, confronted with a given situation, to use a given behaviour, comes down to making an inventory of the set of factors that contribute to determining the probability of this strategy's being put into action, and making an analysis of the brain mechanisms through whose intermediary these factors operate. The factors are many and varied, for they simultaneously relate to a personality rich in experience, a situation that is part of a sociocultural context, and the individual relationship established between them—a relationship which the behaviour expresses and may aim to preserve or modify. The brain's modes of operation themselves have an effect on the way in which lived experience is recorded in the representations which the brain holds, on the way the brain evaluates the significance of a situation by referring to such representations, and on the way it selects and puts into action a strategy which it judges appropriate. This is to say that reality is infinitely complex and its analysis necessarily requires an interdisciplinary approach. Edgar Morin (1973) was indeed right to say, 'The bell is tolling for closed, fragmentary and simplifying theories of man. The era of open, multidimensional and complex theories is beginning.'

Aggression, a means of expression and action

In the various fields of investigation into human behaviour, emphasis is being ever more unambiguously laid on the *function* of this or that aggressive behaviour, on the goal for whose attainment this means of action is put into effect. It is becoming apparent, through converging studies, that an aggression more often than not reflects a given state of affairs at the same time as it is intended to act upon it.

When the development of social behaviour in children is analysed, it is noticed that all children pass through aggressive phases of varying duration between the ages of 15 and 24 months, the period when a child increases its participation in competitions.[6] The child seeks out competition as an opportunity to 'try itself out' in contact with others at a time when it is conquering space by walking, and to 'position itself' with regard to others and the objects it desires. In the adult, numerous forms of behaviour are aimed at providing the 'Ego defences' (which are exercised primarily against an internal danger, anguish) and the 'social defences' (exercised against an external danger),[7] and we shall see later the reasons for the increase in aggressive forms of social defence behaviour that we are witnessing nowadays. Nuttin (1980), for his part, stresses the important role played by actions, which are resources employed in the operation of human motivation and in the process of forming goals and projects. He refers to 'some studies on the harmful effects of a situation in which the subject perceives himself as devoid of behavioural '"resources"'.

As for the act of auto-aggression represented by suicide, it cannot be understood in a perspective limited to the individual and his 'drives'; quite to the contrary, 'the suicidal act is a social fact whose numerous

[6] A. Restoin *et al.* 1984.   [7] See A. Mucchielli 1981.

functions respond to the multiplicity of determinisms'.[8] It is a means of expression for an individual who wants to signal his distress and rage, and is at the same time a means of taking action which, faced with a situation that has become unbearable, constitutes both an attempt to escape from it and one last endeavour to change it.

As for criminology, it may be considered that 'a transgression [of the law] is simply one means among others of satisfying an urgent desire, of resolving a problem, or of reaching one's goals'.[9] If the crime rate varies so much over space and time, it is because of the extremely large number of factors that contribute to determining the probability of this or that mode of action's being put into operation. This means that we explain nothing by simply putting crime down to man's biological inheritance (his 'natural aggressiveness') or to the 'aggressogenic' influences of society. To do that is to deprive oneself of any real possibility of conceiving and executing a relevant analysis.

For the sociologist who is interested in the origin and development of conflict, it also appears that, with rare exceptions, 'a conflict is not launched for its own sake, but with a view to a goal', and that the relationship here is therefore not a cause-and-effect but a means-to-an-end one.[10] Particular aims which one seeks to attain by means of the conflict come to be grafted on to the general and formal aim of all conflict, namely the wish to impose one's will on the other party. These aims are as various as there are 'desires, ambitions and projects capable of inflaming a community, large or small'. Pondering the true significance of 'social movements' and the violent forms they can take (in particular, the demonstrations in the first semester of 1979 at Longwy), Christian de Montlibert (1984) concludes that, here too, nothing much is explained by blaming violence, considered as a unitary dimension supposedly inscribed in human nature and sure to reappear regularly in the most varied forms. In this case too, acts of violence are in fact a means of expression and action: expressing the realization of identity and of a particular state of affairs (more concretely, sending out a 'cry for help' when faced with 'imminent death'), and taking action in an attempt to change the state of affairs, which, for a social grouping, comes further down to carrying out a certain amount of work on oneself.

As for terrorism, the feeling that resorting to this kind of violence can further the most varied causes is expressed—provocatively, to be sure, but tellingly—in the comparison made by Raufer (1984) between terrorism and the state lottery: 'it is a tool that is easy to use and that often gives gigantic profits compared with the initial investment. We are inevitably reminded of state lottery advertisements: 'It's easy, doesn't cost much, and could get you loads of money.' In a study of the relations between terrorism and its sociocultural context, Jean-Paul Charnay (1981) explains that, in the light of a theory of 'social offences',

---

[8] H. Chabrol 1984.　　　　[9] M. Cusson 1983*a*.　　　　[10] J. Freund 1983.

terrorism attacks 'the deep bases of society, the elements that make man a *zoon politikon*' (a political animal); and he emphasizes the fact that, in that quality, man 'kills for ideas'. Clearly, this is not the inevitable overflow of any supposedly innate aggressiveness: it is a deliberate choice that some people make and that one can equally well refuse ever to make. This is the idea expressed by Montaigne in the quotation that closes the study mentioned above, when he recalls, in connection with the most cruel of the wars of religion: 'There are few ideas for which I should willingly die. There are none for which I should kill.'

In an analysis of the 366 major armed conflicts that occurred between 1740 and 1974, Bouthoul and Carrère (1976) follow the movement of 'fronts of collective aggressiveness'. For the authors, the notion of fronts of aggressiveness 'excludes all possibility of inevitability or determinism'; these fronts are 'evidence of (and not an explanation for) the focusing of warlike violence following certain special lines and certain contexts, notably racial, ethnic, religious and ideological ones'. The analysis shows that foreign and civil wars, which fulfil a number of *functions*, 'express the specific natures of the societies involved and contribute to their transformation'. Here once more, the notion of aggressiveness allows behaviour to be described, but its real motives — infinitely complex and changing — are to be sought elsewhere.

Firearms: a protection, or a potential hazard?

Since aggression, whether individual or collective, constitutes a means of expression and action that can be put into operation in the most diverse circumstances and contexts, it is important to emphasize — from now on — the dangers created by the proliferation, in our homes as well as in the arsenals, of weapons capable of causing death. It is obvious that, in the (imaginary!) situation envisaged above, I run the risk of doing away with my wife with the aid of a .22 carbine only if this firearm is within my reach. In *La révolution conservatrice américaine*, Sorman (1983) quotes the opposing attitudes adopted by two town councils, ascribing them to two contrary ideological positions: in Kennesaw, a small town in Georgia, the council (conservative, i.e. right-wing) has decided that each family is obliged to possess a firearm, whereas in Morton Grove, in the Chicago suburbs, the council (liberal, i.e. left-wing) forbids the possession of firearms. Without wishing to minimize the influence of diverging ideologies, it is hardly necessary to point out that the possession of firearms, for reasons that are not just ideological, does not necessarily have the same consequences in a Breton village as it would in the suburbs of Marseilles.

On the subject of the proliferation of nuclear arms, the author begs to offer a personal memoir (this one is quite genuine). At a dinner given on the occasion of a general assembly of the European Science Foundation, I was seated opposite the Foundation's President (Hubert Curien), who was flanked by his guests of honour, namely the President of the Academy of Sciences of the USA and one of the vice-presidents of the Academy of Sciences of the USSR. Chance had it that these two colleagues were advisers 'for nuclear affairs' to their respective heads of state (President Carter and General Secretary Brezhnev). To our question about the risks of a nuclear war, both of them gave the same answer: 'The American (or the Soviet) people desire peace; but the very fact that such an accumulation of nuclear arms is being pursued, and such a large part of our national product is invested in them, means that the use of these weapons—within what period we certainly cannot say—is almost inevitable!'

All of this initial descriptive analysis of various forms of aggression and the conditions in which they occur leads to the conclusion that aggressive behaviour is in fact a means of action that may be employed in pursuit of the most variegated goals, and is not by any means a simple outward projection of some 'aggressiveness' supposed to be generated by the brain. If this (absolutely essential) conclusion is to be the starting point of all our subsequent examination of the question, we should verify its well-foundedness by considering the interactions of the brain and behaviour, such as they are shaped by their common history, and the origins of the 'motivations' which, at any given moment of this history, underlie the individual's behaviour. These large themes are tackled in the two chapters which follow.

3 Interactions between the brain and behaviour: a common history

Separating aggressive behaviour and its determinants from all the other aspects and dimensions of behaviour, in a totally artificial way, would be a fairly unilluminating way to proceed. And it would be equally wrong when dealing with the brain, a dynamic whole, to consider in isolation those cerebral mechanisms thought to be responsible for the control of such behaviour. Furthermore, since there can be no behaviour without the brain and the latter's functioning has full meaning only when considered in the light of the dialogue conducted by the living being with its environment, it is important to examine the brain and behaviour within the context of the relations that exist between them. And we should emphasize the notion of 'interaction', so as to stress from the outset the bidirectional, reciprocal character of these relations.

In reality, the point of extracting an overall view of the interactions between brain and behaviour goes well beyond a didactic wish to place a particular case in a more general context. For this overall view has important repercussions in that it affects profoundly our way of understanding man's nature and destiny and, thereby, a whole series of practical measures that follow from and reflect that understanding.

In this connection, the main question concerns the complementary notions of self-awareness, free will, and individual responsibility. Man wants to find out to what extent his brain gives him the basis — potentially, at least — for true individual liberty or, on the contrary, the constraints of a narrow and rigid determinism. How could we, after all, carry on a free dialogue with our environment, how could our freedom be other than the most treacherous illusion, if our behaviour — including our verbal behaviour — were nothing more than the outward projection, both inevitable and irreversible, of this or that aspect of cerebral functioning, itself the fruit of the genetic lottery which holds a 'special draw' on the occasion of every fertilization? It is therefore important to consider whether any 'natural laws' exist that would permit the definition and prediction of behavioural 'effects' on the basis of cerebral 'causes', not only when one juxtaposes (in a sort of 'snapshot', momentarily freezing things) this behavioural event with that cerebral event, but also when the relations between the brain and behaviour are reconsidered within the dynamic of their common history. It goes without saying that political philosophy (with its notions of democracy, rights of man, individual freedoms, and so forth) does not remain unchanged whether one believes in (to give only the extreme positions) a rigid and absolute determinism, or a 'free undetermined movement' in individual history, as in history *tout court*. In addition to this, it is quite obvious that there can be no question of the neurobiologist providing *the* answer to these fundamental questions, but rather elements of an answer, such as to enrich and direct the thought that is led to consider them.

The notions which emerge from this reflection do not concern merely a small number of thinkers. They are in fact picked up by the decision-makers and, by virtue of this, they concern each one of us. Indeed, the way that behavioural problems and mental illnesses are explained and treated, the educational system and its pedagogical methods considered, criminal acts judged, responsibilities attributed and penal sanctions applied, is heavily influenced by prevailing 'psychological theories' which are based, explicitly or not, on a certain view of the relations between the brain and behaviour. Given the scope of this statement, it is best to establish it on a corpus of concrete facts.

Psychological theories and practical approaches

The determinants of behaviour: two distinct approaches

If theories about the determinants of behaviour and the genesis of mental activities are many and subtly varied, we can all the same distinguish two fundamentally different approaches. It is permissible deliberately to stress the distinctions between these approaches, even if it means looking at their points of view and methods schematically, so as to get a better grasp of the practical consequences they can have.

Dispositional factors . . .

Adherents of the first approach put the emphasis on 'personality', on a set of 'dispositional factors'. They study the way in which dispositions are expressed in behaviour and how they permit us to foresee that behaviour. In their attempts to identify the dispositional factors with a view to establishing some correlation matrix between them, their aim is to define the various structures or configurations of personality traits. Thus different personality types are characterized, with their aptitudes and their limits.

Within this first approach two main variants can be distinguished, equally schematically. The first emphasizes analysis, the individualization of the traits (with the possible use of factor analysis) and the measurement of the phenomena that express these traits using psychometric methods. This approach often leads to an emphasis on the essential role played by genetic factors, as shown in the ideas of Eysenck (1980) on the nature and measurement of intelligence. The second approach gives more attention to the correlations between traits, the dynamic of their interactions, the 'psychodynamics' of the person.

Psychoanalysis, with its own theoretic structure, falls within this variant.

The neurobiological version of this approach consists in brain research into the different 'generators' of dispositions, in the form of neuron systems characterized by a distinct anatomical topography and/or distinct neurochemical properties. Thus it is that one will describe, for example, different 'motivational systems', implying that the degree of activation of such a system determines the individual's propensity to 'output' the corresponding behaviour. All observable behavioural phenomena, in other words, are related in some way to the brain's very 'matter'.

. . . or situational factors?

A second school insists on the role of the 'situational factors', on interactions with the environment. Behaviour is considered mainly in relation to lived experience, those experiences acquired in a defined sociocultural context that have made a large contribution to shaping a mentality, attitudes, and goals. More attention is devoted to the *function* that a piece of behaviour provides in dialogue with the environment than to the way in which it constitutes the *expression* of a personality trait.

Within this second approach one can again distinguish two main variants which, once more, reflect a concern mainly with either analysis or synthesis. The 'behaviourist' approach looks on a living creature, in essence, as a machine that responds — and learns to respond — to the promptings of the environment. Since these promptings arrive in a haphazard manner without any obvious interrelations, the individual will construct by simple addition a repertoire of responses, and his behaviour will therefore be investigated in an analytical and mechanistic way. A more synthetic approach stresses the person, no longer conceived as a set of dispositions but as a social being whose goals, and the behaviour aimed at bringing them about, are largely shaped by the experience acquired in a given sociocultural context. Investigations will therefore be aimed more particularly at 'social representations', cognitive structures which allow one to apprehend and interpret situations while undergoing an appropriate 'updating', through the very fact of being in dialogue with the environment; for these structures tend to have an internal coherence, necessary to the genesis and recognition of an 'identity'.

Because of the set of concepts and methodologies at his disposal, the neurobiologist is not greatly drawn to this approach as long as it is formulated in its most synthetic version. For the brain is no longer seen as a generator of dispositions and behaviour expressing them, but as an instrument for the analysis and interpretation of situations, as a 'mediator' between the living being and its environment. As we shall see below, it is not easy to tackle by the methods and techniques currently in use in neurobiology the various information-processing operations used by the individual brain to ascribe meaning to the promptings it receives, leading it to generate certain expectations, to aim for certain goals, and to select appropriate strategies for their attainment.

The influence of theories on practical approaches

It is easy to see that practical approaches based on these theoretical conceptions can be very different, according to whether the accent is put on dispositional factors or, instead, on situational ones. This is clearly apparent when one considers the choice of therapies for behavioural problems, the selection of psychopedagogical approaches, or, again, the ascription of responsibilities in criminology.

The choice of a therapy when faced with deviant behaviour will not be the same depending on whether that behaviour is seen as the expression of a certain type of psychopathic personality or, rather, as a reactive response to conflict situations resulting from group constraints. As Leyens (1982) stresses, the therapeutic solution envisaged will generally be 'heavier' if the dispositional factors are emphasized; for the therapist will have to 'change the patient, rather than the disturbing situation'. Furthermore, in a sort of vicious circle, the therapist, if he 'believes that the only existing therapeutic solutions are those aimed at changing the person', runs the risk of being 'quite naturally led to exaggerate the importance of dispositional factors in the etiology of his patient's behavioural problems'. We shall come back later to the reasons reinforcing the tendency to emphasize these factors.

The main objective of psychopedagogic approaches may consist of discovering in each person the dispositions and talents with which his genes have endowed him, and to do everything possible so that these talents can be fully developed and expressed, so that everyone occupies within the group the place for which 'Nature' has predestined him. If, on the other hand, the role played by 'social representations' and 'sense of identity' in the structuring of behaviour is emphasized, one will attempt rather to help each person to develop the awareness he has of himself, his knowledge of the world, and a critical reflection on his own relationship with the world. This aim implies the use of an essential tool, namely a nuanced and authentic language.

One concrete illustration among others: criminology

In the domain of criminology, the attribution of responsibilities will obviously not be carried out in the same way whether stress is laid on the criminal's personality or on his actions, more precisely on the situations in which he 'took action'. If such concepts as 'born criminal', 'cruel, cold psychopath' or 'fou moral' are made use of,[1] the criminal bears the responsibility for his actions, with possibly some diminution of his moral and criminal responsibility on the grounds of pathological traits in his personality. If one believes that a 'chromosome for crime' exists in those people, one is bound to speculate on the chances of genetic engineering being able one day to rid us of this error of nature. But where importance is attached to situational factors, to the action of the environment, one will instead try to involve a welfare service to change the relations between the person and certain situations, or even make a bold attempt to change some aspects of social organization, relying more on social than on genetic engineering.

The reasons for a choice . . .

For each of the alternatives contemplated, the terms have been presented quite deliberately in a highly contrasted manner. It goes without saying that positions normally adopted have a more finely differentiated character, though still with a perceptible tendency towards one or other of the extreme conceptions. And it is not without interest to ponder the reasons that may affect the choice made. In this regard, Leyens (1982) stresses the paradoxical attitude of a number of psychologists who, true to their convictions, claim to be 'agents of social change' and who use as working tools classification systems, constructions of organized data for decoding the person, which lead them to consider the personality in isolation, ignoring the problems of society and so becoming 'the unconscious apostles of the status quo'. There are, says Leyens, two main reasons for this. In the first place, the use of stable and significant classification systems, and easily accessible tables of organized data for decoding the personality and behaviour of others and of ourselves, is a quick and efficient approach. In the second place, the psychologist's context (a fixed location) and working method (a standard interview) often lead him to look on his clients as 'asitituational beings', i.e. as 'immutable personality traits'. It may be added here that certain errors of perspective in behavioural neurobiology also spring from the fact that attempts made to analyse the effects on animal behaviour of some brain operation ignore the fact that the animals are being studied in a situation which, devoid of meaning for them, does not bring out inter-individual differences resulting from factors in their lived experience.

But it has to be said that the tendency to create categories and classify our fellow creatures is a much more profound and widespread one concerning us all. Each person tries to situate himself in relation to others so as to form a 'social identity' and have it recognized. Now, since it is difficult to apprehend others in all the richness of their diversity and the mystery of their singularity simultaneously, there is a great tempta-

---

[1] We shall return to these concepts in Chapters 7 and 9.

tion to standardize by 'pigeon-holing' them in some broad category (sex, race, social class, character type, etc.), not to mention the distinction made between the 'good' and the 'bad'. This is well expressed by Israel (1984) when he writes that 'the same concern with standardisation has haunted taxonomists (classification experts) since the Age of Enlightenment, an age in which all those whose behaviour injured the harmony of civilisation were locked up'.

. . . and its inanity

Now that we have briefly sketched these contrasting ideas, it is important to stress as clearly as possible that there is no good reason to make a choice between a predominance of either dispositional factors or situational ones. Quite the contrary, these two categories of factors interact in such a closely coupled and complex way that it is completely artificial, and therefore useless, to try to decide globally and generally the weighting due to each one.

In this domain, the attempt to achieve a synthesis does not just reflect one's possible preoccupation with a worthy 'ecumenism': it alone is capable of apprehending reality without unduly distorting it.

The human brain's role as 'mediator'

The neurobiologist, for his part, considers that, on the evidence, it is in the brain that the individual, the person, and the social being meet and carry on their mutual dialogue. The cognitive structures with which a person is endowed and which apprehend the situations encountered by the social being are intimately linked to the brain and cannot therefore be independent of the way in which this organ of the biological individual functions. Furthermore, the dialogue carried on by the social being with his environment can only be well adapted and effective in satisfying the needs of the individual and the desires of the person to the extent to which his brain is able to gather and exploit the fruits of experience. But, in return, the brain is ceaselessly changing by virtue of the very fact that it carries out this 'mediating' role in a ceaselessly changing dialogue. The neurobiologist is thus led, in his attempt at a synthesis, to view the human brain as a place of convergence, interaction, and reciprocal structuring of biological, psychological, and sociological systems.

A working hypothesis: the notions of interaction and reciprocal structuring

These interactions and this reciprocal structuring do not compel our recognition as pieces of primary evidence. They are, rather, part of a *working hypothesis* which we must therefore verify by putting it to the test of factuality. The processes we have in mind necessarily imply the intervention of a certain dynamic which can be described only by apprehending it over time. An important *moment* in this time — for each birth of a 'trinity' of individual, person, and social being — is quite obviously the constitution, at the moment of fertilization, of the gene pool bearing the necessary though insufficient information for the development of a human brain worthy of the name. But the genetic reproduction of brain structures and functions is only intended to ensure — in a rigid, but therefore reliable way — that an individual shall belong to a species, by endowing him with potentialities corresponding to the means of action appropriate to that species. We therefore have to go 'upstream' in this genetic reproduction, i.e. all along the phylogenic history of the human species, and 'downstream', i.e. all the way along the ontogenesis of the individual, to investigate the existence of a dynamic for auto-organization and autoregulation and, where appropriate, to describe its essential traits.

Two closely complementary questions arise in relation to the human brain's evolutionary and individual history (its phylogenesis and ontogenesis). Have the changing constraints of the dialogue with the environment, in the course of the biological history of the species, caused adaptive changes in the brain? Is the 'plasticity' of the human brain such as to permit it to adapt itself, in the course of ontogenesis, to social interactions that have been so singularly enriched by the fact that a cultural history has been grafted on to the biological history of the species? These two questions implicitly contain a third: is there, between the animal brain and the human one, a distinct qualitative discontinuity? And, if so, in what does it consist? When we have some elements of a reply to these questions we shall be able to return to the working hypothesis stated above.

Phylogenesis of the human brain

A respect for chronological order is not the only reason leading us to consider the phylogenesis of the human brain before the ontogenesis of the individual brain. In fact, the degree of 'plasticity' with which the brain and the behaviour of a given species are endowed is a function of the evolutionary level, the level of complexity and organization, reached by the species in the course of its phylogenic development. The more complex a system becomes, the more will the organizing influence of history prevail over the internal constraints of the system. It is not, therefore, just in terms of time, but also in terms of the impact of the organizing contingencies that come to be included, that ontogenesis starts where phylogenesis (provisionally) ends.

Since the brain is the 'mediating' organ in the dialogue that the living being carries on with its environment, it is in the context of this dialogue that its evolution is best appreciated. We shall therefore consider the nature of the pertinent information which the brain receives and processes, as well as the nature of the responses it programmes and causes to be carried out by its executive agents, the muscles. But, above all, we shall consider the processes of integration which are interposed between the brain's 'inputs' and 'outputs'.

And we shall be led to put the emphasis not on the way that behavioural responses (the 'outputs') are programmed but on the kind of elaboration that the sensory data (the 'inputs') are subjected to, and the kind of reference system that is used and that permits all this information to be given its full meaning.

Illustration of our approach by an example: postural muscular activity

The pre-eminence just stated for the treatment and evaluation of information received justifies our pausing here for a moment. It would be stating the obvious to say that an example of behaviour cannot always be defined, either in its determinants or in its aim, by the kinds of muscle or type of muscular activity involved. Let us take a very simple example, what is known as postural muscular activity (because it establishes or re-establishes a certain 'posture', i.e. a certain position in space or a particular manner of holding one's body). Sets of muscular contractions that are very similar may none the less have very different significance. Postural activity may be aimed at preserving or restoring equilibrium; this is reflex activity not requiring the use of a system of reference formed and updated by real-life acquired experience. Or it may be that of an athlete taking up a throwing position: the attention he must devote to regular training clearly shows that, in this case, references to a 'body scheme' and to complex sensorimotor representations — which have to be created and maintained — play an important role. It could, again, be the postural activity of someone who has just been told to 'stand up straight' and is therefore 'correcting' his attitude; this correction is not made in a purely reflex way, nor is it done by simple reference to the body scheme, but rather by simultaneous reference to the 'body image' and to social representations linked to a particular sociocultural context. So it becomes obvious that, in the cases mentioned, the information at the origin of the muscular activity undergoes ever more elaborate processing which brings in, on the part of the brain, ever 'higher' levels of integration and organization.

Levels of integration and organization

The idea of levels of integration and organization, inseparable from a certain evolutionary level, should hold our attention for a moment. It may be illustrated with the help of two examples. As regards the structuring of extra-personal space, both perceptual and motor, two levels of integration can be clearly distinguished. In the mesencephalon, that is, in the upper part of the brain stem, the superior colliculus (anterior corpora quadragemina) is a place of convergence for visual, auditory, and tactile information coming from some region of space, and this sensory topography is matched by a motor topography, i.e. a topography of the control exercised over movements directed towards those same regions of space. This overlap of the sensory and motor representations of space has an obvious adaptive value, and it is not surprising that, in the cat as in the monkey, a unilateral destruction of the superior colliculus leads to the animal's losing interest in half of its extracorporeal space.[2] In the cerebral cortex, a highly complex neural network[3] integrates not only the perceptual and motor characteristics of the different sectors of extra-personal space, but also the distribution in that space of the 'motivational valence' and the corresponding expectation. And, in the monkey as in man, certain lesions in the cerebral cortex give rise to a profound 'neglect' limited to one half of the extracorporeal space. Lastly, in man, and probably only in man, the interacting cerebral hemispheres carry out the integration of iconic representations (in the form of 'images') and logical representations (involving the use of a language) of this same space.

The second example concerns the association of an affective connotation with the objective data of sensory information. Here again we find a completely analogous 'hierarchy' to the one we have just seen. In the mesencephalon, the periaqueductal grey matter (located just below the superior colliculus) intervenes in processes that determine an attitude of 'appetence' or its opposite, 'aversion', towards particular stimuli, without any reference being made at this level to traces left by past experience. On a higher level, cortical structures (particularly the cingulate cortex) and sub-cortical structures (particularly the amygdala) intervene in the genesis of the affective connotations whenever an affective *significance* relating to experience is to be integrated into sensory information. The most elaborate level of integration and organization is employed in the intersubjectivity of man's socio-affective exchanges, in the relations created by cognitions and emotions, 'reason' and 'desire', within the social representations that are his alone.

When studying vertebrate evolution, we find that as it progresses the degree of 'encephalization' becomes greater and greater. This is not simply a question of the degree of development of the encephalic structures as such, but also, and above all, the closer and closer control exercised by these structures on the operations of the spinal chord, i.e. on the spinal reflexes. This growing ascendancy shows clearly in the fact that, after a section isolating the spinal chord from the supraspinal structures, the duration of the 'spinal shock' (characterized by an absence of the spinal reflexes) is all the longer according to the degree to

---

[2] See B. E. Stein 1984.
[3] This neural network will be examined in more detail in Chapter 5.

which encephalization has advanced: some minutes in batrachians, some hours in carnivores, weeks and even months in primates. In mammalian evolution we further observe a progressive 'corticalization' of certain brain functions, i.e. a progressive accentuation of the role played in these functions by the cerebral cortex and the type of elaboration undergone by information at cortical level. Thus it is that, of the two cerebral structures (the cingulate cortex and the amygdala) which, as mentioned above, intervene in the genesis of the affective connotations, the cingulate cortex plays a distinctly more important role in the monkey than in the rat. A lesion or dysfunction at cortical level can then bring about regression, as far as behavioural determinism is concerned, to a less highly developed level of integration and organization.

The notion of a 'triune brain': its use and abuse

A theoretical notion can be fruitful if used with discernment, but it ceases to be so when it is mistakenly taken as sacred. The notion of the progressive superposition of ever more highly developed functional levels has led MacLean to formulate highly interesting ideas. In his conception of man's 'triune brain' (that is, his reptilian, palaeomammalian, and neomammalian brains), Maclean (1977) assimilates the limbic system[4] to the palaeomammalian brain, assigning to it, as its major function, the genesis of the emotions (the 'emotional mind'), and contends that the development of the cognitive functions (the 'rational mind') was made possible by that of the neomammalian brain. But if the distinction between the 'emotional mind' and the 'rational mind' has given rise to investigations that have proved fruitful, it has also led to speculations that look on these two brains as concrete entities, well delimited and to some extent autonomous, carrying on 'conflictual' relations of 'domination' and 'revolt'. We must stress that, in reality, at each evolutionary level the brain constitutes a functional entity endowed with its own dynamic, and is not the result of a simple addition of some 'new acquisition' to an otherwise unchanged brain. As we shall see later, the development of new motor activities in the primates and, above all, in man, was made possible not only by the development of the sensorimotor cortex itself, but also by the establishment — through the intermediary of the pyramidal tract, which makes a direct connection between the cortex and the spinal cord — of a new cortical ascendancy at the very level of the spinal cord's 'motor machinery'. If the sensorimotor cortex and the pyramidal tract thus contribute to the evolution of motor functions, it would nevertheless be incorrect to consider that, because of its phylogenetically older character, the limbic system (palaeomammalian brain) had ceased to evolve and made no contribution to hominization. For instance, while the number of fibres in

---

[4] This system will be studied in Chapter 5.

the pyramidal tract doubles as we move from monkey to man, the number of fibres in the fornix — the main efferent pathway of the hippo-campus (which is an important structure in the limbic system) — is multi-plied fivefold.[5] And Livingston adds that the development of each of these two pathways (pyramidal tract and fornix) is linked with that of 'physical dexterity' and 'social dexterity' (hardly a negligible part of hominization) respectively. It also turns out that other structures in the 'emotional brain' (in particular, the anterior nuclei of the thalamus) are especially well developed in man.[6] Furthermore, it will be seen later[7] that the brain mechanisms underlying the most highly developed cogni-tive activities function normally only if they are activated and modulated by chemical substances released by nerve fibres, and the cell bodies where these originate are located in the brain stem, that is, right inside the 'reptilian' brain.

From the monkey to man: hominization

It is of course the phylogenic period closest to man that interests us most, and that for two complementary reasons. First, because of its very nearness, it is the one we know best, in spite of the numerous questions still awaiting answers. And it is obviously during that period, in the course of which the lineage of the hominids separated from that of the great apes before evolving autonomously, that the problem of continuity and discontinuity is most acutely posed. It is within the framework of that period, therefore, that we should trace the simultaneous evolution of the brain and behaviour, particularly that of the mental faculties whose existence is revealed by observable behaviour.

A very close relation . . .

But it is best first of all to 'put ourselves in the picture' in a two-fold way: to situate man with regard to the great apes, in terms of their respective genetic inheritance, and to place matters with respect to time.[8] The joint efforts of cytogenetics, molecular biology, and immunology have *con-firmed* the data of palaeontology, making clear the extreme closeness of man and the great apes, the chimpanzee being incontestably closer to us than are the gorilla and the orang-utan. Cytogeneticists have studied the respective chromosomes of the chimpanzee and man and have come to the conclusion that these two species have genetic inheritances differing little from each other, and that they could therefore have derived from each other by the sole virtue of only a few chromosomal modifications. The study of proteins (which give an indirect indication of the informa-tion contained in the genes) has revealed that those of the chimpanzee

---

[5] R. B. Livingston 1978.   [6] E. Armstrong 1986.   [7] In Chapter 5.
[8] See R. E. Passingham 1982; Y. Coppens 1983, 1984; M. Blanc 1984.

and of man are 99 per cent identical. If these proteins, the fruit of the decoding of genetic information, are so similar in the two species, it must be accepted that the observed differences in their morphology — particularly the difference in size of their respective brains—are due to modifications that have affected 'control genes' and, therefore, the regulation of certain dynamic aspects of development. Man and chimpanzee are made of the same 'dough', to pick up the expression used by Blanc (1984), but it apparently does not 'rise' in the same way.

... but a consummate separation

Whereas the world's oldest known set of tools, found in Kenya, dates back some 14 million years, the hominid–chimpanzee divergence very probably took place about seven and a half million years ago. Coppens (1983, 1984) places this event in East Africa, correlating it with geological and climatological data. The collapse of the Rift Valley, a great East African geological cleft, is supposed to have cut our ancestral population in two while at the same time altering the precipitation rate with the consequence that, in the east, forest was replaced by savannah. The 'Westerners' of this ancestral population, and their descendants (gorillas and chimpanzees), are supposed to have remained thereafter in a moist, forested environment, whereas the 'Easterners' and their descendants (the hominid lineage) were obliged to adapt themselves to an ever drier, deforested environment. The hominids, first australopithecus and then homo habilis, gradually become erect and their brain evolves. The australopithecines have a brain whose organization is of hominid type, as shown by a mould taken of the inside of their cranium, but its volume is still small.

They are essentially vegetarians and their tools, of wood and shaped bone—tools therefore which they adapt before use—lead one to suppose that they are already leading a relatively complex social life. Homo habilis then appears, 4 to 5 million years ago. His brain size has increased and the brain's blood circulation has developed. His dentition has evolved, he hunts, and his feeding has become largely omnivorous. He has a more erect posture and his biped walk is very close to our own. According to Coppens, language will have developed progressively from the homo habilis stage onward. In fact, on the one hand, a study of the internal surface of his cranium reveals preferential development of the parietal and temporal regions of the brain (which are largely concerned with the processes of sensorimotor integration and of memorization), together with the probable development of Broca's area, whose role is essential in the production of spoken language. At the same time, the anatomy of the base of the skull is modified in such a way as to facilitate speech. On the other hand, a complementary view might be that the progressive diversification of the activities of homo habilis (a diversification suggested by traces identifying areas used for carving meat or cutting wood, and others resembling true habitats) had the effect of promoting the development of ever more complex modes of communication.

From homo erectus to homo sapiens

Approximately two million years ago, homo erectus, taller and heavier and with a more capacious brain, appeared not only in East Africa but also in the Far East. In fact, the remains found in China are older than those found in East Africa, and it may be that men in the habilis form, or a transitional habilis/erectus one, left to explore other areas of the world in search of new hunting territories. Homo erectus improves his habitat and diversifies his techniques while, at the same time, managing the space around him and organizing social life. Evolution then proceeds very gradually towards homo sapiens, so that it is difficult, and a matter of relatively arbitrary decision, to pinpoint the moment at which the latter began his existence (depending on the criteria used, this could be either several hundred thousand years ago or less than one hundred thousand years ago). In Europe this evolution gives rise, some 75 000 to 80 000 years ago, of Neanderthal man (homo sapiens neandertalensis), a subspecies which, over a period of 50 millennia, will develop very rich regional cultures, with the first 'industries', and using rituals that are already complex. This subspecies will at last give way to ours (homo sapiens sapiens), which seems to have migrated from the Near East towards Western Europe.

The two great phases of hominization

This brief overview is enough to show that hominization happened in two phases with a very gradual transition between them.[9] In the first phase, there was biological evolution linked with ecological constraints, so involving evolutionary processes analogous to those which were at work over earlier periods of the evolutionary history of the animal kingdom. Constraining our ancestors to adopt a new mode of life, to invent a new way of handling space, using its resources, and adapting social organization to it, this first phase laid the foundations of a culture whose driving force was to be affirmed more and more throughout the subsequent phase, which continues into our own day. The diversification of activities that took place, and the growing complexity of social life, with its requirement for cooperation and sharing, must have encouraged the development of new means of communication and memorization, going beyond sign language and permitting the transmission of 'lessons' drawn from experience acquired by the group and thought useful for, or even necessary to, its survival. We may suppose that the language initially reflected the concrete phenomena of daily life

---

[9] See V. Reynolds 1980.

and was later enriched with symbols arising from the development of ritual while, at the same time, its essentially 'descriptive' function was supplemented by an 'evaluative' one. This cultural evolution went hand in hand with continuing biological evolution, probably accompanied by complex reciprocal interactions, to such an extent that we cannot disentangle from this co-evolution the nature and direction of the causal relationships. It is none the less possible to emphasize two complementary ideas, namely the very gradual transformation undergone by the genus homo and the very homogeneous character of our present-day human species.

Speaking of the gradual evolution of the genus homo, Coppens (1983) rightly stresses that 'as the particular and essential characteristic of man is his cultural development, it is logical to suppose that it is henceforth the latter which acts on biological evolution'. As for the homogeneous character of our species, different groups of researchers have produced mutually corroborative data clearly showing that the human species is very little differentiated in genetic terms and that the idea of the 'great races' of mankind has no foundation in biology.[10] Furthermore, differentiation between ethnic groups does not seem in any way due to processes of natural selection, but much more to the consequence of migrations and the differentiated cultural evolution of populations which thus became separated from each other. Reynolds (1980), considering recent cultural and biological co-evolution, and simplifying things, distinguishes three great regions of the world (Africa and South America, Asia, Europe and the Near East). He stresses the fact that aggressive behaviour has undergone highly specialized development within the context of the sociocultural systems characterizing the third of these regions. We shall come back to this later.[11]

Specificity of the human brain

Since our knowledge does not allow us to put together a serious picture of the simultaneous evolution of the brain and behaviour in terms of precise causal relations, we are led to assemble, so as to compare them, the major aspects characterizing respectively the development of the brain and that of behaviour, together with the faculties whose existence is revealed by behaviour. Looking at the brain, one should first stress that its absolute weight is not our focus of interest, for it varies with the weight of the body. But if the present human brain is compared with that of a hypothetical primate having the same weight as us, our brain turns out to weigh three times as much as his. And if the brain volume is compared to the volume of the spinal bulb, or medulla oblongata (which contains all the nerve fibres making connections in both directions between the spinal chord and the brain, the brain-to-bulb ratio thus

[10] See M. Blanc 1984.   [11] See Chapter 7.

giving a good index of the mass of cerebral tissue in excess of that strictly necessary for analysing 'ascending' sensory messages and generating 'descending' motor messages), it further appears that the difference between man and the chimpanzee is greater than that separating the chimpanzee from an insectivore such as the shrew.[12] Simply from the point of view of volume, there is no doubt that the human brain is like no other. But it has preserved the pattern of the primate brain, and its cortex has the same cellular density and cellular architectonics as would a sub-human primate brain hypothetically increased to the same volume. If, then, there is no radically 'new' brain, the great quantitative increase achieved while following an already relatively ancient pattern has been reflected particularly in the development of the associative areas of the cerebral cortex, permitting the elaboration of new gnoses ('perceptual schemes') and praxes ('motor schemes' or 'kinetic melodies'). The development of the frontal pole of the brain should be picked out for special attention, for this region seems to play an essential role in the operations of anticipatory simulation, with the faculties of attention, concentration, judgement, and initiative that these imply. Lesions in the frontal pole of the human brain alter these faculties and, with them, the individual's autonomy (that is, the ability he normally possesses to explore his environment actively in response to his own motivations, to disengage himself from the primacy of the urges of the moment, and to project himself deliberately into the future). In subjects who had suffered frontal injuries, François Lhermitte (1983) has described 'utilization behaviour', that is, stereotyped behaviour consisting of the utilization of objects that are presented to them; everything seems to show that these subjects could not prevent themselves responding to the stimulus of the presentation of an object and were thus 'forced' to take hold of and use it.

In the control of motor functions, the pyramidal system came to play a greater role. Because of the direct control that it exercises over the motor neurons of the spinal cord, i.e. over the 'spinal keyboard', the pyramidal system can short-circuit the constraints of the programmes that are prewired at brainstem level, and can tell the spinal cord's motor mechanisms about new programmes of action not contained in the primitive, genetically laid-down repertoire. This is particularly useful for the motor functions of the hand, which is also progressively liberated from the constraints imposed by quadruped locomotion: by permitting independent use of the fingers it has led to the full flourishing of manipulative skills.

The brain and language

Given the essential role played in hominization by language, the question arises of how far and in what way the cerebral structures necessary

---

[12] See R. E. Passingham 1982.

for the production and comprehension of language are unique to man. Data furnished by the various methods of investigation applied have shown clearly and convergently that pre-eminent roles are played by Broca's area in the frontal lobe (for language production) and Wernicke's area in the temporal and parietal lobes (for language comprehension); further, the anatomical and functional organization is asymmetrical, with a marked dominance of the left hemisphere in right-handed persons.[13] This functional asymmetry of the two cerebral hemispheres, resulting from the association of certain functions with one side or the other, may be outlined as follows: the left hemisphere (in right-handed persons) is believed to be involved mainly in logical, abstract thought, processing information in an analytical, sequential way, whereas the right hemisphere is concerned with more intuitive and emotionally loaded thought, processing information in a synthetic, simultaneous way.

If we now consider the brain of the chimpanzee and, in particular, the regions of the cerebral cortex corresponding to Wernicke's area in the human brain, the comparison reveals a double convergence: the chimpanzee's brain shows a similar anatomical asymmetry, and the cortical areas involved have the same cell architectonics (only the 'magnopyramidal' area of the planum temporale has yet to be investigated in the chimpanzee: we still do not know whether it exists in that animal). On a more functional level it has been possible to show that, even in the macaque, the TA area of the superior temporal gyrus (which, in man, is part of Wernicke's area) plays an important role in the identification of sounds and the perception and recognition of their sequential order, which means that it is 'equipped with at least some of the necessary mechanisms for comprehension of language'.[14] To the extent that Broca's area seems to play an essential role in the control of the series of movements carried out by the vocal cords and in the adaptation of that series to a more global context, it is interesting to note that the corresponding region of the premotor cortex controls in the monkey, too, the adaptation of a series of motor acts to a given context (here we are no longer dealing with 'language acts' but with hand movements).[15] It is also interesting to note in passing that the one hemisphere, in man, controls both language and the gestures made with the right hand. Study of the coordination between emission and reception of messages exchanged in social interaction showed that primates have a mechanism that—even before the actual emission of a sound signal—inhibits the neurons of the temporal cortex which are normally activated by that signal when it is sent by a fellow-creature or when, having been sent by the subject himself, it is 'played back' to him.[16]

[13] See R. E. Passingham 1982; A. R. Damasio and N. Geschwind 1984.
[14] R. E. Passingham 1982.
[15] U. Halsband and R. Passingham 1982.
[16] D. Ploog 1981.

Given the nature and complexity of the mechanisms brought to light in the brains of sub-human primates, it really seems that what essentially distinguishes the brain of man from any other—apart from its volume, i.e. the number of neural circuits that can be established—is the particularly striking functional dominance of one of the two hemispheres. We may suppose that programming, like execution, gains in rapidity and precision when it does not have to integrate processes going on simultaneously on both sides. But we must add that the dominance of the left hemisphere (in right-handed persons) is not accompanied by the development within it of structures that are missing from the other side. Indeed, when the left hemisphere is injured in a young child, the right hemisphere shows itself perfectly capable of looking after the production and comprehension of language. Even in adults, when injury to the left hemisphere is strictly limited to Broca's area the production of language is recovered remarkably well, after an initial period of aphasia.

Social behaviour of primates: elements for a comparison

The evolution of behaviour and the faculties underlying it lead one also to compare man, not directly with his ancestors (the available data are insufficient), but with primates which, though starting out from our common ancestors, did not follow the path of hominization. And in our field of interest, i.e. aggressive interactions, the emphasis will quite naturally be on the dynamic of socio-affective exchanges, the means of social communication, and the emergence of a culture which, supplementing the transmission of genetic information, provides both a new system for transmitting information and the driving force for a new kind of evolution.

Social communication: the role of sound signals . . .

The social communication of primates is characterized by strongly developed sound signals and visual signals—a development that will be further accentuated in man—at the expense of the olfactory signals which are a more primitive means for recognizing one's fellow-creatures.[17] Where vocal communication is concerned, we must stress its close relations with certain social functions and with the degree of socialization of the species. The range and detail of the signals transmitted go hand in hand with the richness of inter-individual exchanges. This 'vocal behaviour' fulfils various functions: it protects the social unit against predators and keeps neighbouring social units at a distance; by maintaining continual sound contact between group members, it contributes to preserving group cohesion. If the vocal behaviour of a group of

---

[17] See J.-P. Gautier and B. Deputte 1983; J.-J. Petter 1984.

cercopithecinae (which have a strongly hierarchical social structure) is compared with that of a group of chimpanzees (whose social structure is quite loose), it is seen that, vocally, cercopithecinae grow up to be very specialized (thus simplifying the communication system they use) whereas adult chimpanzees continue to produce the full range of calls in the repertoire of their species. Gautier and Deputte conclude from these observations that in the current state of our knowledge 'it seems that hierarchical social systems, in which the roles of individuals are clearly defined, have correspondingly simpler communication systems'. This social determinism is also expressed in the observation that the pitch of the call varies as a function of the animal's age, so allowing its age group to be recognized. It appears that 'an alarm signal sent by a young creature will be taken 'less seriously' than that of an adult, to which its fellow-creatures give more credence'.[18]

. . . and of visual signals

Visual signals, which often reflect the animal's emotions, correspond to a posture adopted by the individual or to the position of the head and the facial expression.[19] In the most fully evolved primates, ever more expressive sign language comes to be associated with their postures and calls. The significance of this sign language can be interpreted only if one knows the history of the group and the way in which inter-individual relations became established. The most evolved monkeys also use gestures that have precise meanings, and the chimpanzee has proved capable of learning and using a true gestural language, similar to that used by deaf-mutes. The famous female chimpanzee, Washoe and, later, the male, Nim, were thus able to acquire a 'vocabulary' of some 150 to 200 signs which they learned spontaneously to combine into brief sequences.[20] It is possible, as has been suggested, that the first hominids communicated with the aid of gestures, and that spoken language developed later from this mode of expression. Given that certain lesions in the left hemisphere disturb both language production (causing aphasia) and gesticulation (causing ideomotor apraxia), it has been supposed that a close relationship exists between the production of a word and of a gesture, and the nature of the concept expressed by each. But detailed analysis of difficulties in gestural communication, in subjects who had suffered localized brain injuries, shows the multifarious nature of the levels of information processing involved and, consequently, the complexity of the relationships (for verbal as for non-verbal communication) between conceptualization, motor representations, and perceptual representations.[21]

[18] J.-P. Gautier and B. Deputte 1983.
[19] See J.-J. Petter 1984.
[20] See R. E. Passingham 1982; J.-J. Petter 1984.
[21] P. Feyereisen and X. Seron 1984.

Capacities for learning and symbolization . . .

Gautier and Deputte (1983) conclude their paper with the finding that 'if it is certain that monkeys do not "talk", they have none the less many things to say to each other'. In such a case should we not ask, if they have so many things to say, why they do not talk? We have already seen that the chimpanzee is perfectly capable of learning and using a gestural language. But its cognitive and 'linguistic' abilities go further yet, as is shown by corroboratory experimental data, even if the interpretation of some of the data remains controversial.[22] The chimpanzee has shown itself able to read and write, using 'lexigrams' which represent, not letters, but words (nouns and verbs). These lexigrams are plastic shapes that in no way resemble the things they represent; in some experiments they are cast on the keys of a keyboard which the animal learns to use. This is why the chimpanzee Elisabeth presents an apple to her 'teacher', Amy, when the latter displays the following four-symbol sequence: Elisabeth—give—apple—Amy. The chimpanzee Lana learned to use the keyboard to ask for food or drink, for a window to be opened, or, again, for her 'teacher' to come and keep her company; and, in order to do this, she correctly used the rules of an elementary syntax. Given that the chimpanzee is highly capable of learning in a mechanical way, one does ask, of course, whether the animal really was capable of understanding the significance of a symbol, i.e. the nature of the exact relationship existing between a symbol and its referent (the object it refers to). After having learned to sort food and tools and to make two distinct heaps of these two *categories* of objects, chimpanzees succeeded in doing the same thing not only with photographs but also with symbols representing all these objects. They must therefore be able to refer to an internal representation of the real object when they are considering the significance of a symbol and the category to which the object indicated by that symbol belongs. When a symbol causes the evocation of an internal representation of the object it indicates, it is no longer a case of simple associative learning carried out in a mechanical way. Furthermore, experiments on 'transfer' (the use in a new context of a rule learned in another one) do seem to show that the chimpanzee is capable of realizing that the serial order of the symbols is itself a bearer of meaning, so that, again, it is not using elementary syntax in a purely mechanical way.

. . . or the illustration of a dilemma

Further experiments showed that chimpanzees could use symbols not only as naming tools (showing the 'banana' symbol in order to get a banana), but also as implements allowing a 'reward' other than the one

[22] See R. E. Passingham 1982; C. A. Ristau and D. Robbins 1982; J.-J. Petter 1984; A. Vloebergh 1984.

indicated by the symbol to be obtained (when one presents it with a banana it shows the 'banana' symbol in order to receive a biscuit or warm congratulations), or again as instruments serving to state the kind of food the animal intends to fetch from the neighbouring room (and which it will be allowed to eat only if it corresponds to the displayed 'intention'). It is therefore difficult to deny the chimpanzee all capacity for symbolization. But why, then, does it not talk? First, the fact must be emphasized that, contrary to what occurs in the young child, a chimpanzee does not spontaneously acquire the 'linguistic' tools that he is actually capable of acquiring and using. It is not enough to expose him repeatedly to symbols for him then to start using them spontaneously: quite the reverse, they have to be 'inculcated' into him with the application of much insistence and patience. The chimpanzee therefore has abilities which, in natural conditions, remain 'latent', i.e. they are not expressed in daily life.

If the constraints of the physical environment and social life did not bring out its abilities more clearly into the open in daily life, we may suppose that neither were they such as to promote the development of spoken language and its corresponding cerebral structures. This way of looking at things leads to a dilemma: on the one hand, some specialists in language acquisition consider that a syntax can be acquired only if the brain contains an innate mechanism specialized for the purpose.[23] But, on the other hand, if the chimpanzee really is capable of acquiring and using an elementary syntax, how can his brain have been endowed with a mechanism he does not use in natural conditions, and which therefore could not give him any advantage that might be 'selected for' by natural selection? We can add that as far as the emergence of spoken language in man is concerned, we have no knowledge of the point during phylogenesis at which it can have taken place or what the volume of the brain was at that point. As for the precise circumstances in which it emerged, we are reduced to speculation. What seems certain, on the other hand, is that the inherent constraints of a new mode of social organization (cooperation, sharing, rudimentary forms of 'gossip' and 'accounting') and the advantages conferred in facing these constraints by the possession of a spoken and written language, played an essential role. From that moment on, the processes of biological evolution and those of cultural evolution have made up an indissociable tangle.

If the role played by language in the development of any culture is obviously an essential one, we should none the less make it clear that a transmission from generation to generation of certain cultural 'traditions' does already exist in the monkey. Social learning through observation and imitation is distinctly more developed in monkeys, especially the great apes, than in less evolved mammals; and numerous observations have shown the transmission within a given group of food prefer-

---

[23] N. Chomsky 1980.

ences or the techniques used for various purposes.[24] The most convincing cases are those in which one is able to observe the moment at which 'invention' occurs—whether fortuitous or prompted by the experimenter—and gives birth to a 'tradition' that is passed to subsequent generations. On the other hand, if a mother ape sometimes appears to be encouraging her little one to observe and imitate, nothing justifies the assertion that she is really 'teaching' it. This is where language comes in in human mother–child relationships, and we see once more the essential role played by this means of communication in the transmission of any culture.

The various forms of 'intelligence'

The acquisition and use of language are not linked only to the existence of linguistic faculties proper but also to that of a set of faculties normally grouped under the name of 'intelligence' (a rather vague notion, difficult to define other than operationally). Apart from practical intelligence, expressed in the use of particular objects as tools, in modifications made to those tools to make them more efficient, and in the invention of particular strategies for using them, the apes give evidence of a highly developed 'social intelligence'.

They know how to put together information about inter-individual relationships within the group, about the 'balance of power' obtaining at a given moment, and about the moods and intentions of the group's various members.[25] They can, furthermore, use the knowledge thus acquired to manipulate for their own benefit the group's internal dynamics, to such an extent that F. de Waal (1982) published his observations under the title *Chimpanzee politics*. Confronted by much more abstract data, the chimpanzee very clearly shows that he can also 'reason'. In fact, he proves capable of proceeding by both induction (analogical reasoning: for example, completing a relationship to make it equivalent to another one) and deduction (for example, a transitive inference of the type 'A larger than B, B larger than C, therefore A larger than C').[26]

If we consider all the cognitive faculties that have successfully been detected in the chimpanzee, there is no doubt that the differences separating it from man are more of *degree* than of kind (the thing being compared is, in each case, 'reasoning intelligence'; fortunately, man's interior life is not limited solely to this). Again, then, we can underline *the* discontinuity that distinguishes man from the rest of the animal world: spoken and written language, a human property, by the richness and diversity of what it allows to be expressed and transmitted, contributes greatly to structuring the person and the social being throughout their common ontogenesis.

---

[24] See R. E. Passingham 1982.   [25] See W. A. Mason 1982.   [26] See D. J. Gillan 1982.

Ontogenesis of the human brain

Every human brain develops on the basis of the information furnished by a set of genes contained in the fertilized egg. This set of genes is the fruit of the evolutionary history of the species, throughout which nature—confronted by the changing constraints of interaction with various ecological niches—has been 'tinkering', slowly reorganizing its work, 'retouching it ceaselessly . . . seizing every opportunity to adjust, transform, create'.[27] Genetic information is only *one* of the sources of the specificity of the individual brain which will develop from the fertilized egg and which will be unlike any other. For what the genetic programme puts into place are only 'reception structures' and it is the many structuring influences of experience, i.e. interactions with the environment, that will cause an original personality to arise in the context and within the limits of the 'play of possibilities' (to pick up the expression used by François Jacob (1981)). The handiwork of phylogenesis is thus followed by that of ontogenesis, with the big difference that from this point onward it is man himself who—in large measure—plays the role of handyman (with the opportunities and the risks that this entails!). If it is obvious, as we shall see in later chapters, that certain aspects of behaviour—normal or pathological—can in part be ascribed to genetic factors, two facts should be emphasized: first, many intermediate events separate the direct biochemical effect of a gene from the impact that it may have on behaviour; and, second, any given aspect of behaviour is modulated by the joint effects of several genes.[28] And we should emphasize even more clearly that, because of the peculiar characteristics of behaviour (it operates principally in relational life; it involves the individual in his totality; it provides a regulating function), the joint development of the brain and of behaviour results from the play of complex and indissociable interactions between genetic factors and factors of experience acquired in dialogue with the environment.[29]

Brain and behaviour: a shared plasticity

If the brain and behaviour are equally 'receptive' to the structuring influences of environment and experience, this is because each is blessed with great 'plasticity'. And it is on all organizational levels that the establishment of structures and functions may be modulated in this way. This plasticity is at its peak during the early phases of ontogenesis, but it persists to some extent throughout life.

[27] F. Jacob 1981.
[28] See J. Médioni and G. Vaysse 1984.
[29] See G. Vaysse and J. Médioni 1982.

The example of the dopaminergic nerve cells . . .

At cell level one can influence the choices made by developing nerve cells as to the kind of chemical neurotransmitter they will produce and the type of synaptic contacts (those with other nerve cells) they will grow, by appropriately manipulating their environment.[30] Even when these choices have been made and the circuits have been established, plasticity persists. As an example we shall quote the nigrostriatal dopaminergic pathway,[31] which plays an important role in the control of motor functions. In the rat, the operation of this pathway — which exists in each cerebral hemisphere — is asymmetric, and a correlation is observed between the side of the brain on which the release of dopamine is more marked and the side towards which the animal spontaneously turns when moving freely (the 'preferred' side is the opposite to that on which the dopamine is more markedly released). Now, the 'normal' preference of an animal can be inverted, simultaneously with an inversion of the dopamine-release asymmetry, if one teaches it that by turning the other way round from its preferred direction it will be rewarded.[32] Following a unilateral destruction of the substantia nigra, which causes 'circling' (the animal circles around continuously, seriously disturbing any directed behaviour), it is noted that this abnormal motor behaviour disappears in the space of a week as cross-connections develop (the corpus striatum and the thalamus located on the side on which the substantia nigra was destroyed now receive projections from the still-intact substantia nigra on the other side). If, on the other hand, the animal is prevented (by being suspended in a hammock) from learning progressively to counter the motor asymmetry induced by the unilateral destruction of the substantia nigra, on emerging from the hammock it will not show the re-establishment of normal motor behaviour, nor will cross-connections be found to have developed.[33] And, following damage to the nigrostriatal dopaminergic pathway, an implant of fetal substantia nigra — whose dopaminergic neurons will reinnervate the corpus striatum — has the effect of facilitating the restoration of most of the faculties disturbed by the lesion, including the correction of postural and locomotor asymmetries.[34] There is no doubt that the way in which the functioning of a given area of the brain is prompted has a *structuring influence* on that region. For instance, Spinelli and Jensen (1979) observed the *massive neural traces* left in the only regions of the cerebral cortex concerned by simply submitting a young

---

[30] See P. H. Patterson 1978.
[31] This neural pathway links the substantia nigra to the corpus striatum; the fibres making it up release dopamine as a neurotransmitter at the level of the corpus striatum.
[32] B. K. Yamamoto *et al.* 1982.
[33] M. Pritzel and J. P. Huston 1983; J. P. Huston *et al.* 1985.
[34] See A. Björklund and U. Stenevi 1984.

cat to avoidance conditioning involving repetitive processing of visual and tactile information. A similar structuring influence is observed even in the adult mammal: the section of a cutaneous nerve brings about a profound reorganization of the region of the cerebral cortex on to which messages relating to cutaneous sensitivity are projected—suggesting that, under normal conditions, its organization undergoes constant rearrangement as a function of the way in which it is activated.[35]

. . . and that of the visual system

A great number of very elaborate studies have been—and will continue to be—devoted to the development of the visual system and to the role played by the visual environmental conditions in determining this development.[36] Elementary functional properties of the visual system, such as binocular vision or the ability to distinguish the orientation of a light signal in space (properties which can be discerned by recording the bioelectrical responses of the neurons of area 17 of the visual cortex), develop normally only if the young organism is able to explore actively a normally structured visual environment and its visual cortex thus receives not only structured messages coming from the retina but also 'proprioceptive' messages coming from the muscles responsible for eye movements.[37] In other words, the very fact of exploring visual space provides the visual system with the coherent set of information it needs so that its own functional maturation proceeds appropriately and so that the organism can thus behave, in regard to that environment, in a well adapted and efficient way.

The plasticity of social behaviour

Plasticity is equally characteristic of the development of behaviour, in particular that of social behaviour. Mason (1979) thus sees the developing individual as 'an ongoing enterprise . . . in continuous commerce with its surround and undergoing continuous change as a consequence of these transactions'. It is by interaction with its fellow-creatures that the young monkey becomes familiar with the members of the group, forms an appreciation of the strength and temperament of each of them, and positions itself in relation to the others within the group. It is not surprising, therefore, that an animal brought up in total isolation over several months beginning from birth should show itself thereafter as an 'emotional cripple': it is withdrawn, easily terrified, and apparently incapable of establishing social contacts.[38] It goes without saying that, in the socialization of the young child, a code of reference is progressively

---

[35] See J. H. Kaas *et al.* 1983.
[36] See J. Stone *et al.* 1984.
[37] See Y. Frégnac and M. Imbert 1984.
[38] See R. A. Hinde 1982; R. E. Passingham 1982.

put together which both models and is modelled by interactions with the social environment. In a study of 107 abandoned children, most of whom had been adopted and brought up by adoptive parents of varied cultural level, the investigation (made 14 or 15 years after the abandonment) clearly demonstrated two essential facts: unadopted children, who had been 'placed' with many different families or had spent long periods in institutions, were severely deficient in both intellect and personality; the sociocultural milieu of adoption has a noticeable effect on the frequency and nature of backwardness at school, and also on the frequency and nature of difficulties in social behaviour.[39] In a study of a different kind, S. Ekblad (1984) showed by comparing groups of 300 Chinese and 300 Swedish children that certain characteristics peculiar to the family environment had a strong influence on the development of the dominant attitudes observed in the two groups.

Self-awareness: contribution to a re-definition

In the socialization process, important and complementary roles are played by the development of self-awareness, the establishment of emotional ties, and the acquisition and use of language. Given its essentially subjective character, the notion of self-awareness is difficult to tackle and define directly and immediately. But if this mental state is approached by considering the way in which it provides for mediation between the brain's 'inputs' and 'outputs', two observations stand out. On the one hand, observable behaviour is the expression of a set of relations with the world, formed progressively by experience. The richness of these relationships and of their mental representations is a function of the richness of the communication established between the individual and his physical and social environment. Self-awareness thus corresponds to the ability to create internal representations of one's relations with the world, to enrich them by reflective thought and by communication with the environment, and to express them through language and non-verbal behaviour. On the other hand, the neuropsychological data show that this mental state cannot be located in a given area of the brain, for it seems to include various facets each of which could correspond to the emergence of peculiar dynamic properties coming to master more elementary mechanisms within a relatively dispersed neural network (special 'configurations' of the interactive dynamic, characterizing the functioning of this network?). This is how two different 'personality styles', reflected particularly in two different modes of cognitive apprehension of the world (the 'compulsive' style of sequential and additive mode, or the 'hysterical' style of combinatory and multiplicative mode), seem to coexist and alternate in the same brain and correspond to distinct dynamic states of that brain;

---

[39] M. Duyme 1981.

these two states, with their alternation and a possible predominance of one over the other, can be objectivized and distinguished by recording the bioelectrical responses of the cerebral cortex, or on the basis of their differential sensitivity to particular psychoactive substances.[40] On the other hand, several lesions that affect different areas of the cerebral cortex and are reflected in (at least partly) different syndromes, have the common effect of causing a lack of interest ('neglect') for one half of the body surface and the adjacent half of the extrapersonal space, the subject being capable of going as far as denying their existence.[41] Finally, it should be pointed out that the memory, which is largely concerned in the genesis and use of internal codes of reference, is in no way 'monolithic': on the contrary, the different attributes of one and the same 'memory' — of one and the same 'mnesic trace' — are processed by different neural circuits,[42] and their processing involves a whole series of brain structures. It thus appears that the awareness we have of ourselves and of our relations with the world presupposes the appropriate and highly integrated operation of a whole set of cerebral regions, particularly those involved in the processes of perceptual integration and motor integration, forward-looking simulation, memorization, and symbolization. It should not surprise us, then, that this awareness should be intimately linked with communication and action and the knowledge that they generate (which comes to enrich one's mental representations), the faculty of reshaping the representations through reflection, and the ability to refer to them constantly in one's dialogue with the environment. But the content and the objectives of this dialogue are not just *cognitive* in nature but also — and very fortunately — *affective*.

Structuring role of affective ties

The establishment of affective ties, especially in the mother–child relation, plays a major role in modelling social behaviour. We shall see later[43] that the processes of attachment between individuals, and social cohesion, are linked with the normal operation of systems of neurons involving the 'endogenous morphines'. These systems are activated, in particular, by exchanges, tactile and otherwise, between the mother and the child. In the monkey, as in the man, separation from the mother has similar effects: initially the child 'protests', is agitated and cries; then it shows all the symptoms of 'despair' and 'depression'.[44] In the adult human subject, the breaking of an affective tie also causes a 'state of mourning' that is likely to express itself by all the symptoms of a

---

[40] A. J. Mandell 1983.
[41] See M.-M. Mesulam 1981, and Chapter 5.
[42] See P. Karli 1984, and Chapter 5.
[43] In Chapter 5, Role played by humoral factors.
[44] See R. A. Hinde 1982; R. E. Passingham 1982.

'depressive state'.[45] Since man frequently claims—or aspires—to be 'purely rational', we should emphasize that cognition and affectivity, knowledge and emotion, are intimately interwoven with each other. In the elaboration of knowledge about intersubjective relations (the knowledge at the base of 'social representations') the cognitive processes integrate (in the *content* that they apprehend and structure) affective, irrational aspects together with strictly cognitive, rational ones.[46] Conversely, the experience of an emotion remains rarely raw, 'pre-reflective' experience; quite the reverse, it is organized and takes on its full meaning (on the basis of a set of 'social norms') through reflective thought.[47] And thus we return to the role played by language.

Language, an essential factor in development

Debate continues on the nature of the processes involved in the development of language. For some,[48] the brain is innately endowed with a sort of 'universal grammar' in the form of structured sets of representations and formal rules, which may be considered true 'mental organs'. Here, Chomsky makes a distinction in language acquisition between knowledge of the language as such and knowledge of how to use it. Viewing the development of cognitive activity in the very general context of the notion of 'vital adaptation', Piaget (1974) considers that the 'instruments for cognitive assimilation' necessary for the 'reading' of this or that experience evolve progressively by endogenous reconstruction in order to adapt themselves to new situations; in this view, the development of language is difficult to dissociate from the use made of it.

When one studies the child's acquisition of mental representations of the semantic content ('meaning') of particular verbs of action, it is clearly seen that these semantic representations depend not only on the level reached by the development of his cognitive faculties, but also, very much, on the child's experience, on the 'knowledge of the world' he has gained from his own perceptuomotor experiences.[49] Searle (1972) rightly states that 'a theory of language is part of a theory of action'. In 'language acts', indeed, there is not solely a transmission of some word or phrase, but also an intention to act on someone, foreseeing the possible repercussions of this intentional act. Language, initially linked with concrete experiences and their immediate context, later undergoes a double evolution. On the one hand, it effects a certain 'distancing' with regard to concrete experience, so opening the door for processes of generalization and abstraction. On the other hand, it develops a more 'committed' use of particular words, in the sense that this use follows the constituent rules of a given society, and that language thus can serve not only to describe facts but also to formulate moral judgements.

---

[45] See D. Widlöcher 1983*a*, 1983*b*.   [46] See S. Moscovici 1982.
[47] See J. R. Averill 1982.   [48] N. Chomsky 1980.   [49] J. Bernicot 1981.

Whatever the *theoretical* disputes as to language and the way it develops, a distinct convergence can be noted in the *practices* intended to ensure, in particular by the use of language, the development and education of the child.[50] The requirement is to kindle and refine the child's awareness of himself, his person, and his action on others; to increase his capacity to apprehend immediate problems in the light of past experience and to formulate plans. Helping the child more and more to 'interiorize' — i.e. more and more to integrate into his mental representations — both the actions he can take in the physical world and the values and prospects offered to him by the social world, one helps him to stand aloof from immediate experience and thus free himself from its ascendancy.

Back to our initial hypothesis

This overview of the phylogenesis and ontogenesis of the brain and behaviour brings out vividly a fundamental idea: for a living being, the brain constitutes the appropriate instrument that permits him to carry on a dialogue with his environment, because it is precisely this dialogue that gives concrete content to the 'reception structures' provided by genetic inheritance. And it is in fact within the brain that interaction and reciprocal structuring takes place between a number of genuinely bio-logical factors and processes, processes linked with the configuration and dynamics of some 'psychological structures', and some determinants of a sociological nature which emerge from the confrontation between the data of immediate experience and a set of 'social representations'. Due to their complexity and diversity, these interactions hardly allow themselves to be apprehended and described in a general and global way; we should attempt to disentangle the skein case by case, in con-crete, well-defined situations, before going on to make, in the form of hypotheses, some generalizations that will run the risk of being excess-ive. None the less one very general idea flows from the mere existence of these interactions: if one (deliberately or otherwise) abstracts whole chunks from a highly complex reality, some speculations about the remaining chunk may be only brilliant ideas, not devoid of charm, whose gratuitous character consists in the fact that they proceed *in vacuo*. Let us take the concrete example of some speculations tending to liken the brain to a computer. Fodor (1980) considers that, because of purely formal properties he finds it necessary to attribute to them, mental processes cannot have access to the semantic content of repre-sentations formed in a dialogue with the environment; in these circum-stances, for the thinking subject, no other reality than himself can exist. Pylyshyn (1980), for his part, considers that some predetermined facul-ties of thought result from a stable 'functional architecture' provided by

---

[50] See D. J. Wood 1980.

the 'hardware' of the neural circuits, totally neglecting the fact that, within the brain, every instance of information reception, followed by processing and storage, sets in motion a reorganization on the anatomical, biochemical, and functional planes.

Fruit of phylogenesis and ontogenesis: freedom

It is appropriate to consider at the close of this chapter the notion of 'freedom', from which that of 'responsibility' logically follows. We are not called on to 'explain' this freedom but we should take note of the fact that the phylogenesis and ontogenesis of man create the conditions necessary for its emergence. Both phylogenesis and ontogenesis are characterized by a double distancing, in space and in time. In mammalian evolutionary history, immediate contact with the environment through the senses of touch and smell gave way progressively to detection and recognition at a distance, by means of sight and hearing, of more highly developed social signals. In the course of ontogenesis, the infant makes extensive use of the senses of touch and smell (he recognizes his mother's aroma from the first days following birth); in the adult it is the *teledetectors* of the eye and the inner ear which pick up the messages whose role is essential in verbal and non-verbal communications. A 'private' space is thus created between the individual and his environment, simultaneously distancing the world and providing a privileged area for dialogue to bring the world closer. At the same time, the characteristics of this private space are progressively ever more thoroughly interiorized, i.e. integrated into internal representations. Bringing these representations into play leads to a less direct, less immediate recognition and ascendancy of signals and situations; as opposed to this, cognitive operations, which act on the content of these representations, permit the external world to be interpreted, and this interpretation is a source of expectations and plans. For, 'distancing oneself' in the spatial sense is accompanied by a similar effect in the temporal one. The development of powers of memory and the ability to refer to the traces left by lived experience have the effect of placing the individual in a segment of time which is simultaneously rich in its past and pregnant with its future. Owing to the more marked dissociation inside the human brain of the processing that is undergone respectively by the objective parameters of sensory information and the affective connotations associated with that information, man is better placed than an animal to stand back a little from the emotion aroused by some event or evocation.

This double distancing—in space and in time—characterizing the dialogue that the individual carries on with his environment permits the development within the individual himself of another, to some extent 'distanced' dialogue between an 'I' and a 'social Ego'. This distancing is necessary if the I is to be able to reflect on and evaluate the social Ego, the fruit of an interiorization of the role played by each of us in the 'shadow theatre' where the scene is set by a given sociocultural context. It looks as though it is here, in the context of this interactive dynamic, both *inter*subjective and *intra*subjective, that the self-awareness is gradually born that will be operating from then on. Sperry is right to state that self-awareness should be considered a 'causal reality on its own' within the complex determinism of the relations between the brain and behaviour; he adopts a resolutely 'monist' perspective (there is no mental activity independent of all material substrates), while keeping its distance from a purely mechanistic and reductionist conception.[51] It is the very complexity of these interactions, and the emergence of new dynamic properties, that create this share of indeterminacy and un-foreseeability which leaves room for the exercise of freedom.

A free dialogue cannot take place in the private space extending between the biological body, with its elementary needs, and the social body, with its unavoidable constraints, unless the space remains avail-able and open. By this I mean that it must not be 'invaded' by either side. One must, on the one hand, prevent 'the body swelling up until it fills the universe', to use Orwell's words in *1984*. Hunger or sharp pain invades the whole field of awareness, by this means giving the biological constraints an excessive importance and 'presence'. On the other hand, it is just as important that the constraints imposed by the social corpus — and by those who make decisions and take actions in its name — should not be too onerous, too confining, too intrusive, to the point of abolish-ing this private space that every individual should be able to enjoy.

If a certain freedom in one's dialogue with the environment is a necessary condition for the free flowering of the inner life, it is still not a sufficient one. First of all, the interior dialogue between the I and the social Ego should be balanced, unlike that between the *pot de terre* and the *pot de fer*; in other words, the I must be able to oppose the demands and temptations of the social Ego with the demands of a personal ethic and a personal discipline. Furthermore, the space separating the two protagonists should also remain available and open, without being 'invaded' by one or other of them. If the social Ego swells up excessively, the I runs a serious risk of being reduced to silence; if the I swells up with a devouring passion, it will either withdraw from the world or, on the other hand, try with the aid of the social Ego to conquer, convert or destroy it.

One very important aspect of these matters remains to be stressed. Creating distances — in the dialogue with one's environment as in the inner life — has meaning only if one always has at one's disposal the instrument allowing them to be crossed. That instrument is language. But, in order to conduct a lucid and enriching dialogue with men and

[51] See R. W. Sperry 1980, 1981.

with things, and above all with oneself, it is essential to possess a language both genuine and rich in nuances. Is there any need to stress that a concern to develop such a language and to use it towards oneself and others is more a matter of a personal ethic than of a high-powered intellect? Furthermore, since our own language cannot avoid being influenced by the surrounding discourse, it is highly desirable that the dominant discourse (in politics, advertising, and the media) should prove its equal concern with truth and shades of meaning.

4   Behaviour and its motivations

An analysis of behaviour aiming at perfect objectivity will naturally tend to focus its attention on the *phenomena* that can be seen and described by an outside observer. But such an approach can obviously register only a partial, even if important, aspect of a much more complex reality. For behaviour can rarely be reduced simply to its observable phenomena. A form of behaviour is not an end in itself: its *raison d'être* is not entirely contained in the behaviour nor exhausted by its execution.

Behaviour is usually an action charged with meaning and also, by virtue of the consequences it produces and their evaluation by the brain, a generator of meaning. All behaviour is in fact deployed in a space and inscribed in a history, both of which are fraught with meaning for the living creature in dialogue with them. For behaviour is not merely action but also event, both the fruit of a history and a generator of history, by contributing to directing its course. And the latter unfolds in a space that is not simply that which could be described by any external observer, but is experienced and shaped by an individual and is his alone. With this in mind it is easy to see that the referents of behaviour — the stimuli evoking it and the objectives it aims for — are constituted not so much by the objects of the surrounding world as such, but rather by the individual's relation to each of these objects. In language, too, which is a form of behaviour, the referent of the signifier is not simply some object but also the speaking subject's relation to that object. It is therefore a complex dynamic interaction — characterized by the forces brought into play and the information whose exchange is mediated — which gives full meaning to the behaviour, the space in which it occurs, and the history in which it comes to be inscribed.

This implies that behaviour and the way it is generated by the brain can be truly understood only if they are seen in the context of this dynamic interaction, in the context of the interactions between an individual rich in experience and an environment that he has progressively shaped and appropriated. From this perspective, behaviour cannot be reduced to a simple expression of cerebral operation, to the outward projection of the modalities of cerebral operation at a given moment. On the contrary, behaviour must be considered an integral part of a process that starts out from the brain and returns there. In fact, behaviour provides the living creature with an essential *function*: not simply the expression of an existing relation but also the creation, preservation, modification, or abolition of a particular relation to the environment. To preserve or modify an individual relation to the environment equates in concrete terms to acting with the purpose of preserving or modifying the way in which some particular aspect of the environment is perceived, evaluated, lived through (together with all the affective experiences connected with it). In other words, then, one essential function of behaviour (an 'output' of the brain) will be to act indirectly on the 'inputs' to that same brain.

The role played by the brain can therefore be fully defined only by starting with this function that behaviour provides in the establishment, expression and evolution of the relations between the individual and his surroundings. The brain assumes a three-fold mediating function in 'administering' these relations, these transactions which lie within time and space:

- faced with promptings from its surroundings, it not only generates objectives and the behaviour intended to realize them, but also records and evaluates the ensuing consequences (costs and benefits), subsequently taking them into account;
- it inscribes this behaviour into its own history, since it is the depository of this history and the place where it undergoes elaboration of some complexity;
- it positions the behaviour within a space of its own, for it carries representations of the familiar physical and social environment, representations which are structured and updated by and for interactions with the environment.

Brain functions and brain mechanisms

Since the brain assumes these mediating functions in managing the dialogue conducted by the individual with his environment, one is led to wonder how these functions are linked with the mechanisms within the brain that are analysed by neurobiology. Given that this kind of mediating function comprises a full set of more elementary functions which form sequences and interact with each other, we are naturally led, in theory and even more so in experimental procedure, to isolate one or other of these more elementary components from the global function. But this step, though necessary and perfectly legitimate, must not be allowed to lead to a misconception of the relationship between mechanisms and functions.

In the case of an elementary component, mechanism and function largely overlap with each other: their relations are both immediate and local. Let us take a very simple example: the flexion and extension movements of the forearm in relation to the upper arm. This displacement function of one part of the arm with respect to another can be described in terms of articular mechanics, of the contraction of an agonist muscle and the relaxation of an antagonist, or, again, of activation of the motor neurons innervating the agonist and inhibition of those innervating the antagonist. There is little likelihood of logical error in stating that it is the activation of the motor neurons of the brachial biceps muscle that explains in essence the flexion of the forearm with respect to the upper arm. There is in fact a linear causality running from the activation of the motor neurons to the contraction of a muscle and the movement of the forearm.

The organizing action of a global function

When elementary functions and the mechanisms effecting them are integrated into a more global function, the latter is not created or explained by their mere addition: on the contrary, it is the global function that 'subordinates' the elementary functions and permits an understanding of how they are formed into chains and the modalities by which they are brought into play. Let us return to the example of flexor and extensor muscles. When certain muscles are paralysed because a peripheral nerve has been cut, a therapeutic 'transposition' may be carried out of other muscles which have retained the ability to contract. After a transposition of this kind a muscle can act differently from normal, even inversely (e.g. a flexor muscle can contribute to an extensor movement). A functional reorganization takes place in which an elementary mechanism (activation of the motor neurons and contraction of the transposed muscle) is subordinated to some new function. Such a reorganization is effected at a 'high' level, for it appears that it requires the involvement of the pyramidal tract, through which the cerebral cortex is able to impose new programmes on the motor machinery of the spinal cord.[1]

The organizing effect of a function to be provided for a living organism is even more clearly apparent in the coordination of elementary functions within an organism which is passing from the sleeping to the waking state and which, therefore, is about to interact with its environment. We observe vasodilation within the muscles, acceleration of the heartbeat, vasoconstriction at skin level, speeding-up of respiratory movements, relaxation of the smooth muscles of the bronchial tubes, and cessation of the motor and secretory activities of the digestive tract. This sequence of physiological changes, the activation of a number of very varied mechanisms, can be understood only when we see that this is all coordinated, integrated, by the 'ergotropic' action (that preparatory for muscular effort) of the orthosympathetic component of the vegetative nervous system. The muscles that are to contract need more nutrients and oxygen, i.e. they need to have a markedly increased flow of blood passed through them, a need responded to by an increase in arterial pressure at the entry to the muscle (consequent on acceleration of the heartbeat and cutaneous vasoconstriction), together with the

---

[1] See J. Paillard 1978.

opening of the arterioles within it. The increase in pulmonary ventilation, facilitated by relaxation of the bronchial musculature, will allow the blood to take up within the lungs an increased quantity of oxygen. Since the organism has at its disposal only a limited volume of blood, it directs less of it to the digestive tract, which, in any case, is in a state of rest. Clearly, the 'ergotropic' global function subordinates a whole series of elementary physiological mechanisms. Furthermore, the reticular activating system (located in the brainstem), which is responsible for cortical and behavioural arousal, will induce a change in the connections of the spinal cord's motor machinery: the motor neurons will become less sensitive to messages originating in the segment(s) (capable of playing an important role in the sleeping animal's defence reactions) and more sensitive to messages coming from the cerebral cortex (which correspond to intentional and directed movements), facilitating the transition from a sort of 'automatic pilot' to voluntary control of the executive agents, the muscles.

The organizing action of the ergotropic function is, of course, genetically preprogrammed and is exerted through the intermediary of 'prewired' circuits. But it was this global function, generating autonomy and efficacy in the dialogue with the environment, that was the driving force behind one particular aspect of evolution, namely the progressive build-up of genetic information that now forms the basis for the positioning of these 'prewired' circuits. Once the subordination of a certain number of elementary functions and mechanisms to a more global function is inscribed in the genome, genetic reproduction faithfully ensures that the same subordination operates in every member of the species concerned.

Let us now consider the brain's mediating functions in managing the individual's relations with his environment. It is obvious that, particularly in man, most of the pertinent information is not inscribed in the genome. It is inscribed in the brain—in the representations formed within it—by experiences that are encountered in dialogue with the environment (but it should be added that this information could be neither retained nor expressed by the brain without the intervention of structures and mechanisms which themselves depend on information of genetic origin). Given that it is through subsequent reference to these internal representations that current sensory information acquires its basic motivating properties and an appropriate behavioural strategy is selected, it appears that these representations, together with the information accumulated and built up within them, are both the fruit of and the driving force behind interactions with the environment. Since the representations are formed within the brain by the very circumstance that it plays the mediating role in these interactions, we can say that it is the brain's mediating function that determines and explains action, particularly the involvement of the cerebral mechanisms underlying observable behaviour.

Reorganization of the injured brain and restoration of its functions

At any time of life, a lesion of the brain causes a reorganization of certain circuits and elementary mechanisms so as to permit the restoration of certain of the functions which were initially to some extent disrupted.[2] It is in this restoration of function that the pre-eminence of the function over the circuits and elementary mechanisms that serve it is most clearly shown. If a function can promote and direct processes of auto-organization (and auto-reorganization) in this way, it is because it is an integral part of a dynamic entity, the brain in action in its role as mediator of the dynamic interaction that characterizes the relations between an individual and his environment. The restoration of a particular function, which contributes to restoring some faculty of interaction with the environment and achieving some objective within it, may operate (schematically speaking) in either of two ways: either the same strategy is produced by a 'vicarious' reorganization of the remainder of the brain, or a different strategy is employed thenceforth in order to achieve the same objective. If the human brain has at its disposal a great potential for auto-organization (and auto-reorganization), this is because it is characterized, above all in the early stages of ontogenesis, by a marked redundancy—non-differentiation—in both structure and operation. Since the restoration of a function *lies within* the (more global) restoration of the interaction dynamics of the dialogue with the environment, it is not surprising that it is affected by the environmental conditions under which the injured brain has to play its mediating role.[3] This point is significant when we are considering what procedures should be carried out for functional re-education following cranial trauma, if we believe that a more global re-education of the person and his interactions with the environment is likely to facilitate the restoration of some more partial function.

The 'specification schedule' of the brain and behaviour

The experimental method of the neurobiologist, which is in essence an analytical one, permits a description of the neuronal circuits and a definition of the elementary mechanisms whose 'substrate' they form. It further attempts to form more complex ensembles from the elements supplied by the analysis, by making an inventory of the various controls exerted on these circuits and elementary mechanisms. This method is quite similar to many others we employ in daily life. Let us take the analogy of a motor vehicle which, in a similar manner to the brain, plays a mediating role in managing some particular relations we have with the

[2] See M. Jeannerod and H. Hécaen 1979; M. W. van Hof and G. Mohn 1981; S. Finger and D. G. Stein 1982; S. Finger and C. R. Almli 1984; B. Will *et al.* 1985.
[3] F. Eclancher and P. Karli 1981; B. Will and F. Eclancher 1984.

space within which we are moving about. We should be quite justified in saying that the combination of a system to generate mechanical energy, and systems for braking and steering, 'explains' the fact that this vehicle lets us move forward, stop, and change direction. But it is equally obvious that this combination did not come about by chance nor, moreover, would just any braking system be associated with just any system for generating mechanical energy. For, at the beginning, there is the 'specification sheet' drawn up by the engineer who designed the vehicle, that is, the set of performances (the global 'behaviour') expected of this means of locomotion. And it is precisely these performances, necessary to the realization of some objectives, which determine a whole series of constraints imposed on the vehicle and 'explain' the selection of the systems to be put together and the exact way in which they are to function and interact. Each mechanism taken in isolation can of course be analysed in terms of linear causality, for there is a direct cause-and-effect relationship, but as soon as the same mechanism is integrated into the dynamic entity of which it is a part, it becomes a 'means' contributing to the attainment of a certain 'end'.

The 'specification schedule' for the brain and the behaviours it makes use of (objectives aimed at, performances required, constraints imposed) has the peculiarity that some of its specifications, after evolving throughout phylogenesis, have been inscribed in the genome, whereas the others will progressively form part of the internal representations built up within the brain. Above all, however, it is interaction with the environment that will both 'reveal' the evolved specifications (by ensuring the complete phenotypic expression of this part of the genome) and provide the main 'source' of the others. In other words, the specifications which structure the dialogue between the individual and his environment are themselves the fruit of that dialogue.

What determines behaviour?

A behaviour can therefore hardly be reduced to a simple effect whose cause is the activation of a given cerebral mechanism. In reality, behaviour is a means of taking action, the agent of a strategy directed towards an objective. The brain makes use of a strategy which, through an object in the environment, is often in fact directed at the brain itself (or, more precisely, the brain's own relation with the object). There are no rigid relationships between strategies and objectives: the pursuit of a given objective does not necessarily imply the implementation of a given strategy. Quite the reverse: one objective may be attained by very diverse pathways, and one strategy may be used—in different contexts—for very diverse ends. Besides, given that many factors are usually involved in the elaboration of an objective and in the choice of the strategy that appears appropriate, it is easy to see that a behaviour can rarely be imputed to one simple cause. Confronted with a given situation, the probability that a given behaviour will be put to use will be determined by the configuration of a 'causal field' of some degree of complexity. In other words, finding the 'motive', the 'reason' for the use of this behaviour boils down to examining the *set of factors* which contribute to the determination of its probability. These factors do not correspond in an exclusive way with cerebral mechanisms, or with promptings emanating from the environment, but with the complex modalities of the processing that these promptings undergo at the hands of those mechanisms. Psychologists, who ponder the way in which the 'motivations' arise that underlie our behaviour, are therefore right to reject both exclusively 'innatist' hypotheses (according to which our behaviour is determined by fundamental needs, the constituents of 'human nature') and exclusively 'situationist' ones (according to which our behaviour is determined by environmental constraints), emphasizing the importance of an 'interactionist' hypothesis.[4] According to the latter, 'motivation arises in the encounter between subject and object, each of which has characteristics interacting with one another', and—in their concrete forms—motivations are therefore 'types of habitual interactions constructed by the individual with his surroundings'. In his work devoted to the *Theory of human motivation* (*Théorie de la motivation humaine*), Nuttin (1980) insists on the fact that 'the individual is defined as a 'subject in a situation' and the environment as a situation constructed by the subject' and that 'rather than being simply interdependent or in interaction, the active subject and the world of action exist only as a function of each other'.

Before looking at the concrete forms of motivations, it would not be without interest to examine briefly the kind of realities brought to light by the behavioural neurobiologist in order to point out both their interest and their partial character. By operating on the brain[5] one can elicit, modify, or eliminate—in perfectly reproducible fashion—this or that type of behaviour. The experimental data thus obtained are of the greatest interest, for they allow us both to achieve a better understanding of the roots of certain behavioural disorders and to operate on behaviour in an efficacious and predictable manner. If a certain reaction can be produced by operating on the brain with an electrical or chemical stimulation, this is because, at that particular moment, the brain contains all the necessary information—and all the mechanisms required to process that information—for the reaction to be initiated and completed. In other words, the observable phenomenon thus triggered can, at least in principle, be described entirely in terms of neuronal activities, and the latter are the immediate cause of the phenomenon as it is observed from

[4] A. Mucchielli 1981.
[5] An electrical or chemical stimulation, a localized destruction or a selective blockage of the transmission of certain messages is carried out in a given area of the brain.

outside. It is in this regard—but only in this regard—that one can say that 'all behaviour mobilises defined sets of nerve cells and [that] it is at their level that the explanation of conduct and behaviour is to be sought'.[6] For our experimental methods carve out—necessarily, but artificially—a 'snapshot' from the (reciprocally structuring) common history of the brain and behaviour. They mobilize, here and now, information and processing mechanisms, without telling us about the origin of the information thus mobilized (since it is not wholly contained within the genome), or about the nature of all that information which—in a brain fully open to a familiar environment and to its own history—contributes to determining the probability that this very behaviour will be triggered.

Exemplary value of an analogy

Let us take an analogy while remaining aware of the limitations inherent in the one we select for consideration. If I place the pick-up of my record-player on a precise point in a recording of Beethoven's Ninth Symphony, I know that I shall hear the sounds of the 'Ode to Joy'. I cause the restitution of a structured set of information which was recorded there earlier (besides, the same information can be recorded on a different medium and with a different coding system). I can describe the sounds of the 'Ode to Joy' in terms of mechanical vibrations and in terms of mechanical-to-electrical and electrical-to-mechanical transduction. And there is no doubt that the mechanical vibrations and variations of electrical potential that I analyse give a complete explanation (i.e. a full account) of what I can hear. But it is just as obvious that the genius of Beethoven, and the expertise of the conductor, the orchestral performers and the choir, have something to do with what I can hear. The disc, by recording vibrations and restoring them, carries out the office of 'mediator' between the performers and myself. When we turn to the brain, which carries out much more complex mediating functions in interactions between the individual and his environment, an analysis of the sole restoration of just one part of the information at a given moment, and of the diverse ways in which this restoration can be manipulated, allow us to apprehend only a part of the reality. It is self-evident that, in a human brain, rich in experiences and engaged in dialogue with its familiar environment, the representations which make up important reference points for behaviour derive neither their origin nor their meaning from the functional properties alone of the neural networks. This is what we shall see in more concrete form by examining the genesis of certain motives for action, or motivations.

[6] J.-P. Changeux 1983.

Some concrete forms of motivations

Motives for action arise out of the fact that, in the dialogue that he carries on with his environment, the living being pursues a certain number of goals of a biological and psychobiological nature. For the pursuit of these ends is in some way marked out by set points and frames of reference. As soon as the brain detects, in the information that it receives from the external world and also from the interior of the organism, a difference measured against a set point or a discordance in relation to a frame of reference, there is a high probability that it will operate a behaviour aimed at correcting the difference or reducing the discordance. Certainly it is always tempting—and often dangerous—to generalize. But there is no doubt that many motives for action are linked, directly or otherwise, to the preservation or the restoration of an equilibrium or a consistency.

Adopting the point of view we have just outlined, we can envisage various categories of ends which are distinguished by the nature of the information processed, the type of elaboration the information undergoes, and the degree of directness and rigidity in the links established between the brain's 'inputs' and 'outputs'. A dialogue between the individual and his environment presupposes the existence of a certain autonomy, which implies the preservation of the individual's physical integrity, the maintenance of constancy in his internal milieu ('homeostasis'), and the ability to act against the force of gravity. It is on the basis of this autonomy that the individual can then structure his extra-personal space and the relationships—of whatever degree of complexity—he forms within it. And the interaction dynamics of these relationships reflect in many respects the quest for a kind of relational and affective homeostasis. When information originating in biological history is combined with that furnished by a socio-cultural context, 'desires' mix with 'needs' and 'projects' are born which aim at new equilibria and new consistencies.

The living being who maintains his shape, his internal structures, and his spatial position (by means of defence reactions against harmful agents, ingestive behaviour that furnishes energy and materials, and postural adjustments that combat the force of gravity) is responding to the internal constraints of the system which constitutes him as an individual. Determinism is rigid, and references to past experience are unimportant or even non-existent. The affective connotations (pleasure or pain) that may exist are linked directly and immediately to the sensory information of the moment. Responses are of varying degrees of complexity but by nature they are always reflex, almost automatic, and stereotyped. For they reflect the involvement of links which are very largely genetically preprogrammed and therefore 'pre-wired'. And it is also genetic programmes peculiar to the species which furnish the references, i.e. the various set points. The adaptive character of behaviour is therefore innate: it is the expression of the 'wisdom of the body' (Cannon).

In the dialogue with the environment, socio-affective exchanges correspond to a fundamental need for expression and interaction while at the same time they participate to a great extent in maintaining a certain equilibrium of hedonic nature. The links between the brain's 'inputs' and 'outputs' have a much more diachronic character, for many references are made to individual experience, to the traces experience has left in the internal representations of one's familiar social environment. The affective connotations (pleasant or unpleasant emotions) associated with the objective data of sensory information often arise from this confrontation between current information and the traces left by past experience, and they usually constitute important elements of the 'causal field' which determines the probability that this or that response will be emitted. These responses are also more subtly differentiated, more personalized, for they aim more particularly at the maintenance (for a given individual) of a certain relational and affective homeostasis. Certainly, the genetic programmes largely determine the structure and dynamics of exchanges within the group as well as the means of expression and action (potentially) available to each individual organism. But the adapted manner in which these means are put to use is largely determined by the individual experience that has been acquired in interaction with others through the processes of positive and negative reinforcement that are entailed in the consequences that result—for the individual organism—from such interactions. Behaviour is no longer simply the immediate and momentary expression of internal biological constraints: it belongs within a history. It is an event that both reflects the past and is a source of the future.

In man, the information of the moment, with its initial affective connotations, may undergo a cognitive elaboration of some degree of intensity, affective experiences playing in this regard an important 'energizing' role. This cognitive elaboration, greatly facilitated by language, which particularly enriches the internal 'discourse', is fed by the springs of individual experience and social learning. In the representations that serve as references and which are updated by interactions with 'reality', many markers are furnished by the socio-cultural context, with its value systems and its myths. The degree of adaptation of an individual is then defined by his degree of integration into this socio-cultural system. New constraints are thus created, linked with the quest for a social identity and the defence of the social Ego. But, at the same time, a certain liberty develops in regard to strictly biological constraints. Thus, not only can man write his individual destiny into the course of the history of his species: he can also—for the first time in evolutionary history—'change its course' for the better or, sadly, for the worse.

The concept of a biological or psychobiological 'end'

It may appear contrary to scientific method to talk about *ends* that determine the use by the living creature of *means* appropriate to their realization. For by postulating the existence of and quest for a certain end, one might give the impression that one is postulating the existence of some kind of 'attractor' that a dynamism tends towards, which would imply that the arrow of time might be reversed, that some 'final cause', pre-existing in the future, could attract and direct towards itself actions going on in the present. This would be to forget that 'end' also—and primarily—means 'cessation of a phenomenon in time' (when we speak of the end of the world, we allude to its eventual disappearance, not to some final destination). Now, we have stressed the fact that the brain is the bearer of set points and references and that it detects any difference with respect to a set point and any discordance regarding a reference. 'End-directed' behaviour, then, aims to put an end to a difference or discordance that may have been detected, not to promote some *ideal* perfection composed of equilibrium and consistency. This being so, the objective aimed at (return to a set point, re-establishment of consistency with an internal representation) pre-exists as regards behaviour (which is to have the effect of modifying the brain's 'inputs' in the direction thus predetermined), and the arrow of time therefore points in the correct direction.

We can state briefly some facts which show, convergently and clearly, that it is genuinely the goal to be attained that defines the function to be carried out by behaviour—the means of taking action—and that, therefore, determines the choice and the effective deployment of an appropriate strategy. We should first of all point out that, when the brain programmes a behaviour sequence, it has already developed an 'image' of the objective to be attained; it anticipates the effects of the movements it is programming and it prepares the sensory pathways ('pretuning') to record the results expected of this behaviour.[7] We are thus able to record neuronal activities within the brain, closely linked with a certain expectation, or with registration of the satisfaction, or, in the contrary case, the non-satisfaction of this expectation. Furthermore, the deployment of cerebral mechanisms in order to confront a given situation depends on what knowledge is already possessed, or is not yet possessed, by the individual brain with regard to an effective coping with that situation. The complex humoral response to a situation changes profoundly as it gradually acquires a strategy whose effectiveness it verifies itself.[8] But humoral changes of a different kind become firmly established

---

[7] See J. Paillard 1978, 1982; M. Jeannerod 1983.

[8] It is not simply a matter of the humoral response occurring within the brain: the release of catecholamines and steroid hormones by the suprarenal glands is also involved. See R. Dantzer 1984.

whenever the individual feels incapable of coping with a situation effectively ('learned helplessness').[9]

On the other hand, the 'priorities' which become evident when nervous mechanisms are deployed have more to do with whatever biological or psychobiological goals evolutionary history may have written, by means of the genome, into the structure of the system than with constraints resulting 'mechanically' from the system's elementary properties. Such priorities are already displayed at the simplest level of organization, that of the spinal reflexes. The myotatic reflex (reflex contraction of the muscle in response to its own stretching) is the basic element in the maintenance of any posture, which is itself the starting point for intentional and directed movement. Now, the myotatic reflexes operating in the extensor muscles of a limb will be immediately cancelled whenever a nociceptive stimulation (signalling the initiation of an attack on physical integrity) is applied to the skin of the limb, bringing about the latter's 'withdrawal' by means of a flexion reflex. A myotatic reflex will also be cancelled when the tension developed in the muscle by its stretching and the active contraction in response reaches such a degree that there is a risk of the muscle's being torn or a tendon's being detached. At a completely different level of integration and organization, a hungry individual who is engaged in taking food will cease eating and direct his behaviour towards defence reactions if he senses danger in his environment. It is not as though the hunger motive has disappeared: rather, a 'priority' motivation has come to supplant it in the determination of behaviour. This kind of fundamental biological priority, an expression of evolutionary history and an integral part of genetic information, comes to be joined by priorities of a different sort— the reflections of a socio-cultural context (with its 'scale' of values)— which are recorded in the internal representations of the relations that are developed and expressed by the individual in this context.

Finally, at the risk of re-emphasizing an obvious point, we shall recall that a form of behaviour is a means of expression and action which is capable of being put into action in pursuit of very diverse ends. We cannot therefore consider it an effect springing from a cause that is both necessary and sufficient. Although feeding behaviour is basically caused by a need for energy and materials, and it aims to cover this need while ensuring that balanced body-weight is maintained, we know very well that it also satisfies a whole range of social functions (family meal, wedding dinner, old comrades' banquet, lovers' shared snack, business lunch, etc.). And the sudden recent growth, in the United States, of the number of seriously obese individuals (young women above all) is certainly not linked with the appearance (by mutation) of a gene whose effect would be to disturb one of the mechanisms sharing in the regulation of bodyweight. Conversely, the cessation of all food intake does, it

[9] See P. Willner 1983; R. Dantzer 1984.

is true, indicate the onset of satiety. But this non-ingestion may also be used for ends that have nothing to do with the organism's nutritional state: a child who refuses to eat, or an adult on hunger strike, each of whom is trying to express something and to obtain something; a fast in connection with a religious rite; mental anorexia in the adolescent, more an indication of a psycho-affective disorder rather than one involving the energy metabolism. Similarly, aggressive behaviour—whether offensive or defensive—can in no wise be ascribed to a single motivation. Quite the opposite: as we shall see later, it is a means of expression and action which is subordinated to the most diverse ends; this explains the large number of factors (biological, psychological, and sociological) which—by their interactions—make up the 'causal field' responsible for its observable manifestation.

Set points and internal references

In the case of the homeostasis of the internal milieu, the set points (glycaemic level, osmolarity, central temperature, etc.) correspond to vital necessities, and they are reproduced in an invariant manner thanks to the transmission of the genetic programmes characteristic of the species. The behaviour employed (ingestion of food or water, thermoregulatory behaviour, etc.) is an integral part of the fundamental biological regulations that ensure constancy in the internal milieu. Considered from this viewpoint as elements of a regulated system, such behaviour can be described and explained with the aid of the concepts and methods of systems analysis.[10] In the case of relational and affective homeostasis, in search for equilibria and consistencies in socio-affective exchanges, things become much more complex, for at least three reasons. First, the reference is no longer an equilibrium point but an equilibrium structure. Just as, in the sleeping–waking alternation, equilibrium is not located somewhere between the two states but in the balanced relationship they have with each other, we can conceive of a whole range of equilibria involving contradictory mental states that are either successive or synchronous.[11] Furthermore, since behaviour is part of an individual history, there is no longer necessarily a question of establishing an instantaneous balance or consistency. The brain can integrate experiences over a period whose length varies between individuals and—in one individual—between one period of life and another. It is by such an integration that a feeling of success or failure, of equity or inequity, of satisfaction or dissatisfaction, develops in relation to a particular 'project' or with more general application. Lastly, the

[10] See F. Toates 1980; M. Cabanac and M. Russek 1982.
[11] Successive mental states: for example seeking the exhilaration provided by exploration of the new, then retreating into the soothing calm of the familiar. Synchronous states: for example an anticipated or actual evaluation of the 'benefits' and 'costs' of an action, together with the pleasant and unpleasant affective experiences connected with it.

internal representations—which are structured by lived experience but at the same time furnish references for actions yet to come—evolve over time. This evolution may occur progressively or, on the other hand, through crisis, especially when the adolescent rejects his status as a child and seeks a stable adult status with a new social identity.

The complex role played by the emotions

The *emotions* play a large part in the mediating functions provided by the brain for managing the individual's relations with his environment. The emotion aroused by the perception of an object or a situation is inseparable from the way in which the brain apprehends that aspect of external reality, from the 'reading' the brain makes of it. Consequently, emotion contributes to the definition of the causal field which will determine the choice of an appropriate behavioural response. Likewise, the emotion aroused by evaluation of the consequences that may ensue from this response will form an integral part of the traces left within the brain by the experience thus lived through. Since some emotions originate from the confrontation between current sensory data and the traces left by experience, while others, born of action, have the effect of enriching these traces, it is clear that behaviour is part of a history, for it both draws nourishment from the past and contributes to shaping the future. In the interaction dynamics characterizing one's relations with the environment, these emotions form inexhaustible sources of 'energy'.

Neurobiological data have shown very clearly the important role played by the hypothalamus in the genesis of the emotions. Various types of emotional reactions may be brought about by localized stimulation of the hypothalamus, and several attempts at classifying these reactions have been suggested. One of the most recent and most highly developed of such suggestions is that of Panksepp (1982), who postulates the existence within the class of mammals of at least four neural circuits corresponding to distinct 'emotion generators'. Although it is very interesting to pinpoint within the brain the nervous substrate whose activation engenders this or that 'emotional state', it is important not to forget that a localized electrical stimulation short-circuits the set of factors and mechanisms which are responsible for this activation under normal conditions. So less emphasis should be placed on these emotion-generators and the way—differing from one species to another—in which their activation is expressed, than on the analysis of everything that happens 'upstream' of their deployment. In this regard, the limbic system plays an essential role.

Reducing emotion to a generator and an outward projection does not allow an appreciation of its development and function in their full dimension. For the affective experience generated by an object or by a situation results from both the evocation of previous experiences, positive expectations (or fears) aroused in the current context, and the concrete possibilities that offer themselves for the realization (or avoidance) of these. That is to say that this affective experience is not in any way 'monolithic' but is constantly evolving as a result of a cognitive elaboration of some degree of complexity. In the very course of acting it will further evolve under the influence of the retrospective effects stemming from partial fulfilment. In other words, the configuration of the causal field underlying the action is continually changing. Moreover, an emotion can be experienced in two very different ways: either as an integral part of the action, as a force driving one on and permitting self-affirmation in action, or as a 'passion' that one undergoes. For the individual in dialogue with his environment, the relationship between an emotion and observable behaviour is rarely that between a mental state and its sole outward projection alone. Much more often, the behaviour is aimed at creating, maintaining, modifying, or abolishing an affective experience in the individual who implements it.

Emotions form part of a socio-cultural context

It must be added that emotions also play an important social role whose precise nature depends on the socio-cultural context.[12] Displaying all the signs of burning rage allows one to carry out an act not conforming with society's rules without being held entirely responsible (moreover, it is by reference to social norms that the 'justified nature' of the anger will be assessed). Within a human community, the probability of the implementation of a behaviour as a means of expression and/or action is quite strongly influenced by the general attitude towards it—whether a reproving or, on the other hand, an indulgent one; now, both this attitude and the influence it exercises have a strong emotional component. More generally, emotions may function as 'scapegoats' by allusion to a transitory regression to the state of 'beast', whenever we try to preserve the image we have of man, a being normally free from his 'base instincts' and ruled by 'pure reason'.

Emotion is interpreted through language, itself a reflection of the socio-cultural context. Initially relatively diffuse, until it is named, emotion takes on its full meaning as soon as the internal discourse takes it in hand. The way in which experience is lived through thus depends on the way it is interpreted by language, on the basis of a set of social norms. In turn, by naming an object in a certain way, by speaking of it repeatedly in a certain way, an experience of an emotional kind will be created in regard to that object (as in advertising, propaganda, etc.).

The information furnished respectively by the genome, by individual experience, and by the socio-cultural context is most intimately blended in the elaboration of the social behaviour expressing and creating affec-

---

[12] See J. R. Averill 1982.

tive states (justly referred to, therefore, as 'socio-affective' behaviour). Each of these three types of information is built up progressively into a system having its own structure, content, and dynamics. It is within the brain that these systems interact and shape each other. Since these are dynamic, open, and evolving systems, only a historical and constructivist perspective permits them to be apprehended in their full reality. Given these facts, how is it possible to conceive of and define such a notion as that of a 'relational and affective homeostasis'? Given this notion's abstract nature and the fact that it embraces a reality of extreme complexity, we may think it devoid of interest for anyone trying to clarify and understand some (very partial) aspect of this reality. But it does, in fact, present a major interest: that of underlining the interdependence of the partial equilibria and consistencies that we apprehend in the study of biological systems, psychological structures, and social representations, reminding us that the biological individual, the person, and the social being have at their disposal one and the same brain which endows them — not without some 'conflicts' — with their singular uniqueness.

Socio-affective behaviour

It is obviously only in a very fragmentary way that one can touch here on the historical dimension and the dynamic interaction characterizing this behaviour. The development of the means employed, the evolution of the objectives aimed at through these means of expression and action, the (changing) way in which this behaviour both reflects and inspires a social structure and a social dynamism, can only be appreciated over time. Ethological analyses of ever greater refinement have permitted us to describe the development of communications systems and of social interactions, both in animals[13] and in man.[14] These investigations put stress on the processes of inter-individual attachment and the development of social ties, as well as on the processes of competition and the establishment of relationships of dominance and subordination.

The mother–child relation

In the monkey as in man, the very young creature's need for contact with its mother — or her substitute — corresponds to a basic need for 'comfort' and 'safety'. Placed in a situation which arouses some degree of anxiety, the young monkey will explore it only if his mother or her substitute is present; for the young child, too, the mother is a focus of security to which he returns regularly for reassurance, when he begins

[13] See P. Marler and J. G. Vandenbergh 1979; R. A. Hinde 1982; R. E. Passingham 1982; B. Thierry 1984.
[14] See R. B. Cairns 1979; R. A. Hinde 1982; R. E. Passingham 1982; A. Restoin *et al.* 1984.

exploring the world surrounding him. Even in the adult, physical contacts (particularly those of the hands and the lips) retain a social function
of reassuring and soothing. When the mother–child relations differ in
kind because they have developed in different social systems, the effects
produced by separation also turn out to be different. In certain species
of macaques, mothers are relatively 'permissive': they allow their very
young child to interact with all the group's members; in others, they are
more 'protective': they restrain their young ones, or take them back
frequently.[15] In the latter case, the disappearance of the mother is a
serious event for the young creature which presents a phase of great
agitation before falling into a deep depression. The contrary obtains
with the 'permissive' mother: separation does not cause these extreme
effects—the young creature is little agitated and continues his interactions with the other animals without exhibiting depression.

In our species, too exclusive a relationship between mother and child
can be 'pathogenic'. Indeed, as pointed out by Balleyguier (1981), 'the
mother who devotes herself exclusively to her child over-protects it and
at the same time becomes too demanding, for she expects too much
satisfaction; in this way she prevents the formation of an autonomous
ego, for the latter requires the establishment of multiple relationships
and increasing independence of the first attachment'. It is also interesting to point out that, even in animals, the existence of a social bond has
an effect on the way in which a situation is perceived and experienced.
Thus it is that two animals confronted together with a frustrating situation will or will not present a distinct reaction (secretion of hormones by
the suprarenal cortex and aggressive behaviour), depending on whether
they originate from two different groups (so having had no previous
social interactions) or instead from one and the same group.[16] We shall
see later[17] that, very generally, individuals' familiarity with each other
markedly lowers the likelihood of conflicts. The development of
language, which is the basic aspect of hominization, has permitted the
development of the world of ideas and the symbols referring to them.
These ideas and symbols, together with the exchanges that they
nourish, give rise to new attachments, potential sources of new frustrations and new conflicts, as well as means of settling them.

Competition and dominance

The analysis of situations in which individuals are liable to enter into
competition with each other, and of the forms such competition may
take, is tightly bound up with the study of the different functions
fulfilled by aggressive behaviour.[18] We shall limit ourselves here to
stressing that the evaluation of the respective probability of 'coopera-

---

[15] B. Thierry 1984.   [16] R. Dantzer 1981.   [17] In Chapter 7.
[18] These functions will be examined in Chapter 7.

tion' or 'competition' between individuals in real life cannot be reduced, as in games theory, to the evaluation by each of some kind of balance between expected gains and losses. Since the partner is not perfectly anonymous, considerations arise which are not purely 'economic': one wonders, for example, if a particular profit is justified, deserved, or fair.[19] As for the notion of dominance, its semantic content is ambiguous, for it often confuses an observed position of dominance in a given conflictual interaction with a personality trait supposed to be its cause. While there can be no doubt that a set of faculties, partly genetically determined, contributes to establishing a dominant position in a given situation, we also know that an observed dominance in a familiar environment can disappear in a new one, and that it can even be inverted by the repeated experience of defeat.[20] In relation to rodents, Benton (1982) questions the usefulness of the notion of dominance, stressing the lack of correlation between dominance as it is observed in aggressive interactions and dominance as defined by the degree of facility with which an animal appropriates food, water, or a certain space. In man, even more than in animals, it is vain to hope to evaluate in a general way the respective weight of the genome, individual experience and social constraints in determining behaviour observed in conflict situations. If we are to try to unravel the concrete causal connections, we have to do it case by case.

Structuring influence of the environment

Having said this, it is clear that the environment exerts a strong structuring influence on the development of socio-affective behaviour. The study carried out by Schiff and his collaborators[21] clearly demonstrates the influence of family surroundings on success at school: abandoned children, coming from disadvantaged social strata and taken into a 'comfortably-off' milieu before the age of six months, turned out later to be much closer to the average intellectual level of their adoptive parents than to that of their biological parents. Moreover, on average they perform much better than their brothers and sisters brought up by their biological parents in the disadvantaged milieu. The investigation carried out by Duyme (1981) shows that the adoptive sociocultural milieu has a clear effect not only on the frequency and nature of inadequate school performance but also on the frequency and nature of disturbances in social behaviour: 'anti-social' behaviour, like poor performance at school, is more common in a disadvantaged milieu (that of unskilled workers, for example) than in other social classes. Hypothesizing that the child's character depends not only on his own peculiarities but also on the relational structure of the milieu, Balleyguier (1981) notes that the child's

---

[19] See M. Plon 1972; J. R. Eiser 1978.   [20] See I. S. Bernstein 1981.
[21] Study cited by J. Médioni and G. Vaysse 1984.

character in fact differs with the type of custody he is used to, there being distinct differences between custody within the family and external custody. His character 'is generally modified in the direction of a calmer, more passive, less autonomous attitude, with less well developed relations in these situations of external custody'. In the development of non-verbal communication in the child, the progressive differentiation — during the third year of age — of characteristic behaviour patterns which shrink to a minimum of elementary acts, without loss of function or significance, is effectuated in close relationship with the communication modes that the family gives importance to in regard to the child.[22] Comparing some 300 Swedish children aged between 11 and 12 years and brought up in a 'permissive' environment with a similar group of Chinese children brought up in a 'restrictive' environment, Ekblad (1984) observes that the Chinese children were more capable of controlling the expression of their emotions and that they adapted their behaviour to collective norms, whereas the Swedish children showed distinctly more individualist attitudes.

Social structures are quite rigid, and it is the individual who, because of the 'plasticity' of his brain, will be integrated into them by adapting himself to the constraints of the system. In this adaptation, an important role is played by the reciprocal character of the relations between brain and behaviour: neural and neuro-endocrinal mechanisms underlying the development of socio-affective behaviour undergo, in turn, the structuring influence of the experiences linked with this very behaviour.[23] In the context of cultural evolution which, in man, have come to be grafted on to biological evolution proper, the individual brain can thenceforth itself contribute to the definition of the social norms which it uses as references in its developing behaviour. This means he can make a voluntary contribution to the 'stabilization' or 'destabilization' of the social system into which he has been introduced.

Equilibria, coherences, and their representations

It is in social interactions that a 'balanced' dialogue is constructed between oneself and others, at the same time as a 'coherent' internal discourse is organized in order to interpret and evaluate the ins and outs of this dialogue.

And it is in the internal representations of which the brain is the bearer that the interface is to be found between the processes animating respectively the external dialogue and the internal discourse; an interface through which a certain 'existence' and a certain 'substance', the 'being' and 'becoming' are mutually structured and enriched. For it is within these representations that interaction takes place between the

---

[22] A. Restoin *et al.* 1984.   [23] See Chapter 5.

traces left by concrete experience and the cognitive and affective development of which they are the object.

Body scheme and body image

This being so, it is somewhat artificial to isolate each of the equilibria and coherences relating to the biological individual, the psychological person, or the social being. Let us take an apparently very simple example, that of being introduced into a given space. The young biological individual develops his 'body scheme' progressively; thereby he becomes aware of his own shape, his position in space, the respective positions of the different parts of his body, and the way he can alter these positions and act within the space surrounding him. But this appropriation of space is far from being the simple internalization and organization of purely physical spatial parameters. It is intimately linked in the child to socialization and the formation of the personality, by means of the play activities taking place there. Space is integrated into individual experience, invested with desires and meanings, marked with personal imprints; and there comes to be superimposed on the 'body scheme' a 'body image' that integrates elements whose origins no longer relate only to purely biological structures. It is to be understood that these elements, too, become part of the biological structures of the brain. For, although the content of the internal representations owes much to experience, their physical medium and their coding correspond to structures of a neurobiological nature.

Cognitive coherence of attitude and situational determinants of action

In the dynamic of the human cognitive system, it does seem that there are adaptational reactions which express a tendency to preserve or recover a certain cognitive equilibrium. When a new element is integrated into a given group of elements of the pre-existing cognitive system, it is perceived as *consonant* or *dissonant* with the other elements in the group (if the relation between two cognitive elements is dissonant, this means that one psychologically implies the other's negation). A dissonant relation between elements of a given group creates a state of psychological tension, and adaptational reactions are triggered to try to reduce or eliminate this dissonance. The coherent ensemble made up by the cognitive elements relating to a defined object determines a certain disposition or individual attitude to it. This attitude is expressed in the judgements and actions targeted at that object. But how does it come about that judgements and actions are far from being always in concord? It is because, in the genesis of the action which comes to form part of a concrete context, the basic attitude — that expressed in the formulation of a 'timeless' judgement — is no longer the sole determinant: numerous 'situational' factors can exert their influence and provoke a more or less marked 'change of attitude'.[24] It must be added that the sociocultural context does not just supply this kind of situational determinants for current actions; it contributes in a more general way to the organization of the cognitive system itself. For, as Zimbardo (1972) emphasizes, the individual often rejects 'reality' as it is given, in order to substitute for it a cognitive reality that permits him to retain a satisfactory image of himself while carrying out a cognitive restructuring of the milieu so as to make it tolerable. Now, it is very obvious that the cognitive organization of a satisfactory self-image, like that of a tolerable milieu, is not designed to satisfy simply the demands of formal logic but just as much, if not more, those of a given sociocultural context and, in the final analysis, by means of the internalized social norms, those of affective homeostasis (or, for psychoanalysis, those of the 'libidinal economy').

Social and personal identity

Every individual needs other people, for it is by interacting with others that he forges, expresses, and preserves his social identity and originality. For the individual, 'to be social is to have a personal identity through belonging to a reference group', within which he compares himself with others; and the importance of other people is measured by the fact that, at any age, isolation causes painful and traumatic experiences.[25] Identity, which is located at the junction of the personal with the group's common culture, has to do most of all with the social roles the individual assumes by virtue of the positions he occupies in a social structure. But it also has to do with the individual's awareness of his membership of a certain social group and with the 'emotional and evaluative significance that results from this membership'.[26] Further, identity is linked with the way in which the members of a social group view their own group and other groups. In other words, social identity is linked with social categorization, the relationships between groups, and Deschamps (1982) has shown that differentiation (discrimination, distance) between oneself and others, and differentiation between groups—established by the same subjects—vary concomitantly. While it is legitimate thus to put stress on the social determination of identity, we must not in any way neglect the search, on the part of the individual, for some originality to distinguish himself from the other group members. This is to say that, besides 'social' identity, we should recognize the existence of a more 'personal' identity, corresponding to the concern for constructing and showing to others a 'self-image' that is original (i.e. 'different'), coherent, and permanent.[27] As the individual requires that this personal and original image be recognized and positively evaluated by the group, he will strive to reduce the discordances he perceives in the image of himself that the group returns to him.

[24] See J. M. Nuttin 1972; J. M. F. Jaspars 1982; G. de Montmollin 1984.
[25] J.-P. Leyens 1979.   [26] H. Tajfel 1972.   [27] J.-P. Codol 1982.

The right to promote and express one's 'sense of identity', to be oneself to the fullest extent, to 'live one's difference', comes fatally into collision with the constraints of the sociocultural context, constraints which often go well beyond the *necessary* limitations. While the individual's membership of a group and the internalization of social norms in his representations correspond to a necessity, and they can be very enriching, we must none the less stress the fragility of the human person confronted by the constraining influences of the group. Paradoxically, it is language, this instrument of liberation and progress, that also conveys the most telling and harmful influences. The most insidious danger lies in the use of verbal labels to separate and characterize others. For these stereotyped labels, with the equally stereotyped images and slogans associated with them, have depersonalizing and dehumanizing effects: in a group that is so well labelled, the individuals are 'interchangeable' and attitudes and behaviour towards them will be determined by their membership of the group, not by each one's personality characteristics.[28] The word, in such conditions, becomes a powerful determinant of behaviour, and this verbal determination, as Zimbardo with great justness emphasizes, is a threat to 'human compassion and comprehension'. The tendency to categorize, to exclude, and to reject is particularly marked in the subject endowed with an 'authoritarian character' who converts his relation with the people around him into a relationship of subordination and exclusion.[29] In his case, a genuine mutual reinforcement occurs between the person and the group which is constituted by opposition to 'outsiders'. For 'grouping permits the authoritarian character to pass, without apprehension or anguish, a pejorative judgement on others, a judgement authorising any hostility and any injustice'.[30]

It is clear that personal thought and personal conduct are intimately linked with the conditions obtaining in the sociocultural context within which they are carried on. To the extent that we consider that the brain's essential function is that of an organ 'generating sense' in the management of the individual's relations with his environment, the 'social' dimension is important in the representations that he constructs and uses. In this view, the social character of a representation 'flows from the use of systems for coding and interpretation furnished by society, or from the projection of social values and aspirations'.[31] Given the 'creative' role of the social representation, it will not be simply the *reflection* of certain sociocultural conditions but will in turn be the *source* of certain social attitudes and conducts. Everyone is influenced by the images—formed in a collective manner—that he has of 'human dignity', of 'illness', of 'madness', of 'violence', and of many other equally fundamental notions. These representations, which are more concerned

---

[28] See P. G. Zimbardo 1972; H. Tajfel 1982. [29] See H. Tajfel 1978; L. Israël 1984.
[30] L. Israël 1984. [31] D. Jodelet 1984.

with intersubjective relations than with relations with some concrete object, are not of a purely cognitive nature; they also have unconscious and irrational aspects.[32] It goes without saying that demonstrating one's lucidity consists not in ignoring these aspects but, on the contrary, in becoming fully aware of them and examining them in a critical way in order to be able to master them.

[32] S. Moscovici 1982.

5 The neurobiology of the processes of motivation and decision

Behavioural neurobiology attempts to establish the existence of close and stable correlations between well defined neurophysiological mechanisms and particular behavioural phenomena or processes. To test the general validity of correlations thus brought to light, a comparison is made of experimental data obtained in different animal species (particularly rats, cats, and monkeys). Where appropriate, this comparison is extended to neuropsychological data obtained in man concerning the correlations between a localized lesion of the brain and a particular disorder affecting a specific conduct or a given mental faculty. But it must be said that, when using these methods, one comes up against a two-fold difficulty. On the one hand, inter-species comparisons and extrapolations from species to species have full significance only insofar as mutually homologous terms are considered from the neurobiological as well as from the behavioural point of view in a series of correlations. Now, it is not easy to define or agree on the pertinent criteria for such homology. Two examples of aggression, in two different species, may take the same *form* while carrying out different *functions*, and vice versa. What criterion, then, should be used in order to decide that these two forms of aggression are, or are not, homologous: their form or their function? When it comes down to destroying 'homologous' brain regions in species located on very different levels of phylogenetic evolution, comparative neuroanatomy does, it is true, supply us with interesting materials for an appraisal, but it gives us no absolute criteria. It is, therefore, not always possible to avoid circular reasoning: one attempts to ascertain whether two structures considered homologous (on the basis of neuroanatomical criteria) are involved, in each of two species, in one and the same given function; and, if such is the case, one will be led to believe that this 'proves' that they are in fact homologous. The second difficulty arises from the fact that, in each correlation thus examined, terms are often taken in artificial isolation from the dynamic ensembles of which each forms an integral part: whether we are considering the global functioning of the brain on which the functional role of each particular region depends, or the global behaviour of the individual as displayed within a familiar social structure and familiar social dynamics. In some cases, the homology of the terms being compared will be more apparent than real, because, in the experimental conditions employed, they will have been 'stripped' of some attributes which are expressed only within the ensemble of which they normally form part. These difficulties do not in any way diminish the interest of the experimental methods of the neurobiologist. On the other hand, the risk exists that they could, if not constantly borne in mind, falsify interpretation of the data obtained.

We must also recall an obvious fact: the data we have at our disposal reflect the questions that have occurred to us. Now, it is not certain that, in the area of interest to us here, all the 'right' questions have already been concretely asked. Some have been able to germinate without ever leading to any experimental verification, quite simply because the appropriate methodological and technical tools do not yet exist. But others may very well be floating around in the limbo of human thought without ever emerging, because they are too distant from currently received ideas.

In this regard, one last preliminary remark is called for. It is in the context of a consistent synthesis of data already brought to light that the researcher conceives his own working hypotheses and then interprets the results he obtains. In behavioural neurobiology, these data are still quite fragmentary and often contradictory (partly because of the difficulties stated above). To give his personal synthesis the consistency it requires, the researcher will necessarily be led, consciously or not, to make choices: he will give more attention and more weight to some facts than to some others, while accepting — perhaps with difficulty — that at any time the well-foundedness of the choices provisionally made may be put into question.

Mastery over relations with the environment

The behavioural repertoire, as we have said, endows every individual with means of action that permit him to enter actively into his environment, to establish relationships within it, and to act on it in order to preserve or modify those relationships. Employing of the various forms of behaviour requires a set of faculties and performances without which the individual would be unable to master the dialogue that he conducts with his environment. But it is quite obvious that this mastery depends every bit as much on the actual conditions of the dialogue, which must not be such as to prevent the individual from foreseeing the consequences of his actions and choosing the means of action that appear pertinent to him, so as to deprive him of the possibility of exerting real control, through his actions, over his relations with the environment.

Before examining the neurobiological foundations of the faculties and performances required by the individual in order to interact with his environment, it is worth pausing briefly on the notion of 'mastery'. The fact of being able to control by one's actions this or that aspect of the dialogue, and the awareness the individual has of this mastery, do indeed play an important role in the preservation of mental equilibrium.

If given a choice, an animal, too, prefers a situation in which it controls the events that occur in its environment, rather than one where it is not given this opportunity.[1] To obtain food, nearly 100 per cent of rats will choose a compartment where they have learned to obtain it by pressing a lever (reinforcement closely linked with a behaviour) in preference to one where there is free access to food. When, in a T-shaped maze, rats have to choose between an arm in which they receive an electric shock at the instant they enter it, and another in which the same shock is given only after a 30-second delay, they regularly choose the side where the shock is directly linked with their behaviour and where, therefore, they themselves determine the moment of its application. These observations, and a number of others, show that the affective significance of many situations changes by virtue of the mere fact of their being controllable. Moreover, repeated experience of the uncontrollability of a situation has repercussions on the subsequent acquisition of adapted behaviour to meet other situations, leading to a deterioration of the faculties for learning and adaptation. As for the cognitive and affective development of the infant, emphasis has progressively been placed not only on the necessity of an environment providing numerous and varied stimuli, but also on the importance of stimuli over which the infant actively gains mastery by virtue of the fact that he elicits them with his own behaviour.[2] In this regard, a number of observations appear to indicate that controllable and controlled stimuli have the effect of promoting development of the cognitive faculties as well as exploration of the environment, while creating positive experiences on the affective level. At this age, such stimuli are essentially provided by the mother–child relationship.

Let us make further mention of a phenomenon that is induced experimentally and is very interesting. When animals have been subjected to shocks which arrive unpredictably and which cannot be avoided by any form of behaviour, they are thereafter incapable of learning an avoidance response which is acquired easily by control animals.[3] Everything happens as though these animals, because of the uncontrollable character of the shocks to which they were first exposed, had acquired the 'conviction' that action is utterly without effect. They no longer attempt to escape shocks when, later, learning an avoidance response really would allow them to escape them. They show 'learned helplessness', and this state is accompanied by modifications of a neurochemical nature within the brain.[4] We may add that the behavioural depression thus provoked experimentally, by manipulating the animal's previous experience, shows analogies with certain forms of human depression.[5]

If it is to be able to carry on a dialogue with its environment, the living creature must be endowed with 'reactivity', that is, it must be capable of

[1] See J. B. Overmier *et al.* 1980.   [2] See M. R. Gunnar 1980.
[3] See R. Dantzer 1984.   [4] J. M. Weiss *et al.* 1981; P. Willner 1983.
[5] See R. J. Katz 1981.

responding to the stimuli it receives from that environment. But more than this, its faculty of reacting must be able to evolve over time so as to remain at all times adapted to the changing conditions of the organism's internal state and its interactions with the external milieu. And, in fact, reactivity is the object of numerous modulating controls. Some modulate the intensity of reactivity: this quantitative modulation can affect the individual's responses either globally or, on the other hand, more selectively. The others are concerned with sensitivity to the qualitative, hedonic aspects of the stimuli and, therefore, the choice between two basic attitudes in regard to them: to 'approach' or, instead, to 'escape'.

But a living creature is not a simple reaction machine. In the course of evolutionary history, 'spontaneous activities' develop and their 'generators' become parts of integrated systems of ever increasing complexity. For, if these spontaneous activities have at first a tendency to be directed anywhere and everywhere in the course of the exploration of the environment, they later become more clearly canalized and orientated, as an integral part of a more controlled dialogue.

The distinction between reactivity and spontaneous activity is, in certain respects, quite artificial. On the one hand, indeed, we may feel that some apparently spontaneous actions are in reality delayed reactions in response to a signal or a situation held in memory. On the other, and of the greatest importance, reactivity and spontaneous action both depend on the nervous system's degree of 'activation', which, in turn, is determined by the totality of the messages of endogenic and exogenic origin which converge on the ascending reticular activating system of the brainstem.

Here we are concerned with some very simple but absolutely fundamental ideas, which should therefore be examined first of all.

Some basic notions (activation, reactivity, spontaneous activity). Ontogenesis of the mechanisms brought into play

Every reaction or action takes place against the background of a certain level of 'behavioural arousal', that is, a certain degree of openness to the environment and availability to it. This behavioural arousal is a function of an 'activation state' that involves the totality of the nervous structures, this state being itself linked with the activity of several neuronal systems in the reticular formation of the brainstem.[6] In this reticular formation, which stretches from the bulb to the most anterior region of the mesencephalon, it has been possible to pick out an 'ascending reticular activating system': under the effect of an electrical stimulation of this same system, the animal straightens up to adopt a posture testifying to a state of attention and investigation; following an extended lesion of this

[6] See P. Dell 1976; A. Hugelin 1976.

same system the animal displays, on the contrary, a marked state of 'adynamia' (absence of motor initiative; the animal maintains uncomfortable positions in which it has been placed; there is little response to stimuli).

Under normal conditions, the activation state is maintained by sensory messages which, in parallel with their transmission along the specific sensory pathways, converge on the reticular neurons; moreover, the latter are excited by humoral factors, particularly by the catecholamines (adrenaline and noradrenaline) which are released into the circulating blood by the medulla of the suprarenal glands, especially in the context of emotional reactions. The role played by the activating system is not limited to the genesis of behavioural arousal. Indeed, it is by means of this system's multiple 'outputs' and its highly developed integrating action that the waking organism's global behaviour is a consistent whole and not a simple collection of mental, motor, and vegetative activities, totally out of phase with each other. In the alternation of sleep and waking, the lowering of the level of activation consequent on a reduction of sensory messages (absence of light and sound; reduction of messages originating in the muscles and tendons, due to the lying position) does, it is true, ease the process of falling asleep, but is not sufficient to cause it. An active disactivation is required, in addition to the purely passive disactivation ensuing from sensory deprivation. 'Classically' it was accepted that serotonin, released by the neurons of the raphe nuclei, was responsible for the process of falling asleep. But more recent research has shown that serotonin, which is in fact released during wakefulness, induces the biosynthesis of the real 'hypnogenic' factors at the level of the posterior hypothalamus.[7]

Pathology of activation

Some believe that excessive activation—excessive in intensity and duration—is responsible for the development of various so-called psychosomatic illnesses, the type of activation itself being a function of certain personality traits.[8] In animals, experimentally produced intense and prolonged activation can cause ulcerations of the gastric wall, hypertension, or a myocardial infarction; it seems likely that, in man, too, excessive activation plays an important role in the pathogenesis of similar illnesses. The level of activation depends on the way in which the individual perceives the situation with which he finds himself confronted, the strategies he has at his disposal for handling it, and the experience he has already acquired of the efficacy of these strategies in the same situation. Activation can be attenuated by psychological 'defence' mechanisms which, modifying his perception of the situation, reduce the latter's impact, or by efficacious action directed at the situa-

[7] M. Jouvet 1984.   [8] H. Ursin 1980.

tion itself, i.e. coping. We may add that some studies have shown that different capacities for mastering difficult situations, and more or less marked predispositions for psychosomatic illnesses, are correlated on the neuro-endocrinian level with a predominant responding of either the suprarenal cortex (release of corticosterone) or the suprarenal medulla (release of catecholamines).

Detection of a biological need and behavioural arousal

The fact that reactivity and spontaneous activity are linked to the state of activation is clearly shown by the following observation, which provides a valuable example. If the hypothalamic glucoreceptors signal the existence of a state of need (in the present case, a glucose deprivation of the neurons due to a drop in the level of glucose in the blood), this detection of an energy need will activate complex interactions between the lateral hypothalamus and the reticular activating system, so as to bring about both an increase in diffuse vigilance and a greater openness to the environment; a raising of locomotor activity, which presents an obvious interest for the so-called appetitive phase of behaviour (the search for food); a general facilitation of the spinal reflexes and the simple motor responses, which presents an obvious interest for the so-called consummatory phase of behaviour (ingestion, mastication, and swallowing of food).[9]

Processes that moderate behavioural arousal

It is interesting to see how the ontogenesis of behavioural arousal, thus caused by the detection of a biological need, is effected. In rats, this arousal is not very pronounced in the first ten days of life, but it becomes more marked later, reaching, towards the age of 20 days, a level nearly ten times greater than that seen in adult rats; between 20 and 28 days, i.e. during the course of the fourth week of life (the weaning period), the intensity of behavioural arousal diminishes sharply, falling at the age of 28 days to the level seen in mature rats.[10] This very strongly marked attenuation of the activation state reflects the maturation over this same period of moderating processes which involve telencephalic structures, especially the hippocampus and the frontal cortex.[11]

The moderating processes, which are put in place when, after weaning, the young rat enters into competition and conflict with its fellow creatures, cannot be reduced to a simple attenuation of behavioural arousal, whether endogenous or exogenous in origin. For it is also during the fourth week of life that the reactions of 'behavioural inhibition' are developed — against the background of a general attitude of appetence and approach — in response to stimuli having an aversive

---

[9] See P. Karli 1976.   [10] W. H. Moorcroft *et al.* 1971.   [11] W. H. Moorcroft 1971.

character, i.e. those generating unpleasant affective experiences.[12] By means of processes that carry out more selective moderations and inhibitions, the individual experience acquired in social interactions progressively shapes forms of behaviour that are more nuanced, more 'personalized', starting with fairly stereotyped behaviour of a reflex type. In this important phase of ontogenesis, several structures of the limbic system (hippocampus, amygdala, and septum) participate very largely in putting the moderating processes into place. This can be seen from the fact that a destruction of the septum, carried out in the seven-day-old rat, brings about the development of a marked hyperreactivity that persists into adult age;[13] similarly, lasting hyperreactivity is seen following early bilateral lesion of the amygdala.[14] The moderating influences exercised on the activating reticular structures of the brainstem by this forebrain control system make use of the descending pathways running through the hypothalamus and the periaqueductal grey matter, structures which are capable of relaying messages. Lesions carried out at the level of the hypothalamus, median or lateral, lead to an accentuation of the rat's reactivity in regard to electrical or thermal nociceptive stimuli.[15] Given that the periaqueductal grey matter is involved in the genesis of attitudes of 'withdrawal', we can understand that lesions of this periventricular structure can facilitate certain forms of 'approach' behaviour, and that they cause in particular a lasting increase in food intake.[16]

Structuring influence of experience

We have already emphasized in the preceding chapter the fact that some nervous and neuroendocrinal mechanisms which underlie the development of socio-affective behaviour in return undergo the structuring influence of the experiences linked with this very behaviour. This very general idea of the reciprocity of brain–behaviour relationships finds a concrete illustration in the following observation: major alterations in behaviour (markedly increased reactivity, stereotyped, and poorly adapted behaviour) resulting from a lesion of the septum or the hippocampus are also the main distinguishing features of the behaviour of animals brought up in social isolation.[17] Generally speaking, animals brought up in an 'impoverished' environment present a whole range of behavioural anomalies, consisting mainly in the exaggeration of oral activities and the appearance of stereotyped behaviour, as well as deficiencies in learning strategies for adapting to danger.[18] Furthermore, it is not surprising that the environmental conditions in which

[12] P. M. Bronstein and S. M. Hirsch 1976.
[13] F. Eclancher and P. Karli 1979a.   [14] F. Eclancher and P. Karli 1979b.
[15] G. Sandner *et al.* 1985.   [16] J. P. Chaurand *et al.* 1972.
[17] P. F. Brain and D. Benton 1979; D. G. Jones and B. J. Smith 1980.
[18] R. Dantzer 1981.

animals are brought up affect their sensitivity to various drugs acting on
the nervous system[19] if one remembers that, even in adult animals, a
period of social isolation lasting several weeks brings about neurochemical modifications within the brain.[20] It is clear that the development of those cerebral mechanisms by which social interactions shape
some of an individual's behavioural traits requires that the mechanisms
be actually brought into play by virtue of the individual's interactions
with a sufficiently diversified environment.

Modulated processing of sensory messages

Besides the modulating influences which globally affect behavioural
arousal and the level of reactivity, some mechanisms modulate much
more selectively the organism's reactions to stimuli from the environment. We may mention first of all the processes and mechanisms whose
involvement has the effect of selectively modifying the processing of a
given category of sensory messages. We saw earlier that the detection by
specialized hypothalamic receptors of a need for energy induces behavioural arousal. But it also induces activation of a hypothalamo-
olfactory circuit that will allow food odours to undergo privileged processing in a starving subject. It has, indeed, been noted in rats that the
responses of the olfactory bulb to a food odour are markedly modulated
by the animals' nutritional state (that of hunger or satiety), while a
similar modulation was practically never observed for responses elicited
by an olfactory stimulus devoid of any feeding significance. If the average amplitude of multi-unit activity at the mitral-cell layer is recorded, it
is found that the odour of customary food has effects that are approximately inverse depending on whether the animal is starving (usually an
increase in basic activity), or, on the contrary, satisfied (basic activity
usually decreased or unchanged); in the case of olfactory stimuli that are
devoid of feeding significance, on the other hand, the response of the
olfactory bulb is the same (basic activity decreased or unchanged) whatever the animal's nutritional state.[21] Another example is that of the
descending control exercised — at the level of the first spinal relay — over
the ascending transmission of pain messages. Analgesia can be induced
by 'stressful' stimuli or situations. And it is interesting to emphasize
that, here again, we find the notion of the 'controllability' of the stimulus or the situation. In fact, when rats are subjected to electric shocks,
the analgesia that develops will be distinctly more pronounced if the
animals have learned that they have no control over the application of
the shocks.[22] Similarly, a mouse which penetrates as an intruder into
the cage of a fellow creature and which is attacked by it will develop an
analgesia which becomes particularly noticeable at the very moment

[19] J. M. Juraska *et al.* 1983.   [20] J. Glowinski *et al.* 1984.
[21] J. Pager *et al.* 1972.   [22] See A. I. Basbaum and H. L. Fields 1984.

when, adopting the characteristic 'defeat' posture, it appears to be signalling that it has lost all hope of mastering the situation.[23] Very generally, 'centrifugal' controls are exercised over the transmission of sensory messages and over the processing these messages undergo at the level of the relays punctuating the centripetal pathways. It must be added that this processing is also influenced by numerous humoral factors. This is why, by recording neuronal activities, we note that, in the hamster, under the effect of the female sexual hormones that facilitate the induction of lordosis (the characteristic posture of the sexually receptive female) by the appropriate stimuli, there is a genuine reorganization of a particular sensorimotor integration within the mesencephalic tectum.[24]

Modulatory controls exercised over responses

Other modulatory controls are selectively concerned, not with some particular stimulus that is being transmitted to the brain, but rather with the response it elicits from the brain. In the newly born infant and during its first months of life, tactile stimulation of the palm of its hand provokes a forced reflex of prehension, a grasping reaction; in a similar way, visual stimulation causes a forced fixation, a 'magnetization' of the gaze. As the maturation of the cerebral cortex proceeds (more precisely, that of cortical areas 6 and 8), the hand and eye free themselves from this ascendancy: the hand can accept or reject; the eyes can carry out a voluntary exploration of the visual environment.[25] In adult subjects, a lesion of area 6 leads to the reappearance of reactions of forced prehension, and a lesion of area 8 will disturb voluntary control over the gaze. François Lhermitte (1983) has described 'utilization behaviour' in subjects who have suffered a frontal cerebral lesion: they behave as though compelled to seize and make use of any object presented to them. More generally, lesions of the frontal lobe cause the appearance of more impulsive behaviour, less well adapted, less controlled.[26]

Genesis of affective connotations

An essential role is played by the neuronal systems of 'pleasure' and 'aversion' as well as by the limbic system, whenever reference is made to the traces left by past experience in the genesis and modulation of the affective connotations associated with sensory data and thence of the attitudes of 'approach' or 'escape' aroused by these connotations. But quantitative and qualitative control is already effected over reactivity at a relatively elementary level of sensorimotor integration, namely at the mesencephalic level (superior colliculus and the underlying periaque-

[23] K. A. Miczek and M. L. Thompson 1984.   [24] J. D. Rose, 1986.
[25] See J. de Ajuriaguerra 1977.   [26] See B. Milner and M. Petrides 1984.

ductal grey matter). This is clearly shown by the following observations, made in our laboratory by Pierre Schmitt and his collaborators. If a unilateral micro-injection of a GABA agonist (that is to say, of a substance *activating* GABA-ergic transmissions) is made into the periaqueductal grey matter, it is noted that the rat thus treated 'ignores' any stimulus applied to the contralateral side of its body surface (the side opposite that where the micro-injection was carried out); on the ipsilateral side, by contrast, it shows distinctly more reactivity, the latter being of the 'approach' type, i.e. the animal displays a marked tendency to orientate itself towards the stimulus and to interact with it. If, on the other hand, a unilateral micro-injection of a GABA antagonist (i.e. of a substance *blocking* GABAergic transmissions) is carried out, the rat displays ipsilateral 'sensory neglect'; on the other side it presents hyperreactivity of the 'escape' type, i.e. it attempts to avoid the stimulus.[27]

Spontaneous activities and their 'generators'

In the absence of stimuli coming from the external milieu, the living being is far from inert. Quite the reverse: it presents numerous spontaneous activities (locomotion, visual and tactile exploration, vocalizations, etc.) which correspond to a fundamental need to interact with and position itself within extra-personal space. For each of these elementary activities there exists, at the level of the brainstem, a 'generator' whose neurons discharge spontaneously. But the functioning of these generators is subordinated to several levels of control, which permits the brain to integrate each of the elementary activities in behavioural sequences of varying degrees of complexity. These control levels correspond to levels of integration and organization of increasing complexity which developed progressively through phylogenesis and which are put into place, one after the other, in the course of the early phases of ontogenesis. It would, however, be incorrect to consider the evolution of motor functions as a linear one, reflecting a simple juxtaposition of successive and hierarchical functional levels. Gesell[28] considers that it is rather a case of spiral evolution on the principle of 'reciprocal weaving', by the interplay of a process of reincorporation and consolidation rather than one of hierarchical stratification.

Let us take the example of the generator of the rapid eye movements (the 'saccades') which is situated within the reticular formation of the brainstem and which is subjected to two main levels of control: on the one hand, the superior colliculus which receives messages directly from the retina; on the other, the oculomotor area (area 8) of the cerebral cortex which receives the information processed by the visual cortical areas.[29] These two levels do not function independently of each other,

---

[27] P. Schmitt *et al.* 1984.   [28] Quoted by J. de Ajuriaguerra 1977.
[29] See C. J. Bruce and M. E. Goldberg 1984.

although an electrical stimulation of the oculomotor area can still cause eye movements in the absence of the superior colliculus. Under normal conditions, there is interdependence and cooperation. On the one hand, indeed, the maturation of the visual cells within the superior colliculus is connected with that of the visual cells of the cerebral cortex: the development of some complex properties of the receptive field of the collicular cells reflects the intervention of influences originating in the cortex; and the functional anomalies provoked at the level of the visual cells of the cortex by a limitation of visual experience are also reflected at the level of the visual cells of the superior colliculus.[30] Besides, an analysis of visuomotor development in cats has shown that feedback effects arising from the eye movements induced by the visual environment are necessary for the construction of a representation of the visual space, which will be used for a more elaborated voluntary exploration of the environment.[31] Moreover, while an essential role is played by the oculomotor area of the cortex in voluntary command over the gaze, and while this area can act directly on the generator of eye movements, a unilateral destruction of the superior colliculus nevertheless disturbs the exploration of the visual environment and the choice of visual targets for eye movements, with development of a certain unilateral visual 'neglect'.[32]

The production of vocalizations also undergoes hierarchical control. It seems that it is the periaqueductal grey matter that ensures coupling between a given 'motivational state', whether of endogenic or exogenic origin, and the corresponding vocal expression; and this phonatory function of the grey matter, it appears, undergoes three-fold control by the hypothalamus, the amygdala, and the cingular cortex.[33] The control exercised by the hypothalamus permits the momentary physiological state of the organism to be taken into account, while the intervention of the amygdala is necessary so that a signal or a situation can acquire its full significance and so that the vocalization reflects the individual's past experience. Lastly, it is through the cingulate cortex that vocalizations can be used voluntarily and can become fully fledged means of expression and communication.

The activity of the 'mesencephalic locomotor region' is also subjected to a certain number of controls situated at several functional levels.[34] We have already seen that the detection of an elementary biological need has the effect of bringing about behavioural arousal which is particularly reflected in a markedly increased locomotor activity. When a pleasant or unpleasant affective experience determines an attitude of approach or escape, this also has the effect of increasing locomotor activity, while directing it in one or other direction. If aversive effects are induced in rats by carrying out a micro-injection of bicuculline (which blocks

---

[30] See B. E. Stein 1984.   [31] A. Hein *et al.* 1979.   [32] J. E. Albano *et al.* 1982.
[33] U. Jürgens 1983.   [34] See G. J. Mogenson 1984.

GABAergic transmissions locally) into the periaqueductal grey matter or into the medial hypothalamus, an increase in locomotor activity is brought about together with jumps that correspond to escape behaviour.[35] Depending on the site of the injection, these jumps differ sharply in appearance: at the level of the grey matter, the GABA antagonist causes uncoordinated, poorly directed jumps; at the level of the medial hypothalamus, by contrast, the same blockage of GABAergic transmissions causes the appearance of well coordinated escape behaviour, with repeated, well orientated jumps. In social interactions, affective experiences and the direction in which they orientate behaviour arise from a comparison of current sensory information with the traces left by lived experience, which implies that this information is processed by the structures of the limbic system. Influences of limbic origin may then be exercised on locomotor activity through the intermediary of the nucleus accumbens and the sub-pallidal region,[36] or else—in the case of more advanced elaborations—through the intermediary of the prefrontal cortex and the corpus striatum.

Let us stress that the information processing that goes on at the level of the nucleus accumbens or the prefrontal cortex is facilitated by the 'permissive' or 'enabling' action of the dopamine released in each of these structures. Here, dopamine plays the role of a 'neuromodulator' rather than of a 'neurotransmitter'. The action of the neuromodulators produced by the brain itself is similar to that of certain hormones brought to it by the circulating blood. For this reason, we shall later consider both of these actions.[37]

We should now look in more detail at three topics whose importance has been indicated throughout the preceding paragraphs. First of all, the *neuronal systems of 'reward' and 'aversion'* which play an essential role in the genesis of the affective connotations as well as in the processes of positive and negative reinforcement of behavioural responses. Next, the *limbic system,* by means of which the emotions generated as much by the evocation of the past as by the evaluation of the present as a function of the past provide the dynamogenesis for socio-affective exchanges, which are generators of new significances and sources of new emotions. Lastly, the neuronal systems of the cerebral cortex making up the *medium for the most complex internal representations*—iconic or logical in type—on the basis of which the brain carries out its operations of anticipatory simulation.

The neuronal systems of 'reward' and 'aversion'

It was in 1954 that papers were published whose complementary data laid the experimental foundations of a very large number of research

---

[35] G. Di Scala *et al.* 1984.   [36] G. J. Mogenson 1984.
[37] See Role played by humoral factors, p. 112.

projects dedicated to studying the neuronal systems of 'reward' and 'aversion'. Olds and Milner (1954) implanted electrodes in various regions of a rat brain and gave the animal the option of stimulating itself ('self-stimulation') by pressing a lever. They noted that the animal's behaviour was very different according to the site of implantation of the active electrode, i.e. according to the neural structure being stimulated. For certain points in the brain the animal stimulated itself practically continuously; for others, by contrast, it carefully avoided touching the lever; and for yet other ('neutral') points, finally, it 'did nothing to obtain or to avoid stimulation'. Delgado and his collaborators (1954) showed that certain brain stimulations (particularly to the mesencephalic tectum) had effects whose motivational values were very markedly expressed in the behaviour of the cat being stimulated: on the one hand, the animal rapidly learns to carry out a manoeuvre (for example, turning a small wheel) that lets it put a stop to the stimulation being imposed on it (the 'switch-off' response); on the other, it learns to avoid food, even when starving, if taking food is regularly 'punished' by the stimulation. The general conclusion that emerges from the body of these experimental data is that certain points within the brain are part of a neuronal system whose activation induces effects which the animal seeks (reward, appetence, or pleasure system) and which positively reinforce any behaviour giving rise to them (positive reinforcement system); whereas others are part of a neuronal system whose activation induces effects which the animal tries to avoid (punishment, aversion, displeasure system) and which negatively reinforce any behaviour giving rise to them (negative reinforcement system), and positively reinforcing any behaviour liable to put an end to them.

The neuronal reward system

Self-stimulating behaviour can be elicited at numerous stimulation points widely distributed in the brain (the neocortex presenting the most noteworthy 'neutral' zones). Furthermore, the numerous experimental lesions performed have shown that it is difficult to cause more than a transitory depression of self-stimulation. It therefore seems that the neuronal reward system is characterized by quite scattered anatomical distribution, highly redundant operation, and a certain plasticity. This having been said, self-stimulation is particularly easy to obtain all along the medial forebrain bundle, not only in the hypothalamic section (in the lateral hypothalamic area) but also at its 'limbic' telencephalic expansions (nucleus accumbens, lateral septum, amygdala, dorso-medial prefrontal cortex) and its caudal expansions (particularly the ventral tegmental area of the mesencephalon and the raphe nuclei). The neuronal aversion and negative reinforcement system corresponds basically to medial and periventricular structures (the medial hypothalamus and the periaqueductal grey matter). But it must be added that the two systems are not sharply dissociated in spatial terms; quite the contrary, in different regions of the brain they present some degree of interpenetration.[38]

Numerous experiments have elucidated the great reward value of the effects induced by activation of the positive-reinforcement system.[39] Thus a starving rat prefers self-stimulation to food, and it is more difficult to inhibit (by repeated 'punishment') operation of the self-stimulation lever than to inhibit operation of the food-obtaining lever. Moreover, a cat learns to take food from a well defined receptacle (even when perfectly sated) if this behaviour lets it cause an electrical stimulation of its positive-reinforcement system; whenever the same behaviour ceases to cause the stimulation and its reward effect, it is rapidly extinguished. When one or more natural reinforcers (intake of food or liquids, with the possible addition of saccharin to the latter) are associated with the 'rewarding' intracerebral stimulus, it is noted that the level of motivation underlying instrumental behaviour (detected by the frequency of operation of a lever, or by the choice the animal makes between two levers giving access to different reinforcers) is determined by the sum of all the reinforcing consequences ensuing from the behaviour. We can therefore assume that the reward effect due to experimental activation of the positive-reinforcement system is of the same kind as that arising in natural conditions; and that natural reinforcers are effected precisely by activating the nervous structures which give rise to self-stimulation under experimental conditions. It is interesting to point out, in this regard, that certain unit neuronal activities modified by the intake (or the simple sight) of food are also modified in the same way by electrical activation of the neuronal reward system.[40]

Behavioural repercussions of reward system activation

Given the 'pleasant' nature of the effects induced by reward system activation, it is not surprising that bringing them into play experimentally can markedly modify the attitude of an individual confronted with certain situations and the observable behaviour by which that attitude is expressed. In macaques, remote stimulation of the reward system attenuates the fear reactions provoked by a snake and markedly accentuates the degree of dominance displayed over a fellow creature.[41] If a rat is made 'ill' by the intake of a toxic solution having a specific flavour it rapidly develops a marked aversion for the latter; but the acquired aversion is markedly less if the animal is allowed an option of self-stimulation in the lateral hypothalamic area at the time it seems to be 'ill'.[42] If lateral hypothalamic self-stimulation is associated with the pre-

[38] P. Schmitt *et al.* 1974; P. Schmitt *et al.* 1979*a*.
[39] See P. Karli 1976.
[40] E. T. Rolls 1976; E. T. Rolls *et al.* 1980; J. P. Kanki *et al.* 1983; J. R. Stellar and E. Stellar 1985.
[41] P. E. Maxim 1972.   [42] B. T. Lett and C. W. Harley 1974.

sentation of a previously unfamiliar flavour, the rat develops a marked preference for it; and resistance to extinction of this acquired preference will be all the greater as the number of associations is increased.[43] In the human clinic it seems that the affective experience of 'pleasure' or of 'well-being' produced by self-stimulation can attenuate a pain of pathological origin.[44] When chronic pain is caused in a rat (through the induction of arthritis by an injection of *Mycobacterium butyricum*) it is noted that the animal spontaneously increases the number and total duration of operations of the self-stimulation lever.[45]

Reward system and appetence behaviour

If the lateral hypothalamic area is a privileged region for self-stimulation, it is also the location where electrical stimulation can induce feeding, water-intake, or copulatory behaviour. And the majority of the points where stimulation causes one or other of these kinds of appetence behaviour also turn out to be self-stimulation points. Moreover, factors controlling the initiation or termination of some given appetence behaviour also have the effect of modulating the intensity of the reward effect — assessed by the level of self-stimulation performance — induced at the very place where electrical stimulation causes the behaviour.[46] Thus, food deprivation causes an increase in the frequency of self-stimulation at the lateral hypothalamus; this effect is only observed with hypothalamic points whose stimulation brings about feeding behaviour. Moreover, neurons activated both by the sight or taste of food and by self-stimulation of the lateral hypothalamic area in a hungry monkey no longer respond when the animal is sated.[47] By contrast, the frequency of self-stimulation diminishes markedly following the intravenous or intragastric administration of a hypertonic solution of glucose. For certain points (located in the posterior hypothalamus) whose stimulation brings about copulatory behaviour in the male rat, it is noted that the frequency of self-stimulation diminishes following castration and increases once more if testosterone is administered. In other words, activation of the lateral hypothalamic area, by lowering the blood glucose level or raising the level of circulating male sexual hormones, causes a selective accentuation of the reward effects initially linked with the anticipation and then with the registration of the behaviour's sensory consequences.

We have previously seen that activation of the lateral hypothalamic area on detection of an elementary biological need caused both a non-specific behavioural arousal and privileged processing of meaningful sensory information. It is the conjunction of this privileged treatment and the selective accentuation of the anticipated reward which will

[43] A. Ettenberg 1979.   [44] R. Heath 1963.   [45] P. De Witte *et al.* 1983.
[46] See P. Karli 1976; J. Le Magnen 1983, 1984; J. R. Stellar and E. Stellar 1985.
[47] E. T. Rolls *et al.* 1980.

orientate the behaviour of the 'aroused' organism towards satisfaction of the need. Hypothalamic activation, whether natural or experimental, therefore induces several complementary processes. This functional heterogeneity is also shown by the fact that aphagia (abolition of feeding behaviour) caused by lateral hypothalamic lesions can be of two very different types, depending on the precise location of the lesions.[48] More posterior lesions cause 'passive' aphagia: the animals react neither to food nor to other exogenous stimuli (marked sensory neglect), because of defective behavioural arousal. More anterior lesions cause the appearance of 'active' aphagia: in regard to both food and various sensory stimuli, animals present reactions of aversion and escape, probably reflecting a tipping of the balance between attitudes of approach and escape in favour of the latter.

Integrating role of the hypothalamus

We should stress here the very important 'crossroads' role played by the hypothalamus, the point where many and diverse processes converge and are integrated. We have already seen that, by means of specialized detectors, it records the fluctuations of various parameters of the internal milieu and participates in the bringing into play of a set of processes aimed at orientating behaviour towards the satisfaction of some detected need. But information originating in the environment and that takes on its full meaning, both cognitive and affective, through the processing carried out by the limbic system, also converges on the hypothalamus. Moreover, because of the control it exercises over the vegetative nervous system and the endocrine system, the hypothalamus coordinates the various aspects of the organism's vegetative life while continually adapting them to the current circumstances of dialogue with the environment. This being so, it is easy to see that this region of the brain plays an essential role in the integration of the somatomotor, visceromotor, and affective components of the emotional reactions.[49] When, following a lateral hypothalamic lesion, an animal is motionless (akinesia) and no longer ingests food (aphagia) or water (adipsia), this absence of all motivated behaviour reflects, not muscular paralysis but an absence of motor initiative, of 'spontaneity'. The rat presenting passive aphagia, discussed earlier, is perfectly capable of masticating and swallowing food that is placed directly in the buccal cavity.[50] In an animal rendered akinetic, progressive recovery of certain movements is witnessed: first, movements to straighten the body, then movements of exploration with the head, and finally locomotion. But all these movements lack true spontaneity: 'the animal seems to be a little robot, lacking all goal-directedness, responding reflexively with a stereotyped response to each configuration of surfaces it encounters'.[51]

[48] T. Schallert and I. Q. Whishaw 1978.
[50] T. Schallert and I. Q. Whishaw 1978.
[49] See O. A. Smith and J. L. De Vito 1984.
[51] P. Teitelbaum *et al.* 1983.

Genesis of the reward effect

Given that self-stimulation within the lateral hypothalamic area activates a functionally heterogeneous nervous substrate, several methods have been developed that permit one to distinguish the 'rewarding' effect proper from the simply 'arousing' effect of a given stimulation.[52] As for the reward effect, a consistent set of experimental data permits at least a partial description of the neural circuit involved in lateral hypothalamic self-stimulation.[53] The stimulation activates fibres descending in the medial forebrain bundle towards the ventral tegmental area of the mesencephalon, which brings about a release of endomorphines at this level (a morphine injection in the ventral tegmental area produces a reward effect and the animal carries out self-injections of morphine if the cannula is implanted there). The endomorphines thus released act on dopaminergic neurons of the ventral tegmental area that project upwards to the nucleus accumbens. The release of dopamine within the nucleus accumbens (without excluding the role that may be played by other projection sites of the dopaminergic fibres) is necessary to the genesis of the reward effect induced by self-stimulation. We may add that by blocking the functioning of the dopaminergic synapses one attenuates or abolishes the rewarding effect due to an intravenous injection of amphetamine or cocaine; and the animal interrupts its self-injections of amphetamine or cocaine if the dopaminergic nerve endings in the nucleus accumbens are destroyed. We shall have occasion at the end of this chapter to return to the complex 'modulating' role of the dopamine released within several structures in the forebrain.

Matters are complicated by the apparent fact that the reward effect can result, at least in part, from the involvement of different neural circuits. In this regard emphasis has been put on certain apparent differences that depend on whether self-stimulation involves the lateral hypothalamic area or the medial prefrontal cortex. First of all, the animal learns less easily to self-stimulate at the level of the prefrontal cortex; and when the self-stimulation behaviour has been acquired, performances are inferior to those seen for the lateral hypothalamus.[54] When one studies the changes induced by self-stimulation in the energy metabolism of the different areas of the brain (by the use of 2-deoxyglucose as marker, a glucose analogue that accumulates in the cell, for the latter cannot metabolize it), one notes that the most markedly activated structures are not the same: ventral tegmental area, nucleus accumbens, and lateral septum, in the case of lateral hypothalamic self-stimulation; dorsomedial thalamus, baso-lateral amygdala, and entorhinal cortex, in the case of self-stimulation at the level of the prefrontal cortex.[55] If an animal may choose between two self-stimulations, one of them affecting the

[52] See J. R. Stellar and E. Stellar 1985.   [53] R. A. Wise and M. A. Bozarth 1984.
[54] See J. R. Stellar and E. Stellar 1985.   [55] E. Yadin *et al.* 1983.

lateral hypothalamus and the other the ventral tegmental area of the mesencephalon, an injection of amphetamine (which activates dopaminergic transmissions) has no effect on the preference it has previously shown; by contrast, if the choice is between the lateral hypothalamus and the prefrontal cortex, the same amphetamine injection accentuates preferentially the reward effect induced at the level of the lateral hypothalamus, for the animal now shows a marked preference for lateral hypothalamic self-stimulation.[56] These observations—and some others—lead certain authors to consider the existence of several distinct neuronal reward systems, with interactions that have yet to be analysed.[57]

The neuronal aversion system

Following the initial observations made by Delgado and his collaborators (1954), other experimental investigations have confirmed the existence of a periventricular 'aversion' system including, basically, the periaqueductal grey matter and the medial hypothalamus. A cat or a rat rapidly learns instrumental behaviour permitting it to switch off electrical stimulation applied to the dorsal region of the periaqueductal grey matter[58] or to the medial hypothalamus.[59] An affective experience of aversive nature, reflected by flight or defence behaviour, may also be induced by neurochemical manipulations carried out on this periventricular system: local injection of excitatory amino acids—such as glutamic acid or aspartic acid—to cause a neuronal activation[60] or local injection of a GABA antagonist which may release a GABAergic inhibition normally exercised on the neuronal substrate involved in the genesis of aversive effects.[61]

The neuronal aversion system has undergone much less study than the reward system, and we shall simply state some data obtained in our laboratory. If an electrical stimulation is applied to the system within the central grey matter, the rat quickly learns to interrupt the stimulation by, for example, pressing a lever or running towards a photoelectric cell (placing itself in front of this cell, the animal interrupts a beam of infrared light), its 'escape speed' being directly proportional to the intensity of the stimulation applied. If two different points in this periventricular structure are simultaneously stimulated, the aversive effects so induced are additive, and the escape speed corresponds approximately to the sum of the speeds induced by the isolated stimulation of each of the two points.[62] When an aversive experience and an escape response are thus induced experimentally, it is noted that the 'escape speed' is closely correlated both with the frequency of discharge of certain neurons with-

[56] T. H. Hand and K. B. J. Franklin 1983.   [57] A. G. Phillips 1984.
[58] See P. Schmitt *et al*. 1974.   [59] See P. Schmitt *et al*. 1979a.
[60] R. Bandler *et al*. 1985.   [61] G. Di Scala *et al*. 1984; P. Schmitt *et al*. 1984.
[62] P. Schmitt and P. Karli 1980.

in the aversion system and with the degree of mydriasis (dilatation of the pupils, a vegetative sign of emotion) presented by the animal. The neuronal responses recorded are genuinely related to the aversion and escape, not simply to the behavioural arousal induced by the electrical stimulation. Indeed, in the case of hypothalamic stimulations that cause both approach and escape responses (the animal seeks the effects of the stimulation and then itself terminates them), neurons are found in the dorsal part of the central grey whose activity is closely correlated with the vigour of the escape responses and—in a more ventral region— other neurons whose activity is closely correlated with the vigour of the approach responses.[63] When two distinct points are stimulated within the aversion system, the rat is perfectly capable of discriminating between the effects induced by each of these stimulations, for it quickly learns to interrupt one of them in one of the two arms of a T-maze and the other one in the maze's other arm. This discrimination is achieved even if the *intensity* of the aversive effects induced (measured by the vigour of the escape responses expressing them) is identical for both stimulations; we therefore have to accept that the *nature* of the induced effects differs, at least in part, between one stimulation point and another.[64]

The behavioural responses thus evoked are in no wise characteristic of quasi-automatic reflex responses; on the contrary, the animal makes a choice which is determined by the nature and intensity of the aversive effects induced by the stimulation. This is shown clearly in the following observation: a rat is trained to interrupt a high-intensity stimulation by pressing on a lever situated in one of the arms of a T-maze and to interrupt a low-intensity stimulation applied to the same point by pressing on a lever situated in the other arm; following a micro-injection of morphine into the periaqueductal grey matter (whose effect is to attenuate the aversive effect induced by an electrical stimulation[65]), it is noted that the animal will now interrupt a high-intensity stimulation by pressing the lever it previously used to interrupt a low-intensity stimulation.[66]

The reward and aversion systems interact

The existence of functional interactions between the reward and aversion systems is well evidenced by the data furnished by lesion experiments and by the use of combined stimulation. The destruction of a region of the brain where self-stimulation is easily obtained (ventral tegmental area of the mesencephalon; the raphe nuclei), brings about a marked facilitation of the switch-off responses elicited by stimulation of the periventricular aversion system; we therefore have to accept that, following damage to the reward system, the activation of the periven-

[63] G. Sandner *et al.* 1982.   [64] R. Lappuke *et al.* 1982.
[65] F. Jenck *et al.* 1983.   [66] F. Jenck *et al.* 1986.

tricular system produces aversive effects of greater intensity. On the other hand, a 'rewarding' stimulation (applied to the lateral hypothalamic area or the dorsal raphe nucleus) has the effect of depressing, attenuating the switch-off responses connected with the aversion system, and positive correlation is observed between the respective amplitudes of this 'aversion-attenuating' effect and the 'appetitive' effect of the same rewarding stimulation.[67] Interactions of another kind are involved when the brain anticipates, and later registers, the positive and negative consequences of some behaviour, and when it calculates the 'cost/benefit' ratio. In this internal algebra relating positive and negative reinforcements, an essential role is played by the limbic system.

The limbic system

Historical précis: the notion of the limbic system

This notion was developed by stages. The adjective 'limbic' appeared in 1878, when Broca described under the name of 'great limbic lobe' the rim (*limbus*, in Latin) running around all sides of the threshold of each of the two cerebral hemispheres, that is, in the region where the brainstem penetrates into the hemisphere's internal or median face. Considering that 'its evolution in the mammal series is closely linked with that of the olfactory lobe', Broca includes this great limbic lobe (made up of two convolutions: that of the corpus callosum and that of the hippocampus) in what he terms the 'olfactory apparatus'. Half a century later, Herrick (1933) was still using the term 'olfactory cortex' for the whole of the archipallial surface (hippocampal cortex) and palaeopallial surface (piriform cortex). But, noting the presence of a well differentiated hippocampal cortex in a totally anosmic animal (i.e. one with no sense of smell) such as the dolphin, he suspects that the 'olfactory cortex' could be the site of a non-specific activity (not linked to the olfactory function) which 'finds expression in global behaviour, in learning and memorization abilities' and which is also exercised on the 'internal system of the organism's general attitude, its dispositions and its affective states'.

A first confirmation of Herrick's purely speculative hypotheses was quickly and almost simultaneously provided by the experimental results of Klüver and Bucy (1937) and by the anatomic-clinical observations of Papez (1937). Klüver and Bucy noted that a bilateral temporal lobectomy led to marked emotional hyporeactivity: after the operation, the animal seemed incapable of recognizing by sight the significance of the objects making up its normal environment ('psychic blindness'). Going on to an attempt at synthesis based on data from comparative anatomy and anatomic-clinical observations collected from the literature, Papez described a functional anatomical circuit (that was to bear his name from

[67] P. Schmitt and P. Karli 1984.

then on) to which he attributed a fundamental role in the elaboration and expression of the emotions and which he hypothesized was made up of the mammillary bodies of the hypothalamus, the anterior nuclei of the thalamus, the cingulate gyrus (cingulate cortex), the hippocampus, and their reciprocal connections.

From 1937 onwards, an ever-increasing amount of experimental data showed that the structures described as olfactory (or rhinencephalic) do participate in functional circuits which well overflow the framework of the merely olfactory function. And an important threshold was crossed when, in 1958, Nauta described in the cat a circuit which puts the hippocampus (and, secondarily, the amygdala) into a reciprocal relationship with well defined mesencephalic structures grouped by the author under the notion of a 'limbic' area of the mesencephalon; the pathways of this circuit run through the diencephalon (thalamus and hypothalamus), relay there and, in passing, send to it and receive from it numerous collateral connections.

Thus, the currently accepted significance of the notion of a limbic system has been gradually built up by a process of moving back from its olfactory origins in two senses, that of anatomical dimension and that of physiological competence: a study of the many anatomical relations maintained by the limbic structures with the hypothalamus, with the 'limbic' area of the mesencephalon, with the 'limbic' nuclei of the thalamus, pushes into the background the interest of the (properly speaking) olfactory afferent connections of these structures; in a parallel sense, in the appreciation of their functional significance, attention is switched from the olfactory afferences to be concentrated on the modulatory action exerted by the limbic structures on the somatomotor and visceromotor efferences. The most recent anatomical data show clearly the two following notions: the limbic system receives all sensory information (already largely pre-processed) from networks of the associative areas of the cerebral cortex; this system can act on somatic motricity through the intermediary of the nucleus accumbens, the substantia nigra, and the neostriatum (caudate nucleus and putamen) and on visceral motricity through the intermediary of the hypothalamus. We may add, generalizing to some extent, that the involvement of the limbic system allows sensory information to be compared with traces left by past experience and that it is enriched both by spatio-temporal references (especially through the intervention of the hippocampus, interacting with the prefrontal cortex) and by affective ones (especially through the intervention of the amygdala, interacting with the cingulate cortex and the orbitofrontal cortex).

The structures making up the limbic system

The limbic system is made up of an ensemble of cortical and subcortical structures which are phylogenetically old telencephalic structures (hence the notion of a paleomammalian brain), that is, their development in the vertebrate series preceded, in the course of evolutionary history, that of more recent telencephalic structures (the neocortex and neostriatum). The cortical structures of the limbic system are located on the inner or median surface of the cerebral hemisphere;[68] they are:

- the *archicortex* with, in particular, the hippocampus or 'horn of Ammon'; the hippocampus, the subiculum, and the gyrus dentatus together make up the hippocampal formation;

- the *paleocortex*, comprising the olfactory or 'rhinencephalic' structures proper: the olfactory bulb, together with a certain number of structures presenting close relations with the latter (olfactory tubercle, septal area, pre-piriform, and periamygdaloid areas);

- the '*transitional*' *cortex*, a transition between the archicortex and the paleocortex, on the one hand, and the neocortex on the other; the transitional cortex includes, among others, the entorhinal, orbital, and cingulate areas.

As for the ganglionic subcortical formations of the limbic system, they correspond to the nuclei of the septum and of the amygdala.

For the amygdala, as for the hippocampus, neural messages arriving from the various sensory receptors reach these limbic structures of the temporal lobe by a two-fold pathway: on the one hand, starting from certain sub-cortical relays of the great afferent systems; and in a much more developed form, on the other hand, from the associative areas of the cerebral cortex. The organization of the projections originating in the cortex seems to be different for the amygdala and for the hippocampus: whereas the various unimodal associative areas project on to distinct regions of the amygdaloid nuclear complex, each of these regions thus seeming to fall under the privileged influence of a given sensory modality, all of these projections converge on to the transitional (perirhinal and prorhinal) cortex, whence they reach the hippocampus.[69] It must be added that the prefrontal cortex (itself a site of convergence of connections arriving from the unimodal associative areas located in the parietal, temporal, and occipital lobes)[70] projects on to the hippocampus through the intermediary of the parahippocampal and presubicular cortex. Now, the latter cortex presents, as does the prefrontal cortex, a 'modular' organization; and it is therefore possible that the various lamellae of the hippocampus (with their functional organization in a plane perpendicular to the major axis of the hippocampus) are selectively associated with the various 'modules' of the prefrontal cortex, with great capacities for processing information.[71] As Swanson (1983) very rightly points out, the functional anatomical relations that obtain between the

---

[68] See P. Karli 1976.   [69] B. H. Turner *et al.* 1980; L. W. Swanson 1983.
[70] P. S. Goldman-Rakic 1984*b*.   [71] P. S. Goldman-Rakic 1984 *a*.

hippocampus and the prefrontal cortex on the one hand, and the unimodal associative areas on the other, plus the high degree of information processing taking place in the hippocampus, reduce the major interest previously thought appropriate to the distinction between a phylogenetically old 'archicortex' (the hippocampus) and the more recent 'neocortex'. We have also pointed out above[72] that it would be wrong to think that because of its phylogenetically greater age the limbic system had ceased to evolve and had made no contribution to hominization: while the number of fibres in the pyramidal tract—the neocortex's 'recent' efferent pathway—doubles as between monkey and man, the fibres of the fornix—the main efferent pathway of the hippocampus— have undergone five-fold multiplication. And, given that the development of the limbic system (the paleomammalian brain or 'emotional brain' of Maclean (1977)) has permitted development of the faculties of memorization and of affective experiences, Livingston is no doubt correct in attributing to it an essential role in the development of 'social dexterity'.[73]

At the prefrontal cortex, the amygdala projects in a privileged, if not an exclusive, manner on to the 'limbic' areas (cingulate cortex and orbitofrontal cortex), both directly and through the intermediary of the magnocellular part of the medio-dorsal nucleus of the thalamus.[74] The amygdala, which is connected to the hippocampus, sends descending projections towards the hypothalamus and towards certain structures of the mesencephalon via the stria terminalis and the ventral amygdalofugal pathway.[75] The hippocampus projects on to these same hypothalamic and mesencephalic regions through the intermediary of its principal efferent pathway, the fornix. The septum may be regarded as essentially a relay station for the pathways joining, in both directions, the hippocampus and the amygdala to a set of diencephalic and mesencephalic structures. The control exerted by the hippocampus and the amygdala respectively over somatomotor activities is made possible by the existence of projections from these limbic structures on to the nucleus accumbens and the neostriatum, i.e. the ensemble of the caudate nucleus and the putamen.[76]

Role played by the limbic system

Since it is through the involvement of the limbic system that cognitive and affective significance is conferred on current sensory information, by reference to the traces left by past experience, and since this significance orientates and nuances behaviour, it is easy to understand the important role played by this system in the adaptation of behaviour to

[72] In Chapter 3. [73] R. B. Livingston 1978.
[74] L. J. Porrino *et al.* 1981; J. L. Price 1981. [75] J. L. Price 1981.
[76] A. E. Kelley and V. B. Domesick 1982; A. E. Kelley *et al.* 1982.

the individual's experience. In an organism carrying on dialogue with its familiar milieu, and so open to its own history, the limbic system is indeed essentially involved in the expression of individual nuances of behaviour, in the expression of a personality that has been shaped by experience. It is not surprising, therefore, that a limbic lesion (especially one affecting the amygdala, the hippocampus, or certain regions of the 'transitional' cortex) should have an effect precisely on acquired behavioural characteristics, and that the lesion's effects should be so much more pronounced that, among the motivating factors concerned, those linked with past individual experience play a more important role. This role is basic in the case of socio-affective behaviour; it is much less so in regard to feeding or sexual behaviour.

Effects of limbic lesions on motivated behaviour

(a) *Feeding behaviour.* Limbic lesions do not in any way prevent energy needs being satisfied by an adequate intake of food. They only modify certain individual attitudes to food. Thus we observe changes of varying degree in feeding habits and preferences; deficiencies in the acquisition of a conditioned aversion for a given food; and a disturbance of processes which, taking account of the quantity and palatability of ingested food and acting through bucco-pharyngeal and gastric afferences, play a part in the onset of satiety.[77]

(b) *Sexual behaviour.* It is classic to place the signs of 'hypersexuality' among those consequences ensuing from bilateral lesions of the temporal lobe in adult males of various species.[78] Besides a general increase in sexual activity, a lack of discernment is noted in the choice of partner: thus, homosexual tendencies are more marked than before the operation; after it, cats mount even inanimate objects. There is also manifested a relative inadaptation to the general situation: cats so operated on will couple anywhere, whereas control animals only carry on sexual activity in a territory to which they have become adapted. The concept of hypersexuality takes in phenomena that would probably be more justly considered as essentially *qualitative* anomalies of sexual behaviour, in keeping with the much more general disturbances that affect the behaviour of animals with temporal lesions. Given that lesions of the amygdala, like those of the hippocampus, seem to depress rather than raise the level of sexual drive (measured by the frequency with which a male rat crosses an electrified grid to join a receptive female, or by the latent period of covering and intromission), it is possible to conclude that the main effect of temporal lesions is to abolish selective inhibitions

---

[77] See P. Karli 1976.
[78] These signs of hypersexuality have been described mainly in cats, but also in monkeys and man.

that were developed in the course of ontogenesis, to disturb the expression of individual behavioural nuances, and to cause a regression of behaviour towards something more stereotyped and automatic.[79] This view of matters is corroborated by data obtained by Aronson and Cooper (1979) who consider that lesions of the amygdala do not interfere at all with a mechanism controlling sexual behaviour as such, but that they cause deficient involvement of selective behavioural inhibitions which normally reflect the individual's past experience.

(c) *Socio-affective behaviour.* Results obtained in two experiments carried out respectively on monkeys living in captivity and others living in the wild give a convincing illustration of the major role played by the amygdala in the development of emotional states and the behaviour that expresses them. In macaques, Downer (1961) carried out both a section of all the inter-hemispheric commissures (a realization of the 'split brain') and a unilateral destruction of the amygdala. Following these operations, animals behave as follows: if the animal sees its habitual environment through the intermediary of that eye linked to the intact hemisphere, the other eye being closed, it presents differentiated, sometimes violent, affective reactions in regard to it depending on its past experience; if only the other eye is open, by contrast the monkey appears perfectly placid and indifferent. In other words, depending on whether the cerebral hemisphere concerned contains an amygdaloid nuclear complex and, consequently, whether visual information is processed by that amygdala, the behaviour of any one organism placed in a given situation can be totally different. If bilateral lesions of the amygdala are carried out in monkeys living in the wild, it is noted that the animals so operated on are incapable of rejoining their group or a neighbouring one.[80] It appears that these animals can no longer adapt their behaviour to that of their fellow-creatures by referring to experience (perhaps because they have become incapable of recognizing the significance of social signals emitted by their fellow-creatures), and their re-socialization therefore becomes impossible. In fact, even if no vegetative function is deeply disturbed and their fellows try to reintegrate them within the group, the monkeys that have been operated on isolate themselves and do not survive very long. It can be added that a similar disturbance of social relations is observed following a lesion to the prefrontal cortex (which is closely linked to the limbic structures of the temporal lobe, directly through the pathway of the uncinate fasciculus and indirectly via the medio-dorsal nucleus of the thalamus): a marked reduction of mimicry, of vocalizations, and of all social interactions is noted; and most of the monkeys that have undergone surgery become isolated and prove incapable of rejoining the social group of which they formed part.[81]

[79] P. Karli 1976.   [80] A. Kling 1972.
[81] E. A. Franzen and R. E. Myers 1973; R. E. Myers *et al.* 1973.

Mode of action of the limbic system

On the basis of observations made in patients and much experimental data obtained in various animal species, we may conclude that the limbic system is basically involved in two sets of closely complementary processes:

- On the one hand, processes by means of which cognitive elements and, above all, a specific affective content are associated with the objective data of current sensory information, by reference to traces left by past individual experience. This association, which gives the information of the moment its full significance, may lead the brain to foresee specific results that may be obtained by responding (or avoided by not responding) in a certain way. The information of the moment can thus acquire motivating virtues, i.e. it may become a motive for action, inciting the individual to respond (or to abstain from responding) by employing one of the means of action at his disposal in the behavioural repertoire.

- On the other hand, processes by means of which the brain records 'successes' or 'failures', when it compares results actually obtained with those anticipated in programming the behavioural response. Depending on whether there is concordance, a pleasant or unpleasant affective experience will be created and will modulate the significance of the sensory information that has just been responded to. And, by virtue of the involvement of the positive or negative reinforcement systems, the probability of initiation of this same response by the information in question will be either increased or reduced.

Experimental confirmation of this mode of action

The fact that the limbic structures of the temporal lobe are indeed involved in the association of an appetitive or aversive significance to a stimulus or situation, in the recording or evocation of the pleasant or unpleasant nature of an experience, is attested to by three sets of data furnished respectively by lesion experiments, localized and transitory functional disturbances, and recordings of unit neuronal activities.[82]

(a) Although section of the connections between the visual cortex and the temporal lobe does not prevent analysis of the physical parameters of the visual stimulus, it does, nevertheless, cause in monkeys a loss of the significance—especially affective significance—first attached to this visual information. A lesion interrupting the connections between the somatic sensory cortex and the temporal lobe, for its part, disturbs recognition of the significance of objects on the basis of tactile affer-ences. A bilateral destruction of the inferior temporal cortex, or the

---

[82] See P. Karli 1976, for details of the experiments and for bibliographical references.

temporal pole and the amygdala, causes a similar deficiency: a monkey that has undergone the operation is incapable of recognizing the positive signal, i.e. the one it had previously learned to associate with obtaining a reward.

(b) Numerous experiments carried out in various species (monkey, cat, rat, mouse) have shown that a transitory experimental disturbance of the functioning of the amygdala and the hippocampus (e.g. by an *epileptogenic* electrical or chemical stimulation, i.e. one which induces a neuronal synchronization that is reflected in a paroxysmal electrical activity) prevented the association of aversive significance with a stimulus and, therefore, the reactions of fear and behavioural inhibition normally initiated by this signal.

(c) At the level of the neuron, certain modifications of the bioelectrical response—of the 'discharge pattern'—reflect the coding of the significance conferred on an initially neutral stimulus. Thus, some neuron in a limbic structure, initially responding in an identical way to two sounds of differing frequency (for example, 1000 and 1500 Hz), progressively presents different characteristic discharge patterns in response to each of these two sounds, as the latter acquire a particular significance for the animal, one announcing the presentation of food, the other presaging the application of a painful electric shock; and these discharge patterns are truly functions of the kind of reinforcement forecast by the sound, for they are inverted as soon as the significance (appetitive or aversive) attached to the individual sounds is inverted. Another example: certain neurons in the infero-temporal cortex of the monkey present a different discharge pattern in response to a visual stimulus, depending on whether the latter (whose physical parameters remain unchanged) has acquired a particular significance by being previously associated with a reward and causes anticipation of that reward. It is quite obvious that modifications like these, recorded at the level of the single nerve cell, are not the only representations of the coding of the significance acquired by a given stimulus: they are only an elementary manifestation of a process that involves a whole neural network of some degree of complexity.

The essential part taken by the limbic system in the evaluation of the consequences of some behaviour, in the genesis of the resulting affective experience, and in the possible modification of the significance—and therefore of the incitatory properties—of the stimulus that caused the behaviour, has also been clearly shown by many experimental investigations. Thus, a bilateral lesion of the amygdala has the effect of reducing the animal's sensitivity to any change in the reward anticipated on receiving a stimulus. In general, behavioural disturbances appear following various limbic lesions, whenever the animal has to learn to refrain from responding to a stimulus, to inhibit a given behavioural response in order to avoid *punishment* in the form of a painful stimulus or *frustration* due to the absence of the expected reward. The

individual seems not to profit from his failures and thus tends to per-
severe in behaviour that is now inadequate.[83] It is therefore genuinely
due to the involvement of the limbic system that complex interactions
between positive and negative reinforcements take place (including a
calculation of the ratio between the 'benefits' to be expected from some
behaviour and the 'cost' that may have to be borne). And it is the result
of this 'interior algebra' that will be reflected in either the facilitation
(and consolidation) or the inhibition (and extinction) of a given behavioural response.

After thus having outlined the limbic system's mode of action, it
would be proper to give a brief account, though in more detail, of recent
ideas about the role played by this system in mnesic processes (the
acquisition, retention, and recall of information) and in the genesis of
the emotions.

Limbic system, mnesic activities, and genesis of the emotions

It is a little artificial to separate memory from the genesis of the emotions, for affective connotations themselves undergo memorization and
it is by reference to its affective attributes, thus memorized, that a piece
of information acquires its full significance and gives rise to a certain
emotional state. Having said this, it must be emphasized that recent
research on the neurobiological bases of memory put the accent on the
fact that memory cannot be considered a monolithic phenomenon and
that it is proper to distinguish at least two main categories of attribute for
any specific memory, each category being the object of separate encod-
ing. It thus seems that the amygdala encodes, conserves, and extracts
affective attributes, whereas the hippocampus encodes the environmental context — particularly spatio-temporal attributes — of the same specific
memory.[84] This functional specialization of the amygdala and the hippocampus explains why a monkey's visual memory is profoundly disturbed by the combined ablation of these two temporal structures,
whereas it is not disturbed by the ablation of one of them in isolation.[85]
A similar distinction can be made between the processing of affective
attributes and that of contextual attributes, when examining the role
played respectively, in the primates, by each of the two parts of the
medio-dorsal nucleus of the thalamus: the medial (magnocellular) part,
with its projection on to the orbitofrontal ('limbic') subdivision of the
prefrontal cortex and the lateral (parvocellular and phylogenetically
more recent) part, with its projection on to the dorsolateral subdivision
of the prefrontal cortex.[86] A distinction of a different kind may be
established between a memory concerning particular items or events

---

[83] See P. Karli 1976; J. A. Gray 1982.
[84] R. P. Kesner 1981; R. P. Kesner and J. D. Hardy 1983.
[85] M. Mishkin 1978.   [86] H. J. Markowitsch 1982.

and a memory dealing with a skill, with perceptuomotor and cognitive capacities: lesions to the medial surface of the temporal lobe noticeably disturb the former while affecting the latter much less.[87] Generally, the existence of correlations between different amnesic syndromes and cerebral lesions with a particular localization confirms both the complex and heterogeneous character of mnesic activities and the functional specialization of the various cerebral structures that make up the memory 'circuits' linking the medial surface of the temporal lobe (together with the hippocampal formation and the amygdala), via thalamic relays, to several regions of the prefrontal cortex.[88]

There is no doubt that the hippocampus, an important link in the Papez circuit, plays a privileged role in learning processes and memory (let us recall that Papez described this circuit, in 1937, as a 'circuit of emotion'). Apart from anatomical–clinical data, a very large amount of experimental research has permitted the establishment of interesting correlations between the bioelectrical activities of this temporal structure or some of its neurochemical characteristics and various phenomena of learning and memory.[89] When a very fine analysis is made of the neural circuits within the hippocampus of the rat or the mouse, it is noted that genetic differences exist between one stock and another. Now, these hereditary differences in hippocampal circuits go together with differences in the capacities to acquire and execute certain tasks, more especially in the faculty for eliminating inappropriate responses, i.e. those found inefficacious, when a certain type of avoidance behaviour is to be acquired.[90]

Role of the amygdala in the genesis of the emotions

The amygdala plays an essential role when, by reference to the traces left by experience, a piece of sensory data gives rise to an emotional experience and an emotional reaction. This fact is clearly shown by numerous observations made in the human clinic and by a large amount of experimental data obtained in animals. It can be said at the outset that electrical stimulation of the amygdala causes emotional reactions, both in man and in various animal species.[91] In the content of hallucinations brought on by stimulation in human subjects, affective elements are closely associated to the cognitive elements. It is likely that this activation of the amygdala—which also involves structures connected to it— initiates processes which, in natural conditions, give affective significance to current sensory information by reference to past experience. Indeed, the content of the induced hallucination and the quality of the affective experience that accompanies it depend largely on the personality of the subject and the concrete psychosocial context in which

[87] L. R. Squire 1982.
[89] See W. Seifert 1983; J. Delacour 1984.
[91] See P. Karli 1976; E. Halgren 1981.
[88] See J.-L. Signoret 1984.
[90] H. Schwegler and H. P. Lipp 1983.

the amygdaloid stimulation takes place. One may therefore consider that this stimulation mainly produces a state of emotional tension whose 'discharge' is organized by the integrated use of mechanisms involved in natural conditions, mechanisms corresponding to certain personality traits and capable of being modulated by the psychosocial context of the moment. Moreover, stimulation of the amygdala has vegetative effects which correspond to the vegetative components of the natural emotional reactions; and a close correlation is noted in animals between the somatomotor and vegetative responses so triggered: a slowing-down of cardiac frequency when the stimulation causes flight, and an acceleration of cardiac contractions when it causes a defensive reaction.

The recording of unit neuronal activities has also shown that the amygdala and the orbitofrontal cortex with which it is interconnected are involved when a stimulus causes the anticipation of certain consequences, and that it thus arouses an expectation which includes positive or negative affective attributes. Certain neurons, for instance, are activated selectively within these 'limbic' structures (at the same time as are others in the lateral hypothalamic area, at locations which also elicit self-stimulation) at the mere sight of a foodstuff for which the animal has developed a strong appetite.[92] Other neurons respond, equally selectively, to objects having aversive significance for the animal; and the experiments carried out show that these neuronal responses are indeed linked to affective attributes, not to the objective sensory characteristics of the objects presented. Given the primary role played in socio-affective exchanges between primates by recognition of faces and their expressions, it is interesting to note that, within the amygdala of monkeys, neurons exist which are activated only by the sight of certain faces of fellow-creatures or human subjects; moreover, these neurons respond differently to different faces, and responses of some of them are particularly marked when the face concerned is one which provokes a clear emotional reaction.[93]

There are reasons to think that the amygdala takes part not only in the mechanisms presiding over the development of an emotional reaction based on the perception or evocation of a given stimulus or situation, but also in those which, by the play of numerous intracentral and peripheral reafferences, are capable of maintaining, or even amplifying, the initial emotional reaction.[94] To cite only one concrete aspect of things: an amygdaloid stimulation which causes a defence reaction in the cat regularly determines a discharge of catecholamines by a set of cerebral structures as well as by the suprarenal medullae; the catecholamines thus released bring about an elevation of muscular tone as well as various vegetative effects whose recording, conscious or otherwise, may in turn maintain the emotional reaction.

[92] E. T. Rolls *et al.* 1980; S. J. Thorpe *et al.* 1983.   [93] C. M. Leonard *et al.* 1985.
[94] See P. Karli 1976.

Effects of a bilateral lesion of the amygdala

If we consider the role played by the amygdala in the genesis and fostering of the emotions, it cannot surprise us that a bilateral lesion of the amygdala leads to a marked diminution of emotional reactivity, both in man and in the various animal species that have been studied in this connection. In every case, this reduction of emotional reactivity is reflected by a certain indifference towards the environment (particularly towards threats that may emanate from it), by clearly attenuated reactions to the various social stimuli. Thus, lesions of the amygdala strongly reduce the probability that an aggressive response will be triggered, and they attenuate reactions of fear, flight, or defence just as much.[95] The behavioural effects of amygdaloid lesions are particularly marked in cases where, before the operation, the level of emotional reactivity was particularly high and was expressed in behavioural responses that were often violent. Wild rats which react violently to the least stimulation and are difficult to handle even with thick protective gloves become placid and easy to handle with bare hands after they have been amygdalectomized.[96] In the same way, the post-operative attenuation of defence reactions is particularly marked in cats which are initially particularly savage.[97] And it is also in 'hyperkinetic' children that the effects of amygdaloid lesions—done for therapeutic purposes—are especially marked: reduction of irritability and of disordered activity, suppression of aggressive impulses.[98]

Some characteristics of amygdaloid mechanisms

The functioning of amygdaloid mechanisms is conditioned both by a set of intrinsic functional characteristics and by humoral factors which come to act on the neurons of the amygdala. These neurons are richly provided with receptors which bind steroid hormones (sexual and cortico-suprarenal hormones) or endorphins (endogenous morphines), both of which are known to modulate the way in which the individual perceives a potentially stressful situation and, consequently, the way he will confront it.[99] In cats, an individual predisposition to display a primarily 'offensive' or primarily 'defensive' attitude to the most various threats is closely correlated with certain neurophysiological characteristics of the amygdala: a predisposition for the 'defensive' attitude goes together with more marked and prolonged neuronal responses within the amygdala, when the animal is faced with a threat; it also goes together with the great ease with which paroxysmal crises are caused within the amyg-

---

[95] See P. Karli 1976; H. Ursin *et al.* 1981; and Chapter 8 of the present work.
[96] P. Karli 1956.   [97] H. Ursin 1965.
[98] H. Narabayashi 1972. It is obvious that this kind of intervention, termed 'psycho-surgical', raises ethical problems. These will be examined in later chapters.
[99] See 'Role played by humoral factors', p.112.

dala by electrical stimulation.[100] This individual predisposition, which is very stable in a given animal and constitutes a true 'personality trait', can be modified experimentally. A repeated electrical stimulation of the amygdala ('kindling') in fact permits the lasting transformation of an initially offensive attitude into a more and more fearful and defensive one; and, in parallel with this, the neurophysiological characteristics of the amygdala become those found in the animal that is spontaneously predisposed to a defensive attitude.

The amygdala does not operate in isolation

These researches on the neurobiological determinants of a certain general attitude to threats have more recently shown that the intrinsic functional characteristics of the amygdala are not the only ones involved. It turns out, in fact, that an important role is played by certain characteristics of the transmission of neural messages between the amygdala, the ventral hippocampus, and the ventro-medial hypothalamus.[101] It is not, therefore, the functioning of a structure taken in isolation that is the determinant but rather the *configuration* and the dynamics of the *interactions* between this structure and several other ones. Given that, at such an early a stage as the mesencephalic tectum (superior colliculus and dorsal region of the periaqueductal grey matter), there are mechanisms which become involved in qualitative modulation of the reactivity of the organism and which orientate its attitude towards 'approach' or 'retreat', it would be of the greatest interest to know the exact nature of the relations between these mesencephalic mechanisms and the mechanisms in which the amygdala participates. In this regard, only some fragmentary data are available. Thus, Jürgens (1982) analysed the mechanisms responsible for the different types of vocalization in monkeys, and was able to show that distinct circuits linking the amygdala to the periaqueductal grey matter underlie the respective emission of vocal messages signalling motivations characterized by confidence and determination, and those expressing 'resentment', fear, and flight.

Role of the septum in moderating emotional reactions

While the amygdala plays an essential role in the genesis and fostering of emotional reactions, it is another limbic structure that is involved in the process moderating those reactions: this is the *septum*. In all species which have been studied in this connection (including man), a destruction of the septum causes the appearance, generally transitory, of all the signs of marked hyperreactivity.[102] The septal lesion seems mainly to accentuate sensitivity and reactivity to stimuli and situations of aversive

[100] R. E. Adamec 1978.   [101] R. E. Adamec and C. Stark-Adamec 1984.
[102] See P. Karli 1976, 1981, 1982.

character. Thus, in rats, the destruction of the septum induces a more marked defensive attitude in response to threatening signals from fellow-creatures. In the opposite direction, an activation of the septum by electrical stimulation has the effect of attenuating behavioural and vegetative reactions by which an aversive emotion is expressed; and the animal will learn behaviour that permits it to obtain this stimulation of the septum if—and only if—an aversive emotion is created by stimulation of the medial hypothalamus.[103]

Reciprocal shaping of septal functioning and social interactions

The way in which the septum is involved in the processes that moderate emotional reactivity at a given point in ontogenesis depends on its own functional characteristics at this time. These characteristics are, it is true, determined by the genome; but they also depend on lived experience, i.e. on the social interactions themselves. Thus social conflict situations affect the activity of an enzyme, tyrosine-hydroxylase (which is involved in the biosynthesis of the catecholamines), at the septum, and this effect is more or less marked according to whether the animal has already learned to face the situation.[104] In other words, the septum is an integral part of a system which, like the brain within its ensemble, is both dynamic and open. The reciprocal character of the relations between the brain and behaviour also appears in the fact (alluded to previously) that a similar hyperreactivity can be brought about by a septal lesion or by social isolation. Under normal conditions, the septum's functioning interacts with social contacts to determine an appropriate level of reactivity. Social interactions cannot have their normal effect on this behavioural dimension without the mediation of the septum; and, vice versa, the septum cannot exert its moderating action if it has not undergone the structuring influence of social interactions.

Cerebral cortex, internal representations and operations of anticipatory simulation

If the living creature must interact in a fashion adapted to its environment, if it must acquire and preserve mastery in this dialogue, it is essential that it be able to interpret the events occurring in the surrounding space. It must, at all times, be able to compare the information emanating from the 'real and present world' with those furnished by the internal representations of the extra-personal space and the familiar social environment. It is this comparison and the ensuing interpretation of events which generate a set of predictions and expectations, in turn leading to the fixing of concrete objectives and the development of

[103] E. Thomas and G. J. Evans 1983.   [104] A. Raab and R. Oswald 1980.

appropriate action projects. The internal representations under whose influence the behaviour employed will consist of actions charged with sense, are continually brought up-to-date by the very fact of retrospective effects of past action. These representations, which are dynamic, open structures, integrate the perceptual characteristics of the environment, the expectations they arouse, and the behavioural strategies likely to achieve them. We may therefore expect the neuronal systems that constitute the medium for these representations to be quite widely distributed within the brain.

Internal representations of extra-personal space

Based on an analysis of syndromes involving 'sensory neglect' brought about by certain lesions to the cerebral cortex, and of the unit neuronal activities recorded at the level of the cortex in well defined experimental conditions, Mesulam (1981) considers that the internal representation of extra-personal space and the appropriate means for its exploration are provided by an integrated neural network comprising three complementary and closely interacting components. On the one hand, a very highly developed sensory representation is constituted within a particular region of the posterior parietal cortex (the PG region, or area 7a), which receives sensory information that has already undergone advanced processing in the unimodal and polymodal association areas. On the other, a sort of 'motivational map' is built up within the cingulate cortex, representing the spatial distribution of 'motivational valence' and the corresponding expectation. And these two representations (one 'sensory', the other 'motivational') interact with certain regions of the frontal cortex (in particular, oculomotor area 8) and structures of the brainstem (superior colliculus and activating reticular formation) which are known to play an essential role in the genesis of behavioural arousal and in the execution of the movements necessary to visual and tactile exploration of extra-personal space. Given the importance of these interactions, it is easy to see that the functional deficiency caused by a lesion affecting one or other of the cortical regions concerned cannot be a purely sensory, motor, or motivational one, but rather somewhat more complex, because more global. For all that, these regions are not equivalent, for a lesion involving them all causes a more pronounced neglect syndrome than a lesion involving only one of them. Unilateral injury to this integrated neural network causes lack of interest in the contralateral half of the body and the adjacent half of extra-personal space and, therefore, difficulties in exploring and acting on this side in a well orientated and effective way. For reasons which are poorly understood, the most marked neglect syndromes are caused in man by posterior parietal lesions but, in monkeys, by frontal lesions.[105] However this may be, the

---

[105] See J. Hyvärinen 1982.

recording of unit neuronal activities has clearly shown that there are neurons, at the level of both the posterior parietal cortex and the prefrontal cortex, which do not respond to a given signal unless the monkey executes towards that signal a specified positive movement that will be rewarded.[106] In other words, similar neuronal activities are indeed involved in the coding of a close integration of perceptual data with motor and motivational data.

Integration of temporal information

While *spatial* data thus come to be inscribed in the internal representations that permit the brain to 'scan' the environment and to localize pertinent information within it, *temporal* data also come to be integrated, relating to the order of events over time and the way in which the partial realizations of a given objective succeed each other. To the extent that the realization of an objective corresponds to the employment of a strategy taking place in space and time, the programming of a component of this strategy is possible only if the brain has retained the memory of what has gone before. When monkeys are subjected to 'delayed response' experiments (the animal can respond to a signal only after a certain delay during which the signal is not present), it is noted—by recording unit neuronal activities—that many neurons are activated within the prefrontal cortex throughout the delay separating the response from the arrival of the signal: some of these neuronal activities are linked to the signal and constitute an *eye on the past*, while others are linked to the response and constitute an *eye to the future*.[107] By thus underlying the processes of memory and anticipation, these activities ensure liaison between the past signal and the action to come. The fact that this act corresponds to a certain motivation and that it is aimed at a specific goal, is attested by the existence of still other neurons whose activity is correlated with a certain expectation, with the recording of reward, or, on the other hand, with the recognition of error and failure, i.e. with motivational aspects. Given all this, it is easy to understand that lesions of the prefrontal cortex, in humans as in monkeys, cause deficiencies of some degree of depth in tasks requiring the faculty either of situating items of information in relation to each other on the time scale or of linking a set of acts into a correct and appropriate sequence, or, again, of correcting a strategy on the basis of the errors recorded.[108]

Interactions between cortical and sub-cortical regions in data processing

The prefrontal cortex is thus involved in the processes that permit behaviour to be inscribed appropriately and efficaciously in space and

---

[106] See J. Hyvärinen 1982; C. J. Bruce and M. E. Goldberg 1984; J. M. Fuster 1984.
[107] J. M. Fuster 1984.   [108] See B. Kolb 1984; B. Milner and M. Petrides 1984.

time. We have already seen that the hippocampal formation, for its part, plays an essential role in processing spatio-temporal attributes in the course of the acquisition and use of memorized information, and that it has 'modular' relationships with the prefrontal cortex. It is, therefore, highly likely that the hippocampal formation and the prefrontal cortex *interact* closely in the processing of spatio-temporal information, when the real and actual world is to be compared with the internal representations of extra-personal space and behavioural sequences that may occur therein. In his analysis of the neuropsychology of anxiety, Gray (1982) considers that the subicular area (close to the hippocampus, on the medial surface of the temporal lobe) could, in respect of all sensory information received and transmitted by the entorhinal cortex, play the role of 'comparator' with the forecasts generated by the cingulate cortex (a 'limbic' region of the prefrontal cortex), the latter having access to the representations of the environment and the relations that the individual maintains with it.

In the processing of information, particularly visual information, relating to socio-affective exchanges and in the structuring of these exchanges by individual experience, the privileged role no longer falls to the connections between the posterior parietal cortex and the cingulate cortex (the parieto-frontal or *dorsal* sensori-limbic system), but to the connections between temporal structures (infero-temporal cortex and amygdala) and the orbito-frontal cortex (temporo-frontal or *ventral* sensori-limbic system).[109] We have already seen that lesions of the amygdala deeply disturb socio-affective exchanges. Lesions of the orbito-frontal cortex, too, cause profound changes in personality and social behaviour: deficient perception of the emotions expressed by others, impoverishment of spontaneous facial expressions, and a tendency to social isolation.[110] In monkeys, these changes have the consequence that a dominant animal loses its status and falls rapidly to the lowest level in the hierarchy, presenting submissive behaviour towards other group members. This change of position in the group's internal dynamics does not in any way result from some change in the animal's physical capacities but rather in its capacity to process correctly the sensory information emanating from others and to respond appropriately.

Right cerebral hemisphere and affective states

It must be added that, in man, it is the right cerebral hemisphere which plays a privileged role in the recognition and expression of the affective states.[111] Subjects who have suffered lesions of the right hemisphere present clearly more marked deficiencies than subjects with lesions to

---

[109] D. M. Bear 1983.   [110] See J. P. C. de Bruin 1981; B. Kolb 1984.
[111] See D. M. Bear 1983; L. I. Benowitz 1983; G. Gainotti 1983; M.-C. Goldblum and A. Tzavaras 1984.

the left side, as regards the ability to evaluate affective states that are expressed by non-verbal means, in particular by facial expressions. In subjects whose two hemispheres have been isolated from each other ('split brain'), the right hemisphere proves more apt than the left hemisphere to recognize individual faces and evaluate the affective significance of facial expressions. It is also in subjects who have suffered lesions to the right hemisphere that errors in the evaluation of the emotional content of a text are particularly numerous. It must be stressed that the observations made in this field favour 'a disturbance of supra-modal nature, and a specialisation of the right hemisphere for processing emotional information *per se*'.[112]

Motivation and decision

When a piece of information, received or recalled, has acquired its full cognitive and affective significance by reference to internal representations, it can give rise to a *behavioural response* reflecting both the individual's personality, the result of experience, and the current psychosocial context. In the choice of an appropriate strategy and its effective utilization, an important role appears to be played by neural circuits which cause the prefrontal cortex and the caudate nucleus to interact, the thalamus acting as a relay.[113] It is interesting to note that, as with the connections between the prefrontal cortex and the hippocampal formation, those linking the prefrontal cortex to the caudate nucleus present a similarly modular organization.[114] It is, therefore, tempting to think that information processing leading to the choice of an appropriate behavioural strategy corresponds essentially to a process of anticipatory simulation which carries out a review of the set of available 'modular units'.

In the title of the present chapter we speak of processes of motivation and decision. In reality, the ideas of 'motive for action' and 'decision to act in a certain fashion' are very largely overlapping. We have, furthermore, previously indicated that, for the neurobiologist, analysing the genesis of a motive for action comes down in specific terms to analysing the various factors that determine the probability of deployment of a given mode of action, and the mechanisms through which they act. In other words, the possible contribution of a factor to the 'motivation' process is judged on the basis of its influence on the probability of the 'decision'. Since the 'motivating properties' of a factor are thus defined operationally by reference to the probability of deployment of a certain mode of behaviour, it is obvious that—for the behaviour in question—the processes of motivation and the processes of decision are, with great exactitude, the same thing. Given that these processes are not in any

---

[112] M.-C. Goldblum and A. Tzavaras 1984.   [113] E. V. Evarts *et al.* 1984.
[114] P. S. Goldman-Rakic 1984*a*.

way organized into a linear sequence but often occur in parallel, and that they interact in ever more complex fashion, it would in any case be perfectly impossible to say where motivation ends and decision begins. Furthermore, the probability of the use of some behaviour and the corresponding 'motivational state' are in continuous evolution over time, even after the behaviour has been initiated, because the retrospective effects arising from partial realization may very well increase or decrease the probability of the action's continuation. At any time, this probability of initiation or continuation of the action is determined by the configuration of a complex causal field. Every pertinent factor whose influence comes to form part of this causal field is, in turn, influenced by it, by its configuration, and by the internal dynamics of the moment. In the development of a given form of behaviour, every particular operation of information processing therefore interacts with a certain 'motivational state' corresponding to the instantaneous configuration of the whole causal field.[115] Among the factors involved, we still have to look at the hormonal substances and neuromodulators which exert a modulating or 'enabling' influence over most of the mechanisms we have reviewed so far in the present chapter.

Role played by humoral factors

There are many hormonal substances (discharged into the circulating blood by the endocrine glands) and neuromodulating substances (produced by the brain itself) that act within the brain throughout its ontogenesis: in the course of development they contribute to promoting morphological and functional differentiation of the nervous system by acting on the establishment of connections; in the adult, they selectively facilitate the involvement of these connections in appropriate spatiotemporal sequences. This is an aspect of neurobiology that has recently undergone significant development. But here we shall limit ourselves to some data more particularly concerning socio-affective behaviour.[116]

Action of the sexual hormones

The sexual hormones (androgens and oestrogens) are involved mainly in the differentiation and subsequent deployment of sexual behaviour: at early stages in development, certain structures of the hypothalamus undergo differentiation under the action of the sexual hormones, an anatomical and functional differentiation which is at the basis of sexual behavioural differentiation; in adults, the action of the sexual hormones

---

[115] See P. Karli 1984.

[116] Some of these data will be discussed in greater detail in Chapters 7 and 8, which deal with the various categories of factors and brain mechanisms underlying aggressive behaviour.

on their receptors—paticularly dense within the hypothalamus—plays an essential role in the control of sexual behaviour.[117] But the role played by these hormones is not in any way limited to the sexual sphere. In rodents, for whom olfactory signals contribute greatly to social interactions, the circulating sexual hormones tightly control both the production of specific olfactory signals ('pheromones') and sensitivity in regard to them. In other species—including humans[118]—in which such signals are far from playing such a preponderant role, it has been found that plasma testosterone levels correlate with individual sensitivity to frustration, threat, and provocation. If male sexual hormone levels are thus capable of modulating the probability of the initiation of an aggressive response, it is interesting to note that, in reverse a conflict situation and aggressive behaviour have an effect on those same levels. In the macaque, the levels drop following an experience of defeat by a fellow-creature.[119] In the rat, too, a marked fall in plasma testosterone level is noted in the 'loser' following an aggressive interaction involving some degree of violence, while the same level in the 'victor' remains unchanged.[120] In man, differences observed between the sexes in the morphology of certain regions of the brain (the planum temporale, frontal operculum, corpus callosum), and in the degree of functional specialization of the two hemispheres, arouse much interest—along with much controversy, not all of it of a scientific nature because, it seems, equality and identity are confused with each other. The deterioration of spoken language following lesions to the left hemisphere, like the deterioration of visuo-spatial faculties following lesions to the right one, is generally more marked in the male than in the female. As regards most verbal or non-verbal tasks, left or right lesions have different effects in the male, while they often have approximately the same effects in the female. These observations have led to the conclusion that a man's brain is more markedly 'lateralized' (that it is characterized by a more marked functional specialization of each of the two hemispheres) than that of a woman, at least as far as language and visuospatial faculties are concerned.[121]

Action of hormones of the suprarenal cortex

The steroid hormones secreted by the suprarenal cortex also contribute to determining the probability of initiation of certain behavioural responses; their secretion is, in turn, modulated by behaviour and by some of its consequences. Confronted by a given situation, the activation of the suprarenal cortex will be more or less intense depending on the factors of novelty, uncertainty, conflict, and frustration which the individual discerns in his environment; activation of the pituitary and

[117] See C. Aron 1984.   [118] D. Olweus *et al.* 1980.   [119] A. F. Dixson 1980.
[120] J. M. Koolhaas *et al.* 1980.   [121] See T. E. Robinson *et al.* 1983.

the suprarenal cortex will be more pronounced for less familiar situations.[122] In the macaque, the probability of the appearance of fearful behaviour is closely correlated with plasma corticosteroid level, but this level itself depends on the nature of the dominance relationships the animal has previously had with its fellow-creatures. When pigs are put into a competitive situation, the plasma corticosteroid level increases more in the dominated animals than in the dominating ones. If human subjects are gathered together and confined in groups of three for several weeks it is noted that, initially, the plasma cortisol level is higher in individuals occupying a lower position in the group hierarchy ('leadership rating'); after three weeks, when stable and foreseeable relationships have become established within the group, these differences in plasma corticosteroid levels have disappeared.[123] The suprarenal hormones thus released exert a feedback action on certain cerebral structures (particularly the hippocampus and the septum), thus contributing to the acquisition of behaviour appropriate to the situation.[124] It is noted, for example, that a high level of plasma corticosteroids (whether naturally high or artificially raised) facilitates the acquisition of an efficacious response to a danger signal, just as it facilitates the extinction of an unrewarded form of behaviour. On the other hand, it is known that suprarenal activation becomes less intense when the animal has learned, by means of appropriate behaviour, to avoid electric shocks given it in a situation known as 'active avoidance'. It does appear that the endorphins, which we shall look at later, act in parallel with the hormones of the suprarenal cortex (as well as with the catecholamines released by the suprarenal medulla), both in the contextual perception of a situation and in the acquisition of an appropriate behavioural response.[125]

Role of cerebral catecholamines

The catecholamines (noradrenaline and dopamine) released within the brain play an important role in many processes of filtering and selective facilitation that are able to modulate one or other of the stages in information processing, leading from reception of signals to initiation of action. The improvement of signal-to-noise ratio observed in the responses of the neurons of the lateral geniculate body and of the primary visual cortex under the effect of spontaneous or elicited arousal can be reproduced by stimulation of the noradrenergic neurons of the locus coeruleus.[126] In the same way, noradrenaline improves the transmission and integration of somaesthetic information in the cortical circuits, and this process could well be responsible for the sharpening of corresponding perceptions under the influence of cortical arousal and attention.[127]

[122] See R. Dantzer 1984.   [123] J. Vernikos-Danellis 1980.   [124] See R. Dantzer 1984.
[125] S. Amir *et al.* 1980; H. Akil *et al.* 1984.
[126] M. S. Livingstone and D. H. Hubel 1981.
[127] B. D. Waterhouse and D. J. Woodward 1980.

Although the activation of the noradrenergic neurons of the locus coeruleus causes a marked improvement in signal-to-noise ratio within the limbic structures on to which these neurons project, it should be emphasized that this improvement is observed only if the activation is induced by pertinent signals charged with a particular significance for the individual;[128] now, we have seen, the limbic structures are broadly involved in the processes through which this significance is associated with the objective parameters of the signal. It must therefore be accepted that filtering requires the involvement of reciprocal relationships between noradrenergic neurons of the locus coeruleus and their sites of projection: the noradrenaline released modulates the functioning of the limbic structures; but retrospective effects of limbic origin in turn modulate the activity of the noradrenergic neurons of the locus coeruleus.

In the ascending dopaminergic projections, three main components are generally distinguished: the projection of the dopaminergic neurons of the substantia nigra (cellular group A9) on to the corpus striatum (nigrostriatal system), and the projections of the dopaminergic neurons of the ventral tegmental area of the mesencephalon (cellular group A10) on to the nucleus accumbens (mesolimbic system) and on to the prefrontal cortex (mesocortical system) respectively. The release of dopamine within the corpus striatum (caudate nucleus and putamen) facilitates the motor initiative, the choice of a suitable strategy and the inhibition of motor activities capable of interfering with its utilization, as well as accentuation of the incentive virtues of the signals under the effect of processes of positive reinforcement.[129] It is not, therefore, surprising that a bilateral interruption of the nigrostriatal pathway in animals causes akinesia (loss of motor initiative) and a reactivity that is strongly attenuated in regard to stimuli emanating from the environment, or that damage to this dopaminergic system in Parkinson's disease is reflected, in the subject so afflicted, by a poverty of spontaneous movements and a fixed facies.

But the nigrostriatal system is not involved solely in the initial choice of a strategy and the motor initiative necessary to its effective application. It is also involved in the commutative processes by means of which the living being changes strategy on the basis of information which may be of internal or external origin. In rats placed in a situation where survival is at stake, an alteration of the dopaminergic mechanisms in the neostriatum[130] brings about the stereotyped maintenance of whatever behaviour was initially chosen, no matter what the efficacy of the strategy thus adopted and maintained, whereas control animals change their behaviour continuously and rapidly develop a fully efficacious

---

[128] See A. Cools 1981*b*.
[129] See R. J. Beninger 1983; J. B. Penney and A. B. Young 1983.
[130] This alteration is caused by very weak doses, administered systemically or locally, of haloperidol, a substance that blocks dopaminergic transmissions.

strategy.[131] In cats, an analysis of the effects produced by bilateral injections of haloperidol into the caudate nucleus has shown clearly that it is above all the 'spontaneous' changes of motor programme that are made deficient, while the changes induced by exteroceptive stimuli are very largely preserved.[132] At the level of the nucleus accumbens, dopamine facilitates filtering of the information which is transmitted — through the intermediary of this nucleus — from the hippocampus and the amygdala to the structures involved on the motor side.[133] It can be added, to emphasize the complexity of the controls provided by dopamine, that the dopaminergic transmissions within the nucleus accumbens are influenced by the regions of the prefrontal cortex which themselves turn out to be dopaminergic projection sites.[134]

As regards the prefrontal cortex, we have already seen that its participation is necessary to the elaboration of 'delayed responses' in monkeys, i.e. in situations in which the animal has to 'hold' a certain piece of information during a short lapse of time and focus its attention on the information in order to obtain a reward. Now, even in the rat, in which the 'telencephalization' of functions is much less marked, a selective lesion of the A10 dopaminergic neurons at the level of their cell bodies, or at the level of their endings in the prefrontal cortex, causes a marked inattentiveness which renders the animal incapable of selecting and holding the information necessary to the execution of, for example, a so-called delayed-alternation task.[135] Since it is believed that the regions of the prefrontal cortex receiving the dopaminergic projections play an important role — in man — in the comparison of the real and actual world with internal representations, and in the complex interactions between cognition and affectivity, one cannot fail to be interested in the hypothesis that a dysfunction of the dopaminergic systems contributes to the pathogeny of the schizophrenias.[136] We shall return in the next chapter to the important question of the highly complex relationships existing between the neurotransmission and neuromodulation systems — together with the psychopharmacology that acts through them — and normal or pathological conduct.

Role of cerebral serotonin

As for the serotonergic projections issuing from the raphe nuclei, their involvement has the effect of attenuating the sensitivity of the organism in regard to stimuli coming from the environment, particularly regarding those capable of causing an aversive affective experience. Conversely, the destruction of the raphe nuclei — or, more generally, a depletion of cerebral serotonin — causes marked hyperactivity and hyper-reactivity.[137] Given that the effects of experimental manipulation of the

[131] A. R. Cools 1980.   [132] R. Jaspers *et al.* 1984.   [133] See G. J. Mogenson 1984.
[134] J. Glowinski *et al.* 1984.   [135] H. Simon 1981.   [136] See J. Glowinski *et al.* 1984.
[137] See S. A. Lorens 1978.

serotonergic transmissions are measured by the degree of facility with which a given stimulus triggers a given behavioural response, it is difficult to distinguish between the modifications that may affect respectively the perception and the evaluation of sensory information, and the triggering of the observable behavioural response to it. The serotonergic neurons project on to both limbic structures (septum, hippocampus, and amygdala) and the nigrostriatal system, and it is thus perfectly conceivable that serotonergic modulations may be involved at various stages in information processing. Having noted that a destruction of the raphe nuclei causes a lasting facilitation of switch-off responses (interruption by the animal of an electrical stimulation activating the neuronal aversion system), we have formed the hypothesis that this facilitation is not due solely to an accentuation of the aversive effects induced by the stimulation, but also to a general facilitation, to whatever degree, of behavioural expression.[138] Recent experimental data permit a better understanding of this increased tendency to active response. Numerous observations having shown that a depletion of cerebral serotonin attenuates behavioural inhibitions due to novelty, punishment, or the absence of a reward, it was generally thought that the 'disinhibitions' thus observed reflected an attenuation of the anxiety caused by the experimental situation. But it does appear that, in these instances of inhibition removal, an important role is played by the serotonergic projections on to the substantia nigra; for the degree of disinhibition of a punished behaviour proves to be correlated with the severity of an experimentally caused injury to these projections.[139] The idea of an inhibiting influence being exerted on behavioural expression by serotonergic mechanisms, and its corollary, the idea of certain inhibitions being removed due to deficient action by serotonin on the substantia nigra, are all the more interesting because certain observations made on human subjects have led to the postulation of the existence of a relation between a low level of serotonergic activity and a quality of 'impulsivity' reflected by frequent 'acting out', particularly in the form of aggressive and auto-aggressive behaviour.[140]

Role of the endorphins

Generally, the release of endorphins within the brain has the effect of attenuating—while the administration of naloxone, which blocks the action of the endorphins, has the effect of accentuating—the aversive character of certain stimulations, whether the latter arise from ingestive[141] behaviour or from social interactions.[142] As regards more particularly the affective experiences linked with social interactions, studies carried out in several animal species have shown that the action of the endorphins is important in the genesis of the phenomena of inter-

[138] P. Schmitt *et al.* 1979*b*.   [139] M.-H. Thiébot *et al.* 1984.
[140] See P. Soubrié 1986.   [141] L. D. Reid and S. M. Siviy 1983.
[142] J. Panksepp *et al.* 1980.

individual attachment and social cohesion. In all these species, naloxone accentuates—and morphine attenuates—the signs of 'distress' presented by an animal separated from its mother or its social group. Panksepp and his collaborators (1980) consider that, in the course of evolutionary history, the opiate systems, which initially played an important role in pain control, were later made use of in the processes controlling the 'social emotions', when it became important for the sake of survival that individuals should be maintained grouped together. The action of these systems is certainly very complex, employing a heterogeneous set of mechanisms; for recent research has shown the existence within the brain of three distinct families of endomorphins and several distinct types of receptors binding these substances.[143] If these receptors are particularly dense at the level of the amygdaloid neurons (which is not surprising, given the role played by the amygdala in the genesis and modulation of the affective connotations), it does appear that they are already involved in the processing of sensory information at the level of the cerebral cortex. Indeed, pondering the possible role of the endogenous morphines within the cerebral cortex in primates, Lewis and his collaborators (1981) studied the topographic distribution of the (type μ) receptors and noted that a very marked density gradient exists between the primary sensory cortex towards higher and more highly developed levels of processing of sensory information. The authors conclude that these 'μ-like opiate receptors' are capable of participating in an 'affective filtering' of sensory messages at the cortical level by being involved in the mechanisms by which the affective states induced through the mediation of the limbic structures influence the nature of the sensory messages attended to. Given the important role which appears to be allotted to the opiate systems in the mechanisms underlying affective exchanges with the environment and the complex interactions between affective states and cognitive processes, it is easy to understand that groups of researchers are investigating the part played by dysfunctions of these systems in the genesis of autism[144] and schizophrenia.[145] It can be added that, if the opiate systems thus contribute to the development of socio-affective exchanges, the latter in turn have an effect on the maturation of the receptors that bind endogenous morphines: it is sufficient to separate new-born rats from their mothers for three successive nights (the fourth, fifth, and sixth nights after birth) to note, some days later, a delay in the maturation of the receptors at the level of the hippocampus and the cerebral cortex.[146]

Of rats and men . . .

Ending this chapter, some readers may perhaps wonder about the

---

[143] See H. Akil *et al.* 1984.   [144] J. Panksepp 1979.   [145] See H. M. Emrich 1981.
[146] V. R. Olgiati and C. B. Pert 1982.

legitimacy of extrapolating to man from data obtained in animals. Here we have a very important question, whose examination is, in principle, a matter for the reason alone. In reality, feelings and value judgements are doubly involved. First, it seems that man, very conscious of the singular position he occupies in the world, has difficulty in accepting that facts experimentally established in animals can contribute to clarifying the bases of his own personality and his own conduct. And this confused prejudice is only reinforced when, in the development of new therapeutic methods, recourse is had to extrapolations which are somewhat hasty or even frankly ill-founded. The neurobiologist is in any case strongly motivated to extrapolate to man those data he may obtain in animal experiments, independently of his purely intellectual motives. Indeed, the concept of 'valorizing' one's results has come to be introduced into scientific research: giving it to be understood that knowledge leading to action has more 'value' than that which leads only to understanding. And the media rarely miss an opportunity to confirm the researcher's belief that his credit (and his budget) will be all the greater if he intimates that the results to be expected from his work will lead rapidly to specific socially and/or economically beneficial actions.

Although the question of the legitimacy of extrapolation of data from animals to man and that of the well-foundedness of action taken for therapeutic purposes have to do with very different topics (the former does not have the ethical implications of the latter), they are nevertheless linked inasmuch as concrete action should be inspired by a well-founded scientific extrapolation. It is this pertinence of the extrapolation—from a purely scientific point of view—that we shall briefly examine here.

We must recall, and emphasize, that in the sphere of relations between brain functioning and behavioural processes there are—as between animal and man—both close similarities and profound differences. This means that extrapolations are possible but also that they must be made under well defined conditions and that there are necessary limits to them. If we consider the afferent side of the system (the genesis and processing of sensory messages, in relation to the perceptual processes) or its efferent side (the involvement, in appropriate spatio-temporal sequences, of the multiple muscular 'effectors'), the similarities are far more striking than the differences. The situation changes when one becomes interested in the states (states of awareness, affective states, motivational states, etc.) and processes (of interpretation, of cognitive or affective elaboration, of anticipatory simulation, etc.) which are interposed between the system's 'inputs' and its 'outputs', ensuring their coherent and effective articulation. It is clear that, in man, the concomitant development of language and of the abilities to memorize, interpret, and imagine have caused a unique increase in the complexity of the representations borne by the brain, the totality of which, for the genesis of attitudes and behaviour, constitutes the central authority for processing information and generating significance. In man, the control exerted over the emotions by the internal discourse considerably increases the richness of cognitive and affective associations and developments.

Moreover, differences between individuals are particularly marked in man, for at least three closely complementary reasons. First, the great 'plasticity' of the human brain makes it a quite remarkable 'learning machine' which both undergoes structuring by the most varied experiences and proves capable of adapting to a large number of situations and mastering them. In the second place, every individual lives his unique history which endows him with a particular set of markers and references in the form of structured mnesic traces. Finally, the roles taken within a given socio-cultural context determine for each individual a particular set of constraints, and performances required for their mastery. This being so, it is easy to see that, in man more than in animals, the repercussions of a specific cerebral lesion (identical as to its location and extent) can vary according to individual experience and the constraints imposed by family and professional surroundings. Let us take two very simple examples. It is quite obvious that difficulties in putting acts together into a correct and appropriate sequence, resulting from a lesion of the prefrontal cortex, will have quite other effects on a fencing master than on a shepherd looking after his flock. And if a lesion of the orbito-frontal cortex disturbs the expression of the emotions and the perception of the affective states expressed by other people, this will not have the same effect on an actor and a post office official.

It is true that, if one thus considers all the factors contributing to man's singularity, one is easily led to doubt the legitimacy and interest of extrapolations from animals to man. But this is because we too often confuse the human 'content' (of memory, of cognitive and affective elaborations, of awareness, etc.) with the 'operations' that act on this content and that are provided by mechanisms which, for the most part, are common to animals and man. If the content of memory differs greatly between the rat and man, this does not change in any way the fact that, in one as in the other, similar processes of immediate memory, of the consolidation of the contents of that memory, of retention and evocation, are necessary so that the mnesic traces can be constituted and used. And a comparison of the data furnished by animal experiments and neuropsychological investigations on man brings out important convergences, by showing that all these processes of memorization and evocation are subtended by mechanisms which—very broadly speaking—are the same. As regards the genesis of affective connotations and their association with the objective data of an experienced or evoked situation, we also note close similarities in many processes and mechanisms. Even in such a 'human' domain as that of consciousness (that which we have of ourselves, of the world, and of our own relations with the world), we cannot fail to be struck by the appearance, in like manner in both man and animals, of a 'unilateral neglect' (total lack of interest for one half of the extra-personal space and for the half of the body which normally interacts with it) following certain cortical lesions. The awareness we have of our body and the relations it weaves with its environment therefore depends closely on the faculty of satisfactory interaction with that environment through the body envelope. This faculty requires, in man as in animals, the normal functioning of a complex and highly integrated neuronal network.

An extrapolation, if it is to form the basis of a fruitful working hypothesis, cannot be based on generalizing and globalizing notions which have probably been abstracted, in animals and in man, from very dissimilar realities. In the area of interest here, noting that the 'aggressiveness' of a rat can be reduced by causing a given cerebral lesion or administering some neuroactive molecule, and hypothesizing that the same might well be possible in man, is not very sensible nor can it give a solid basis for any therapeutic approach. It is necessary, in animals as in man, to consider a well-defined behavioural event (a given instance of aggressive behaviour employed to confront a given situation) and to characterize in each case all the factors which contribute to determining the probability of occurrence—not timeless and universal, but current and contextual—of this precise event. By analysing in each case the factors involved which interact to determine this probability, it will be noted that many are common to both cases. Some of the factors are very directly linked to some given concrete form of cerebral functioning which contributes to the genesis of an elementary dimension of the personality. Others concern 'operations' (for example, a familiarization process) which, it is true, may process very varied 'contents', but whose dysfunction may perfectly well have a similar effect, in animals and in man, on the probability of the appearance of an act of aggression when confronted by a given situation. If analyses of this kind are made, then extrapolations are completely legitimate, and experience shows that working hypotheses founded on them can turn out to be fruitful.

6   How can behaviour be modified?

It is important to recall here that, when faced with a given situation, the individual brain operates in its capacity as the 'executive organ' of a set of relations established between the individual and his environment. The brain can take on this essential function because it is the bearer of internal representations of the relations woven with the environment. And it is as a result of mediation by the internal representations, furnishing markers for dialogue with the environment and being in turn updated by that dialogue, that observable behaviour is both a reflection of and the driving force behind one or another of these relations. In other words, it is in the internal representations that the individual history is inscribed (that events leave traces), and it is as a result of their mediation that this history is expressed in behaviour (which is also a series of events). We should add, using the terms used by Pierre Bourdieu (1985), that the individual history, the history 'made flesh', at every moment meets the history of society and its institutions, i.e. history 'made things'. This is why individual history to a large extent is socially determined, it being specified that the relations between society and its individual actors are reciprocal in their nature.

In the genesis as in the development of internal representations and the behaviour expressing them there is close interdependence between the biological individual, the psychological personality, and the social being. Indeed, whether one considers the individual, the person, or the society of which the social being forms an integral part, these in no way constitute closed systems with narrow and rigid internal specifications, but on the contrary are open systems in dynamic interaction with each other. At each level of organization there are structures and norms making up, to be sure, a system of constraints (a necessary one, moreover), but they also none the less give free play to the 'possibility set'. At all levels, evolution takes place under the pressure of experience, of history, and seeks new equilibria, new coherences. Even if standardization, imitation, and conformism play an important role in social psychology, the majority responsible for such normalization proves perfectly capable of recovering, assimilating, and translating into its own language a large part of the contribution made by the 'active minorities'.[1] Reflecting on the constitution of ideologies, of these organized ensembles of social representations and perceptions, some sociologists consider that it is also appropriate to take an 'empirical and historical view'.[2] And it is not without interest to note that, in order to describe certain aspects of the evolution of ideologies, of sociocultural history, François

[1] See M. Doms and S. Moscovici 1984; S. Moscovici 1985.
[2] F. Bourricaud 1985.

Bourricaud speaks of 'tinkering', as does François Jacob (1981) on the subject of the biological history of our species.

Given the close interdependence and common history of the biological individual, the psychological person, and the social being, it is in many respects artificial, but still necessary, to make a clear distinction among behavioural modifications, between:

- those which result from the complex play of processes of social influence, whether these be social changes, major innovations concerning the whole of a community, or changes of attitude concerning only one or a few individuals;
- those which the various psychotherapeutic approaches attempt to bring about in treating a person 'taken charge of' as an individual;
- those ensuing from interventions aimed at modifying this or that aspect of the brain's functioning (genetic engineering, psychosurgery, psychopharmacology).

If these distinctions are a little artificial, because of the complex interactions involved, they have a certain 'operative' interest all the same, since we are discussing very specific attempts to reduce the probability of occurrence of a given type of behaviour (e.g. aggressive behaviour, as we shall see in the final chapter of this work). We know, in fact, that, confronted with a given situation, the choice of behavioural strategy is determined by the way in which the individual perceives and evaluates that situation, with the expectations and aims that flow from it. This being so, there are two ways in which one can intervene to modify the probability that a given kind of behaviour will be used in response to a given situation: we can act either on the situation itself to change its objective characteristics, or on the individual relationship with the objective situation left unaltered. Now, it is quite obvious that this individual relationship depends on:

- other people's relations to the same situation: for the individual relationship participates in a more general, more collective relationship (the role of social representations, changes of mentality and of attitudes);
- the personality of the subject, his cognitive structures, his affective dynamics, his plans (influence of educative or psychotherapeutic interactions);
- the way in which the individual brain 'managing' this relationship operates, and precisely where interaction takes place between the biological systems, psychological structures, and social influences (these structuring interactions being susceptible of modification by various manipulations of cerebral operation).

We are therefore going to consider, in order and in outline, the various means available for modifying human behaviour: the implementation of changes of a social type; the use of methods aimed at structuring or restructuring the personality; the employment of interventions practised on the brain. But first of all, we should make a preliminary remark with the purpose of enlightening the reader as to the spirit in which the pages that follow have been written. Essentially we have compiled a relatively succinct table of the existing means of action which are actually used or advocated, without (except in passing) going into a question that is none the less crucial, namely their legitimacy. Certainly, the question 'therapies or manipulations of the mind?' will be briefly examined at the conclusion of the present chapter. But, as for going more deeply into these matters, it has been preferred to concentrate our attention on the possible application of these means of action to the concrete case which we are concerned with: that of aggressive conduct. And this we can do only in the last chapter of the work, when the chapters preceding it have supplied us with the necessary elements of information and appreciation.

The effects of social change on behaviour

It is possible to change some types of behaviour by acting on the situations themselves so as to alter this or that objective characteristic. So behaviour that is liable to lead to accidents (in traffic or in industry) may be partially eliminated by getting rid of 'black spots' in the road network and 'hazards' at the workplace. In these specific examples the technical character of the approaches used and the purely material nature of the situational characteristics one is trying to change may cause one to think that the social dimension proper is absent from them. In reality, they are only the visible part of an iceberg whose submerged portion is of social character, because it brings into play the individual responsibility of each one of us at the same time as precise interests and specific ideologies. We should not be surprised then that even in domains where the purely technical aspects are far from negligible, we very soon witness the arrival on the scene of pressure groups and social movements, i.e. the principal agents of social change or of resistance to it.

Since we are talking of a social change and we are emphasizing the fact that a society cannot survive unless it is able to innovate and transform itself, we should recall first of all a primary necessary condition, namely that a society cannot exist and preserve its integrity 'without imposing on all its members some communal rules and well defined conventions, norms which they must share, provisions which they must obey'.[3] It is by reference to these shared norms that the individual is going to structure his own judgement and that, in various circumstances, he will be able to validate his perceptions and his judgements. To define social change we shall accept the definition of it given by Guy

---

[3] M. Doms and S. Moscovici 1984.

Rocher (1968): 'any transformation that may be observed over time which affects, in a manner not simply provisional or ephemeral, the structure or functioning of the social organisation of a given community, and which modifies the course of its history'. These transformations are concerned with the realm of ideas, modes of thought or behaviour, and, more concretely, the way in which wealth, authority, and culture are distributed. Sociology analyses the various factors for change as well as the conditions cooperating with or opposing these strong determinants of social change. As for the agents of change, of innovation, Moscovici emphasizes that two categories of innovation exist: one kind propagated downwards and the other upwards. He paid closest attention to the second category, that due to the efforts of 'active minorities'.[4] It appears that minority individuals or groups can exert an influence on the majority only if they possess a coherent alternative solution and make active efforts to acquire 'visibility' and 'social recognition' by means of 'consistent' behaviour, i.e. 'by presenting their point of view resolutely and with assurance and by entering into it with conviction'. One might add that the majorities exert a powerful public influence, but, on the private plane, the pressure of a minority may prove more efficient than that of the majority.

As regards the change of attitude brought about by a 'persuasive communication', this concerns the internal disposition of an individual towards an object or class of objects, such individual disposition having its cognitive, affective, and conative components. The change of attitude results from the transmission of a message sent by a 'source' and acting on a 'receptor'. Social psychology studies this individual change by proceeding to an analysis of the change factors linked respectively with the source of the message (its credibility, its attractiveness), to the message itself (its content and form) and to the receptor (his personality traits and moods of the moment).[5] Since here we are discussing modifications of behaviour, the question very obviously arises of knowing to what degree attitude is predictive of behaviour. Although the question is very controversial, it seems likely that, in order to predict conduct in a more reliable way, one must take account not only of the individual's attitude regarding the action being considered, but also of his habits (that is, his past behaviour) and his propensity to conform, or not conform, with the social norm regarding this action. If attitude is certainly an individual disposition, it must nevertheless be added that one can hardly dissociate it from the social context in which it arises and develops. On the one hand, indeed, the individual is a part of one or more groups and we note generally that belonging to a group is reflected in quite a wide identity of attitudes towards a certain number of social objects. On the other hand, we often find promptings from the social

---

[4] M. Doms and S. Moscovici 1984; S. Moscovici 1985.
[5] See. J.-P. Leyens 1979; G. de Montmollin 1984.

milieu at the origin of the development of attitudes: it is because the group is faced with a problem that its members are individually led to adopt an attitude towards it.

Theory and practice of psychotherapies

Psychotherapies are numerous and differ from each other as much in the techniques they employ as in their theoretical underpinnings. Before a succinct presentation of the main categories of approach having psychotherapeutic aims, it is not without interest to examine some aspects which they have in common and which they also share, in a more general way, with approaches having educative aims.

In the first place, psychotherapy—like education—is an interaction, an encounter. The personality of the therapist—like that of the educator—therefore has definite importance, as does the recognition of the fact that the patient—like the subject being educated or re-educated—is very much a co-actor in his own change. What counts with the therapist is his degree of receptiveness towards his own experiences and those of his patient, his degree of sincerity and internal coherence. Rogers (1966) puts the accent on the idea of 'congruence', by which he understands the exact correspondence, the close agreement between lived experience, the realization of which it is the object, and the way in which it is communicated to others. It is only on condition that he himself has reached a maximum of 'congruence' that the therapist can help the patient to get there in his turn. Israël (1984), too, states that this requirement for truth is an essential aspect of the ethics of all psychotherapy, which must be 'an ethic of responsibility towards others'.

Secondly, psychotherapy—like education—is a process of development, a setting in motion, an activation of the capacities to change and learn which constitute the beginnings of a permanent evolution. Benoit and Berta (1973) see in this 'activation' an essential aspect of all psychotherapy. Psychotherapy strives to lead the patient towards more autonomy and self-direction along two complementary pathways: by softening the rigidity of his behaviour, by provoking more suppleness and variability ('defixation'); by opposing 'polysemantism' to his semantic unilaterality. This last aspect is important if one agrees with Israël (1984) that 'unilaterality as the ideal for discourse is the greatest menace to human liberty'.

In the third place, psychotherapy—like education—is always underlain by a reference model having a certain image of the 'normal' adult, a model derived from a sociocultural context; by virtue of this, psychotherapeutic and educative interactions acquire, outside their 'private' aspects, an obvious social dimension. The social dimensions and significance of educational practices are manifold,[6] and everyone

[6] See M. Gilly 1984.

knows that this is a social issue of the greatest importance. This explains the often lively controversies concerning the reference model (one model or several? Who chooses: parents or the community?) or the pre-eminence to be accorded in educational practice to the content, the 'knowledge' to be passed on, or to the processes—cognitive and affective—of pedagogy to be used. In the case of psychotherapy, the influence of the sociocultural context is doubly felt. On the one hand, psychotherapeutic interaction has a specific content that essentially concerns, for the patient, his current relationships with others. 'Pathological' behaviour patterns of which he is to be rid often reflect the existence of conflict situations generated by the rules governing group life. On the other hand, psychotherapy aims at a general objective corresponding to a certain vision of man and of human experience; side by side with its 'technical focus' it has a 'philosophical or existential focus'.[7] When one then reads, from the pen of Watzlawick and his collaborators (1975), that 'the problem is not to avoid influence and manipulation, but to understand them better and to use them in the patient's interest', one is tempted to add: may the sociocultural context always permit the therapist and the patient to define this interest liberally, and may they feel entirely free to interact accordingly!

Given the number and the extreme diversity of psychotherapies in use, any classification is bound to be to some extent arbitrary. We shall distinguish, in outline, those psychotherapies that aim to modify:

- some form of behaviour that is causing a problem;[8]

- mental functioning, in a more global sense;

- the course of someone's existence.

(1) In 'behaviour therapies', the therapist attempts to change a form of behaviour that is causing a problem (in particular, phobic or obsessional behaviour) by using techniques based on principles derived from theories of conditioning and learning. Therefore he concentrates his action on the ill-adapted behaviour itself or on the 'mental images' underlying it (with a 'cognitive restructuration' that has some degree of elaboration). In every case, therapy consists of a precise analysis of the problem behaviour that is to be caused to disappear, a 'contractual' definition of the specific objectives aimed at, the use of a technique previously discussed with the patient and an objective evaluation of the results achieved.[9] In the case of a phobia, as in that of an obsession, behaviour therapy aims above all at reducing the anxiety aroused by a

---

[7] J. Guyotat 1978.

[8] Rather than speak of inadequate, inappropriate, or even deviant behaviour (implying a reference to constraints, norms and models of a given socio-cultural context), the specialist in 'behaviour therapies' often prefers to allude to behaviour giving rise to a concrete problem for the patient, or even more simply to 'problem behaviour', for 'behaviour therapy seeks to aid the patient to resolve "his problems" in the here and now' (J. Cottraux 1978).

[9] See J. Cottraux 1978; O. Fontaine 1978.

phobic situation or an obsessing stimulus (idea or feeling). If a phobic situation (being in a lift, in a tunnel, etc.) is no longer anxiogenic, the patient will no longer present the avoidance or flight behaviour constituting the problem. If the obsessing idea (for example, being soiled by contact with others or with unfamiliar objects) loses its anxiogenic value, the patient will no longer perform the conjuratory ritual of washing which was his problem behaviour. We shall cite only two techniques currently in use, applying them to the case of a phobia. In the 'desensitization' technique a counter-conditioning is effected by associating the phobic situation — which the subject is asked to imagine — with a state of well-being created by relaxation. In the 'immersion' technique, the extinction of the problem behaviour is brought about by exposing the patient to the phobic situation, forbidding him to flee and thus demonstrating to him — repeatedly — that nothing happens to him and that his anxiety is therefore without any foundation.

Similar approaches may also be used as a method of social counter-conditioning, so as to free the subject from certain social inhibitions, to carry out self-affirmation training, and to help him to increase his 'personal efficiency'. By means of 'role-playing' in groups and imitating real or imaginary models, the subject progressively acquires appropriate assertive behaviour patterns (to which the group gives positive reinforcement), while losing the anxiety originally associated with these same behaviour patterns. He is thus freed from the terrible choice between resigned passivity and reactive aggressiveness.

(2) Other psychotherapies try to act on the personality of the patient, on his mental functioning in a more global sense, on the basis that problem behaviour is only a symptom, an indicator of something more general affecting the structures and the dynamics of the personality. But here we can introduce a further distinction according to whether stress is put, in theory and in practice, on a certain intrapsychic dynamic — as in psychoanalysis — or on cognitive structures and exchanges of information.

Indeed, certain psychotherapies concentrate on the internal dynamics of a 'psychical apparatus' (with its various 'agencies') as Freud conceived it, that is to say, on a dynamic ensemble of tendencies that are to some extent in conflict with each other. The unconscious infrastructure of the subject is analysed in order to bring out his unconscious desires (by free association and the interpretation of dreams). These unconscious desires, which are sources of energy, enter into conflict with the realities of conscious life, and this conflict is signalled by behaviour which causes problems.

Independently of whatever opinion one may have of Freud's thinking and the various exegeses to which it has given rise, it cannot be doubted that man is a 'desiring machine'. This being so, all psychotherapy — like all education — must help the subject to 'invest' his desires in a way that liberates rather than enslaving him. It must attempt to reorientate towards relational investments, towards the discovery of shared interests and joys, desires that, too often, are misdirected towards the incessant practice—due to incessant suggestion—of the rituals of consumption and play-acting.

Instead of concerning themselves primarily with the 'affective movements' of a patient, some psychotherapies are mainly interested in internal structures of a cognitive kind, in the internal representations of the objects of the environment and the relationships the individual has with them. Behaviour, which is both the reflection of these representations and the driving force behind their development, is then considered as the medium for unceasing exchanges of information. In this cognitivist view of mental functioning, the mediations provided between the brain's inputs and outputs by the internal representations and the operations of thought have more 'density' than do the mental images in the behaviourist conception underlying behaviour therapies. The actions of the therapist are aimed at 'rectifying' the cognitive distortions which affect an ensemble of behaviour patterns and which may be due to mistakes in reasoning, to an over-selective and one-sided method of picking up the messages emanating from the milieu, or again to the use of questionable social or cultural referents. In a more general perspective, the therapist helps the patient to reflect in a lucid and critical way, to think by himself and for himself, to step back a little so as to judge the content of persuasive communications (of advertising, of some sort of propaganda, etc.) which invite him to adopt some comfortable 'ready-made' ideas.

(3) Some practices with psychotherapeutic aims, and they are many and various, claim to derive from the movement of 'humanist psychology' and offer to act on the progress of existence on the basis of a certain existential philosophy.[10] It is the hierarchy of needs, as it has been defined by Maslow, that constitutes the main theoretical foundation for this. Beyond elementary biological needs and the need for security (the quest for familiar things and for stability), beyond the need to exercise a certain mastery over the environment, to belong to a group, to have a place within it and enjoy the esteem of other members of it, new needs or 'meta-needs' appear: self-acceptance and self-respect, and a need for self-realization. Maslow's conception is meant to be fundamentally optimistic as far as the potentialities of man are concerned: placed in the environmental conditions that provide satisfaction for 'lower' needs, the subject displays his need to understand, to discover, to create, to live intensely. The stress is therefore placed on 'the development of human potential' which should lead to autonomy, spontaneity, and creativity, as well as to richness of interpersonal relationships, tolerance, and respect for others. The therapist's general attitude is that advocated by Rogers, which, apart from respect for the potentialities of others,

[10] See C. R. Rogers 1966; J. Furtos 1978

includes the requirement for truth (for 'congruency') spoken of earlier, and the concern to perceive the other person's experience from within ('empathy'). In addition to this single colloquium, like objectives are pursued in 'encounter groups' with various kinds of 'mise en scène'.[11]

Direct operations on the brain

The limits of genetic engineering

We shall not expatiate on genetic engineering, for it appears illusory to wish to modify behaviour in a well defined manner by carrying out a controlled change to the genetic inheritance. Modifying the genetic inheritance of a bacterium so as to cause it to synthesize a hormone or any other biologically active protein is one thing; claiming to 'correct' or 'improve' some form of human behaviour by operating on the genetic inheritance is in many ways something else again. We have to discuss it, however, for, as we have already indicated, some people wonder whether it might not be necessary to extirpate from human nature its 'more harmful and dangerous side' by means of genetic-engineering techniques ('positive genetic engineering').[12] More generally, as Rose and his colleagues (1984) rightly say, manipulation of the genetic inheritance is the ultimate objective of a reductionist determinism to whose attractions biologists often succumb and, even more so, decision-makers who are quick to pick up on solutions they think simple and effective. But it is enough to recall the fundamental difference separating what is currently done from what some dream of doing in the future. In the case of biotechnological methods which permit the production of biologically active proteins, there exists between the gene and the product of its expression a direct, linear, and perfectly predictable relationship. This is absolutely not the case when one examines the genetic determinants of behaviour, since 'aggressiveness' or 'kindness' can hardly be likened to some kind of secretion products of well-defined ensembles of neurons. When people speak of modifying behaviour by means of genetic engineering they are not alluding to behavioural difficulties that are the symptoms of a hereditary disease, of an 'error in metabolism' of genetic origin. For, in that case, it is in principle possible—and certainly desirable—to perfect a genetic therapy that would allow correction of a hereditary disease responsible for these behavioural difficulties. But when one contemplates, for subjects free from all illness of genetic origin, the development of behaviour deemed 'beneficial' (for whom? From what point of view?) and the eradication of that deemed 'harmful' (for whom? From what point of view?) by altering their genotype, one is falling into (bad) science fiction. For one thing, in fact, a whole series of successive levels of organization separates the genes from their final

[11] See J. Furtos 1978.   [12] J. Glover 1984.

expression in the behavioural phenotype; now, on each of these levels, the elementary structures and mechanisms—controlled more or less directly by a given gene—acquire new properties by virtue of their interactions with other elements within a dynamic and open ensemble. For another, the 'motive for action' underlying the use of a given behavioural strategy most often corresponds in man to a complex causal field, to a structured ensemble of determinants; when one modifies one of its constitutive elements, one changes the configuration and the internal dynamics of the ensemble with consequences that are difficult to foresee and hardly to be generalized. Under these conditions, one can certainly play at being the sorcerer's apprentice; but it is doubtful whether one can claim to be using scientifically based methods.

Historical outline of psychosurgery

As for psychosurgery, it can already boast a century of history.[13] It was in 1891 that Gottlieb Burckhardt, a director of the lunatic asylum at Préfargier in Switzerland, reported interventions that he had carried out on six patients who presented with a marked psychomotor agitation. Taking for his basis the earliest data of cortical physiology obtained from animals (inducement of movement by electrical stimulation of the cerebral cortex; sensorimotor defects brought about by cortical excisions), Burckhardt made the assumption that agitation reflected an abnormal activation of motor centres resulting from a pathological excitation of the sensory centres; and with a curette he carried out a series of little cortical resections on either side of the fissure of Rolando. In one of the patients, the psychomotor agitation was noticeably attenuated; but another died of the consequences of the operation and a third became epileptic. These results excited lively opposition on the part of the local medical community and Burckhardt was forced to interrupt these, the first trials of a psychosurgical therapy.

It was not until nearly half a century later that psychosurgery experienced a return to favour and was practised—this time—on a totally different scale. In 1935 the Portuguese neurologist Egas Moniz, taking part in the International Neurological Congress in London, listened with lively interest to Carlyle Jacobsen, who reported the results obtained by causing localized lesions at the level of the prefrontal cortex in monkeys. The monkeys so operated on did admittedly experience some difficulty in resolving problems they were presented with, but the most irritable animals became noticeably calmer. On returning to Lisbon, Moniz carried out the first 'prefrontal lobotomies' in anxious and agitated patients, and he reported the beneficial results of the operation the following year. Because of Moniz's recognized competence (we owe to him cerebral angiography, a method of viewing brain vessels by means

[13] See E. S. Valenstein 1980; J. Talairach 1980.

of substances opaque to X-rays) and his influence (he is a member of
parliament and minister for foreign affairs), his work was to have a great
impact. It was in the USA above all, under the impetus given by Freeman and Watts, that operations on the prefrontal cortex were to multiply, particularly from 1942 when the need arose to respond to the
mental problems presented by soldiers fighting in the Second World
War (in 1943, the Veterans' Administration encouraged neurosurgeons
to acquire additional training in the field of prefrontal lobotomies). In
1949, Egas Moniz received the Nobel prize for his efforts and his
achievements in the field of psychosurgery. When we judge these
lobotomies with some hindsight,[14] we note that, in the description of
their beneficial effects, the stress was placed much more on the elimination of behaviour that was a nuisance to the entourage (family, nursing
staff) than on the 'quality of life' of the patient himself. Now, these
operations have often caused some deterioration in the subject's faculties of judgement and his autonomy, and above all a depression of
emotional tension and the display of a more or less pronounced affective
neutrality. But it must also be emphasized that neuroleptic drugs or
'major tranquillizers', which were to transform psychiatric practice, only
became available in 1952.

A certain 'renewal' of psychosurgery occurred in the early sixties. It
was thought that one could operate on a more rational and precise basis
thanks to new data brought to light by research in behavioural neurobiology. As far as the prefrontal cortex was concerned, it was better
understood that cognitive defects were above all due to lesions involving the dorsolateral convexity, and operations were therefore carried
out, limited to the cingulate cortex (and to the cingulum linking it to the
hippocampus), that is, the circuits involved in socioaffective exchanges.[15]
After an extensive evaluation of the effects obtained by such 'cinguloto-
mies' in 85 patients, Corkin (1980) indicated very frankly his inability to
understand why the operation proved efficacious in some cases but not
in others, and his lack of knowledge of what precise cerebral mechanisms or other factors should be held responsible for the observed im-
provements. In our more particular sphere of interest, that of aggression
and violence, Mark and Ervin (in the seventies) advocated localized de-
structions of the amygdala in subjects presenting violent behaviour.
Even if they have since modified their position, admitting that this
behaviour was not necessarily due to anomalies of a biological nature,[16]
they were the principal promoters of the notion according to which an
instance of violent behaviour was an expression of a defective inhibiting
control over aggressiveness ('dyscontrol syndrome'), because of dysfunction of certain temporal structures, in particular the amygdala.[17]
They even considered that it would be appropriate to examine violent

[14] See E. S. Valenstein 1980; J. Talairach 1980.   [15] See Chapter 5.
[16] See V. H. Mark and W. A. Carnahan 1980.   [17] See E. A. Serafetinides 1980.

subjects, to verify (in particular by recording the bioelectrical activities of the temporal lobe) whether one could not detect this anomaly in cerebral functioning so as to prevent — with a 'preventive treatment' — the repetition of violent acts. We shall have occasion to come back to this point in later chapters. Lesions of the amygdala were also effected in 'hyperkinetic' children who, besides their excessive and disordered activity, suffer attacks of aggressiveness;[18] these operations have not been carried out since it was discovered that, in a rather paradoxical way, these children could be treated by the administration of psychostimulating substances related to amphetamines. As a last example of this more modern psychosurgery, we may cite the operations carried out on the hypothalamus, especially by German neurosurgeons, in cases of deviant sexual behaviour. These 'hypothalamotomies', which were thought to correct behaviour by acting both on the control centres for sexual behaviour and on the 'pleasure centre' of the lateral hypothalamus,[19] also gave rise to some undesirable consequences, such as difficulties of memory, of emotivity, and of feeding behaviour.

The criticisms, often very lively, directed at psychosurgery more often than not mix scientific arguments with considerations of an ethical and ideological nature,[20] which is not surprising since some people advocate the use of psychosurgical operations to resolve the problems society is faced with. But, independently of all ideology, we should stress that psychosurgery reflects an *outmoded* view of the relations between the brain and behaviour, for it is founded, implicitly or explicitly, on some kind of cortical and sub-cortical 'phrenology' that attempts to locate personality traits, attitudes, and behaviour at the level of well delimited cerebral structures. In reality, the nature of the interactions between the brain and behaviour, and the complexity of the dynamic interactions going on within the brain itself, prevent psychosurgery from being able to modify behaviour in a strictly defined and delimited sense, and in a predictable and regular way. Given these facts, we can understand why, in France, psychiatrists and neurosurgeons have a generally negative attitude towards psychosurgical operations.

Progress in psychopharmacology

The initial methods of psychopharmacology were of a purely empirical nature: one limited oneself to noting that such a substance had the effect of modifying behaviour and basically it was a question of determining the active doses as well as individual sensitivity in this regard. Even later on it was not uncommon, as Giurgea (1983) says, for the first medicament of a new class of 'psychotropic' substances to be discovered by accident during clinical observation. But, in its more recent develop-

[18] H. Narabayashi 1972.
[19] See Chapter 5: The neuronal reward system, p. 87.
[20] S. L. Chorover 1980; S. Rose *et al.* 1984; J. R. Durant 1985.

ment, psychopharmacology owes its progress—that already made and that which may reasonably be anticipated—to the fact that it is based on knowledge acquired in two closely complementary areas which are otherwise very different. On the one hand, psychopathology, which analyses behavioural and mental disturbances, defines more precisely the nature and sequence of the processes involved in a given disturbance. On the other hand, thanks to work carried out in neurochemistry and in neuropharmacology, we have a much better knowledge of the mode of action and the location within the brain of the target of this or that active molecule; and, for its part, behavioural neurobiology is attempting to establish bridges between a given cerebral mechanism — whose functioning can be modulated by some active molecule or molecules—and well defined behavioural processes. It is true that a major difficulty is inherent in these parallel methods because of the fact that the psychological type of analysis is made in man whereas neurobiological experiments are carried out on animals. That is why psychopharmacology perfects and uses 'animal models' that are considered to reproduce in animals equivalent syndromes to those described in man, or, at least, some precise symptomatic characteristics of one or other of those syndromes.[21]

The example of depression

To give a better outline of both the very real interest and the non-negligible difficulties characterizing those therapies which make use of psychotropic substances, we may take the concrete case of *depression*. No one denies the efficacy of antidepressants in the treatment of depressive states, not even the anti-psychiatrists who, totally one-sidedly, state the social origin of mental disorders. Moreover, the case of depression is a good illustration of the parallel methods mentioned above, since our knowledge has advanced as much in the psychological and social dimensions of the 'depressive state' as in the biological anomalies of the 'depressed brain'. These parallel methods analyse distinct 'logics' corresponding to distinct levels of analysis and systems of internal causality, as is well shown by the highly illuminating work of Daniel Widlöcher (1983a). Difficulties and causes of error occur as soon as one attempts to establish bridges between these logics or, more modestly, between the concrete data obtained at these distinct analytical levels. For one cannot avoid being strongly influenced by the very general conception one has of the mutual relations entertained between the brain and behaviour. If the human brain is considered simply as the 'generator' of behaviour, as an organ whose functioning (normal or pathological) is projected outwards in the form of a set of behaviours (normal or pathological), one will be tempted to establish some linear and unidirectional

---

[21] See M. Le Moal 1984.

causal relationship between 'the depressive state' of the brain and 'the depressive state' of the patient; it will then suffice to remedy the brain's state in order to normalize the subject's mental state. While a therapy with a similar basis may very well prove efficacious (apparently confirming the correctness of the underlying ideas), it nevertheless rests on an erroneous conception. For one is considering, in cerebral functioning as in mental functioning, stable 'states', artificially cutting out some 'snapshots' (which may of course have a certain duration) in a common becoming, in a common history, in a complex dynamic interaction.

As has already been mentioned several times, behaviour expresses the way in which the individual brain apprehends situations and events by reference to internal representations shaped by his experience, that is by the specific content of the dialogue he conducts with the environment and by the cognitive and affective elaborations to which this content has been subjected. It also expresses a certain way of responding, of facing up to situations and events so as to exert some mastery over them. The representations, the reference points, evolve over time as they interact with the environment and, keeping pace with them, the way of apprehending the world and carrying on a dialogue with it. The behaviour of the depressed subject should therefore be considered in this dynamic and historic perspective. When a person does not manage to overcome the suffering created by a painful event (loss of a beloved, loss of a job and social identity, etc.) or by living conditions that are felt to be unacceptable, when the person considers that there is no longer any point in struggling, he can 'give up' by taking refuge in a resigned passivity which is expressed, in particular, by 'depressive retardation' that is ideational, verbal, and motor.[22] Certain modalities of cerebral functioning take a determining role here, but the determinism is largely circular; it is therefore at an arbitrary point that one opens the loop, either to describe a sequence of processes (using our discourse, which is linear), or to act on it so as to bring about change. A large number of intra-cerebral processes, involving various neurotransmitters and neuromodulators, contribute to determining the way in which situations are perceived and evaluated, with the pleasant or painful emotions associated with them, as well as the way in which the appropriate behavioural strategies are selected and employed. When the painful event or living conditions described above come about, the individual characteristics of these neurobiological processes have already been shaped by the interactions of an individual genotype and all individual life-history; they differ, therefore, from one subject to another and they may be such as to 'predispose' a subject to develop a depression. The animal model of 'learned helplessness',[23] one of the models used to verify the 'antidepressive' efficacy of an active molecule, shows clearly

[22] See D. Widlöcher 1983*a*, 1983*b*.
[23] See R. Dantzer 1984; M. Le Moal 1984; and Chapter 5 of the present work.

that the constraints imposed on behaviour from outside may modify certain neurochemical characteristics of the brain so that it becomes what one terms a 'depressed brain'. When the depressive slow-down has become established, after initially constituting a response of retreat and defence, it may in turn become—together with the reactions it excites in the members of one's circle—an experience that is lived in a traumatic way, reinforcing the vicious circle. The circle may be broken by modifying with an antidepressant some intracerebral process involved in the 'management' of relations with the environment. Given the multiplicity of processes that may be affected in depression, and their possible interactions, it is not surprising that alterations to several types of neurotransmission (involving noradrenaline, dopamine, serotonin, the endogenous morphines) could be hypothesized, nor that several classes of substances should have turned out to be efficacious antidepressants, nor that the depressed subjects should individually be sensitive to one rather than another of these classes of antidepressants. We shall add that a psychotherapeutic approach, too, can contribute to breaking the vicious circle and preventing a relapse, in so far as it succeeds in following up the efficacious action of an antidepressant by reorientating the way in which the patient apprehends the world and carries on dialogue with it.

Complex determinations of aggressive conduct

Even if depression is not something simple, it still presents a certain unity in both its expression and its general significance. Moreover, one may consider that it is characterized, on the neurobiological plane, by a structured set of anomalies, it being understood that the respective importance of the elements of this set and the way in which it has been constituted over time are liable to differ from one case to another. This being so, the generic notion of 'anti-depressants', which groups together the substances proving efficacious in the treatment of depression, has a basis going beyond this purely empirical aspect. The situation is very different when we turn to aggressive behaviour. It is quite improper to speak of 'anti-aggressive' molecules, and it would be quite wrong to group together in a similarly generic notion all the substances that reduce the probability of the unleashing of one or other of the behaviours which, admittedly, are all described as 'aggressive', but which are distinguished from each other by their manner of expression and, more yet, by the most diverse 'motives for action' underlying them. On this account, one might say—exaggerating a little—that curare is the most efficacious anti-aggressive drug, since there is zero probability that the subject paralysed by it will display any aggressive behaviour. We shall see in the next two chapters that the factors and mechanisms contributing to the determination of the probability of aggressive behaviour are many and various, and that these determinants are far from being the same, or playing the same role, from one case to another. It is on the basis of these complex and variable determinations that successive theories of aggression have been developed—the 'noradrenergic', 'serotonergic', 'GABAergic', etc., theories—which, strictly, mean nothing. If the different types of neurotransmission and neuromodulation do admittedly contribute to the elaboration of this or that motive for action, by intervening in the multiple operations of the information processing involved in such elaboration, they are not responsible for the nature of the information thus processed. Giurgea (1984) is therefore right to say that it is not the psychopharmacological drugs of tomorrow that will turn our world into Orwell's *1984* or, on the other hand, the world promised by Antoine de Saint-Exupery's *Le petit prince*.

Therapy or manipulation of minds?

If it is agreed that it is not psychotropic drugs that will make men 'good' or 'bad', it is just as clear that such drugs can be used—in an Orwellian world—to manipulate minds, altering the faculties of judgement, the exercise of the will, the display of autonomy. As soon as the various means of modifying behaviour are considered, the question necessarily arises of the 'manipulation' of another person and the threat of alienation hanging over him. The moralist makes a clear distinction between the educator and the manipulator, between direction with educative aims and manipulation considered as a 'perverted' form of interpersonal influence.[24] If this distinction is satisfying to the mind, the dividing lines are less distinct in practice. As we have already mentioned, Watzlawick and his collaborators (1975) believe that manipulation always occurs, but that the important thing is that it should be done in the patient's interest. This too is a clear and satisfactory position, but the 'patient's interest' remains to be defined. Now, in this regard, everyone applies value judgements that are themselves the fruit of a certain number of manipulations undergone in a given familial and sociocultural context. It is on the basis of these value judgements that we decide—in a way that, viewed objectively, is arbitrary—whether or not we are dealing with a 'perverted' form of interpersonal influence. Let us take two specific examples, provided by a single issue of *Le Monde*, that dated 7–8 July 1985. On the one hand, we learn that a resolution by the regional committee of the Ukrainian CP advocates the forced psychiatric treatment of activists of the Ukrainian Catholic church. On the other, a Japanese cultural historian tells us that the great advertising agencies are a thousand times more powerful than the military police ever were, that they infiltrate everywhere, and that they govern the Japanese nation with kid gloves. The first case, rightly, evokes our condemnation, for we

[24] M. Nédoncelle 1974.

have difficulty in regarding such forced treatment as being in the interest of those 'benefiting' from it. In the second case, our first reaction is rather to conclude that the Japanese specialist is exaggerating and that, anyway, we remain perfectly free either to follow up or to ignore the stimuli of advertising talk. In reality, advertising manipulates our minds and our attitudes by artificially creating needs and priorities, thus fabricating a new system of values. Now, this influence is exerted without our being aware of it, for a real interiorization of this value system, and the attitudes and behaviour reflecting it, takes place.

7 Factors which contribute towards determining the probability of aggression

The general outline provided by the four previous chapters will enable us to examine in greater detail and with more convincing arguments, the basic idea that emerged throughout the second chapter of this work. Aggressive behaviour is a means of action implemented by the individual brain for the purpose of achieving a specific aim. It can therefore be said, rather paradoxically, that the principal cause of this behaviour lies in its anticipated consequences; these consequences consist largely of a change in any of the relationships which the subject maintains with his environment. What matters, when aggression is triggered — or not — is not the event or the situation, considered 'objectively', but how these are interpreted and the affective state which accompanies their perception and interpretation. This being so, aggressive behaviour should not be considered merely as an isolated response to an isolated aspect of the environment, but as a sign of an individual and well entrenched way of grasping and coping with situations and events in order to master them. Of course this complicates considerably the study of what determines aggression, since so many different factors contribute to the complex causal field, the structured set of interactive elements which provide the 'motive for action' which we are dealing with and attempting to analyse. A few preliminary remarks are therefore in order.

If this chapter is to be something other than a long list of factors likely to contribute towards determining the probable use of aggressive behaviour, it is necessary both to highlight the essentials and to group the factors according to certain main lines, while stressing that these groupings, due to the boundaries they define, are necessarily of a somewhat artificial nature. The emotions and affective experiences play a primordial role. When a situation gives rise to aggressive behaviour, this aggression often aims to put an end to, or at least to attenuate, an aversive feeling (anxiety, fear; annoyance, anger) generated by the situation and what it represents. Other situations, which occur by chance or are actively sought after, provide the opportunity to experience pleasant emotions, since they make it possible, through aggressive conduct, to appropriate a desired object or to bolster one's self-esteem and self-confidence. Even in cases where aggression seems to be 'spontaneous', insofar as it is displayed in the absence of any event or situation which might have provoked it, it may express a feeling of uncertainty and dissatisfaction which it aims to attenuate; it could also be the means for seeking stimulation, excitement, or 'thrills'. While the emotions thus play a primordial part, so, to a major extent, does individual experience.

For apart from those which derive from the satisfaction of elementary biological needs and comprise a significant innate component, emotions that appear at a given time in one's life are closely related to certain traits in a personality which has been shaped by experience. And this individual experience constantly draws on the affective events resulting from behaviour to which it contributed essential determinants.

It may seem strange that a neurobiologist working with animals should stress the role played by affectivity and individual experience. For on the one hand, an animal's affective experience cannot be observed directly, and inferences made in this respect are often considered to reflect 'mentalist' views (which for a neurobiologist used to be considered a major 'heresy'). On the other hand, neurobiological experiments have seldom taken into account, in the study of socio-affective behaviour, the possible role of differentiated and controlled experiences. But as we will explain in the next chapter, which is devoted to the neurobiology of aggressive behaviour, we have insisted since the beginning of the seventies on the major role of brain mechanisms underlying an animal's emotional reactivity and those which enable the influences of actual past experience to intervene.[1] This is despite the fact that research has focused on inter-specific rat–mouse aggression, a type of behaviour which is generally considered (wrongly, in our view) as 'predatory' aggression in the determinism of which neither the emotions nor actual experience are supposed to have a marked impact. Ideas have gradually evolved and it is now more widely accepted that emotional reactions are not merely 'epiphenomena' accompanying and distinguishing certain types of aggressive behaviour, but that processes of an affective nature can intervene in the causal chains, playing a determinant role in the choice of an appropriate behavioural strategy.[2] This highlights the need to study in animals all the brain mechanisms which are involved in the genesis of affective experiences and their modulation by actual experience, that is, through dialogue with the environment.

We will use two criteria to group together the factors to be reviewed. On the one hand, we will consider that in a given situation, there is a high probability of aggressive behaviour if:

1. The situation is perceived, interpreted, and experienced in such a way that aggression appears to be the appropriate strategy.

2. Experience acquired in the same or similar situations has shown that this strategy has in reality every likelihood of being effective.

3. Nothing does really stop the potential aggressor.

We will examine each of these three points in turn, together with the factors that are likely to affect them. But a second criterion will be used in analysing these factors, to distinguish between biological factors,

---

[1] P. Karli *et al.* 1974.
[2] See, for example D. C. Blanchard and R. J. Blanchard 1984.

those which form an integral part of the subject's personality and factors pertaining to the situation and, more generally, to its sociocultural context.

Whilst this system of grouping the factors certainly makes it possible to present them in a structured fashion, its artificial aspects should be emphasized. Let us consider, for example, the experiment carried out by Robert Dantzer,[3] who showed that when two pigs were placed together in a frustrating situation, a transient increase in the level of plasma corticosterone and aggressive behaviour towards one another either did or did not occur, depending on whether or not they had had an opportunity to interact and to establish interindividual 'links'. It may be thought that in this case, prior social interaction clearly reduced the probability of aggressive behaviour by exerting a dual preventive action: on the one hand, by attenuating the aversive nature of the situation as assessed by each of the animals; on the other, by reducing, for each animal, the probability of the use of aggression towards a familiar congener. Likewise, the general attitude which a child develops towards his environment (I can trust the world because it is basically kind and reassuring; or, on the contrary, I should be suspicious and defend myself from a world that is fundamentally hostile and unkind) will affect both the way in which he perceives a given situation and how he judges the use of aggression as a means of action, thereby reducing—or not— the likelihood of its actually being used. In other words, a single factor may act at several levels (in our two examples, at levels 1 and 3 mentioned above) which we distinguish in our way of presenting the subject matter.

Distinguishing between factors deriving respectively from an individual's biological characteristics, the psychological structures of a personality, and the structuring and incentive influences of a sociocultural context, is mainly justified by the fact that these different categories of factors correspond to different levels of analysis, with different systems of internal causality. But it is quite obvious, as we have already emphasized, that there are complex interactions between the processes in which these different factors intervene. Therefore behavioural neurobiology should not limit the scope of its research solely to the brain mechanisms involved in the triggering—*hic et nunc*—of a given type of behaviour in response to a given stimulus or situation. It is equally important to study both the mechanisms underlying the development of the personality (particularly those involved in the formation and expression of the internal representations of an individual brain) and the way in which these mechanisms are affected in their turn by the actual experience that shapes this personality. We have not yet reached this point, because we still lack certain operational concepts and because our minds are not yet fully prepared for it.

[3] R. Dantzer 1981. See also Chapter 4, under Socio-affective behaviour, p. 67.

The importance of fully recognizing these interaction dynamics and their historic dimension cannot be stressed enough, whenever efforts are being made to understand the phenomena observed or to have an effective influence on them. Paul Scott, who for a long time studied the phylogenetic development of aggressive behaviour and the evolutionary changes which affected the functions ensured by it, concluded that a true understanding of these processes is only possible within the scope of a 'polysystemic' view which endeavours to apprehend over time the complex interactions between the genotype, the somatic and behavioural characteristics of the phenotype, and society, together with the ecosystem within which it develops.[4] At the stage of evolution reached by our species, whenever an attempt is made to grasp the notion of an 'aggressive personality', it is also obvious that nothing can be understood if this personality is isolated in space and time. For not only do social interactions have repercussions throughout the ontogenesis, but these repercussions are more or less profound according to the subject's degree of biological and psychological maturity, and an influence occurring at a specific moment may enhance, or on the contrary, attenuate, the effects of a subsequent influence. Here again, longitudinal studies should be carried out, taking into account at each stage of the ontogenesis the data provided by biological, psychological, and sociological surveys.[5] Lastly, those who examine the means likely to be implemented to prevent and control aggression, dwell equally on the fact that any effective strategy should necessarily be of a multidisciplinary nature, since it must operate at multiple levels of social life and act in different ways to make the required changes to the many processes that contribute to the development of aggressive behaviour.[6]

Perception and interpretation of a situation

Even though biological factors by no means always play a predominant role, it is understandable that biologists put them—chronologically speaking—in first place.

(A) Biological factors

It is logical to consider firstly the contribution of the *genotype*, since this set of genes—which exist from the stage of the fertilized egg—ensures that the individual belongs to a certain species with its own possibilities and limitations. Of course, the environment will then contribute to a considerable extent in determining how this genotype expresses itself throughout the individual ontogenesis. But this is no

---

[4] See J. P. Scott 1981.   [5] See H.-J. Kornadt 1984.   [6] See A. P. Goldstein 1983*a*.

reason to be unaware of genetic diversity. It is important to find out what this diversity really consists of, what meaning should be attributed to it, and what attitude may legitimately be adopted towards it (from a biological point of view).

Genotype and 'aggressiveness' in animals

Research carried out on animals has mainly concentrated on aggressive behaviour between males of the mouse species.[7] Two complementary methods of investigation were used: on the one hand, assessing the degree of 'aggressiveness' (measured by the frequency and intensity of the attacks) which characterizes different inbred stock, that is, as genetically homogeneous as possible; on the other hand, efforts were made to differentiate, through selective breeding from a single strain, an 'aggressive' line and a 'non-aggressive' line. The data obtained clearly showed that there were notable differences in the degree of aggressiveness observed between one strain of mouse and another. Experiments in crossed adoption (the young of one strain being raised by mothers of the other) showed that in certain cases the differences noted were partly due to maternal factors, but that in other cases they were entirely of genetic origin. Experiments in selective breeding carried out by Lagerspetz and Lagerspetz with a strain of Swiss albino mice, made it possible to differentiate from the second generation an 'aggressive' line and a 'non-aggressive' line, but the difference between them was no further accentuated after the seventh generation. Crossed adoption did not affect this difference, which was therefore clearly of a genetic nature. It should be added that the two lines did not differ only in their degree of aggressiveness; the 'aggressive' mice displayed greater locomotor activity and generally, a higher level of behavioural arousal than the 'less aggressive' mice. Later research showed the role played by genes located in the sexual chromosome Y (which only exists in the male genotype) in determining the degree of aggressiveness expressed between males. These genes, which are correlated with a high degree of aggressiveness, are also correlated with heavier testicles and a higher plasma level of male sexual hormone (testosterone), especially during the period of puberty which is of importance in the development of social interactions.[8]

Since in the development of an individual, the genotype interacts constantly with the environment, it is easy to understand that certain early experiences can modulate to a lesser or a greater degree the expression of a specific genotype, as will be seen further on. But here it should be stressed that certain genotypes (for example that of the mouse strain C57 BL/10) are more sensitive, more 'unstable' than others with regard to early experiences, insofar as the repercussions of these experi-

---

[7] K. M. J. Lagerspetz and K. Y. H. Lagerspetz 1974; N. G. Simon 1979; S. C. Maxson 1981.
[8] S. C. Maxson 1981; M. Selmanoff and B. E. Ginsburg 1981.

ences on the degree of intraspecific aggressiveness are concerned. Furthermore, even in adult animals, learned experience is likely to alter significantly any differences of a genetic origin. Thus differences created by selective breeding between the 'aggressive' and 'less aggressive' lines can be concealed or even reversed, if the aggressive animals are repeatedly made to experience 'defeat' and the less aggressive animals to experience 'victory'.[9]

In man: the polemic concerning the 'crime chromosome'

Research and discussions (often polemical) have mainly concentrated on the effects of the presence in the genotype of certain men, of a super-numerary Y chromosome (the subjects having as their sexual chromosomes the XYY formula instead of the normal XY pair). In 1965 Jacobs and his collaborators reported on observations carried out on a population of subjects who displayed both a certain degree of mental deficiency (mentally subnormal) and a propensity to violence or crime: of 197 subjects treated in institutions, 8 turned out to be carriers of a super-numerary Y chromosome (that is, a proportion of 3.5 per cent compared with the rate of approximately 0.1 per cent found in the general population).[10] But the authors did stress that they were unable to pronounce themselves on the question as to whether this high proportion of XYY subjects had any relation to their violent or criminal behaviour or their mental deficiency, or even a combination of both. Nevertheless this publication aroused a great deal of interest and gave rise to many works on what quickly became, for some, the theme of the 'crime chromosome' or that of 'supermales genetically programmed for violence'. If totally contradictory results — and above all interpretations — were published, it is because this type of research encounters numerous difficulties, especially those of a methodological nature: rather than compare XYY subjects interned in an institution (psychiatric hospital, prison . . .) with XY subjects living freely in their customary environment, it would have been better to compare two groups living in the latter circumstances; and if subjects deprived of their freedom are considered, the different reasons which may have led to their imprisonment should not be overlooked; these reasons should not simply be dismissed as mere 'aggressiveness'. An in-depth study, carried out on all men born in Copenhagen during the four-year period 1944–1947[11] led the twelve participating researchers to conclude that the higher incidence of criminal acts — observed in XYY subjects as compared with XY subjects — was in no way linked to more marked aggressive tendencies, but possibly to certain

[9] K. M. J. Lagerspetz and K. Y. H. Lagerspetz 1974.
[10] P. A. Jacobs *et al.* 1965.
[11] This study only included subjects whose height was amongst the top 15 per cent; XYY subjects are in fact usually tall, and it was important, for purposes of comparison, to exclude any possible influence this factor might have had.

deficiencies of an intellectual nature.[12] It appeared in particular that aggression against persons was no more frequent in XYY subjects than in XY subjects; only crimes against property were more frequent. A more exhaustive psychological study confirmed that the 'aggressiveness' of the XYY population did not differ from that assessed in the XY population.[13] In the Danish survey, XYY subjects were found to have significantly higher plasma levels of male sexual hormone than XY subjects. However, other studies indicated that these levels were widely distributed (both very high and abnormally low levels), or that there was no difference between the two populations.[14] In any case, men who are bearers of a supernumerary Y chromosome often—but not always—display specific morphological characteristics (especially a tall stature, sometimes with acne or other skin problems), more or less pronounced mental retardation, and in certain cases, signs of hyper- or hypo-gonadism. These somatic and mental features may elicit reactions from his entourage which irritate the subject as well as making him particularly sensitive to it.

Nevertheless, it is a big step to make out that the supernumerary Y chromosome is the 'crime chromosome', on the grounds that it 'predisposes' towards crime, as some people have been quick to do. It has been proposed in the popular press that newborn babies should systematically undergo an examination in order to detect those who, without the parents knowing it, might already be 'engaged on the slippery slope of crime'! Lawyers have asked for such an examination to be carried out on their clients, in the hope that they might benefit, in case a supernumerary Y chromosome should be found, from broadly attenuating circumstances (it was this chromosome that led them to commit the crime!). All serious research leads to the conclusion that it is abusive to consider that the supernumerary Y chromosome 'predisposes', or even 'urges' a subject towards crime, and that it is useless or even dangerous (since it generates anxiety and injustice) to proceed with this kind of 'stigmatization'. Certainly this chromosomal anomaly—because of the somatic and mental features it displays—can sometimes make it difficult to hold a dialogue with an entourage that is not particularly disposed to take differences into consideration. But a possible 'acting out' is by no means inevitable; on the contrary, it is the concern of the joint responsibility of the person in question and his close and less close entourage.

The role played by hormonal factors

Given the number and diversity of brain structures and mechanisms the organization and activation of which may be affected by hormonal substances, it will come as no surprise that the role played by this or that *hormonal factor* differs greatly according to the species, age, and above all

---

[12] H. A. Witkin *et al.* 1976   [13] A. Theilgaard 1981.
[14] See H. F. L. Meyer-Bahlburg 1981*a*.

type of aggressive behaviour under consideration.[15] Even if we consider just the sexual hormones of small laboratory rodents (mouse, rat), it can be seen that the way in which they intervene is far from simple. It has certainly been noticed that, generally speaking, the plasma level of male sexual hormone (testosterone) plays an important role in the determinism of intraspecific aggression.[16] This aggression is far more frequent and more intense in males than in females. It develops during the puberty and post-puberty period, that is, at a time when the plasma level of testosterone in males increases strongly. In adult males, castration distinctly reduces the frequency of aggression, and the administration of testosterone has the effect of restoring this frequency to the levels observed before castration. The male sexual hormone in fact intervenes at two different stages of the ontogenesis. Testosterone only exercises its full 'activating' influence in the triggering of intraspecific aggression in adult animals if the brain has been 'sensitized' since before birth by testosterone secreted by the fetal testicle.

One might be tempted to conclude, on the basis of these facts and several others in the same vein, that testosterone is the 'aggression hormone'. In fact it should be stressed that in certain cases (manipulation by the researcher; certain competitive situations), females show themselves to be as aggressive as males. Even in the case of small rodents, testosterone by no means plays an identical role in different types of aggression. Thus our research on interspecific rat–mouse aggressive behaviour[17] has shown that castration in no way changed the behaviour of the 'killer' rat (which quickly attacks and kills any mouse placed in its cage) and that it was not possible to provoke such behaviour in a naturally 'non-killer' rat by injecting repeated doses of testosterone.[18] In the case of aggression triggered towards a pregnant female which entered their cage, only the female mice were aggressive; the males only displayed this aggression if castrated and it vanished again if they were injected with testosterone.[19] In other words, according to the type of aggression studied, testosterone will either be an activator, have no effect, or act as an inhibitor. Moreover, in a given case of aggressive behaviour, the role played by testosterone may depend on environmental conditions. Let us take the case of male rats which aggress each other spontaneously: the aggression disappears very quickly after castration if the castrated animal is confronted with a congener in an unusual environment, whereas it may remain unchanged for months if the same animal is observed in its own cage, in a familiar environment.[20] It is also true that, very generally speaking, the

[15] See P. F. Brain 1981.
[16] J. M. Koolhaas *et al.* 1980; R. Gandelman 1981; D. Benton 1981; R. J. Barfield 1984.
[17] See Chapter 8.
[18] P. Karli *et al.* 1969.
[19] M. Haug *et al.* 1981.
[20] J. M. Koolhaas *et al.* 1980; T. Schuurman 1980; R. J. Barfield 1984.

action exerted by a hormonal substance interacts with factors pertaining to the situation and the individual experience.

In female rats, aggressiveness towards a male congener placed in the cage develops during the sexual cycle, with the lowest level coinciding with the oestrus period, when the female also displays maximum sexual receptiveness.[21] The attenuation of aggressive reactions during oestrus is believed to be due to follicular progesterone. This hypothesis would appear to be confirmed by the fact that in the case of castrated female rats, a dose of progesterone has the effect of decreasing the level of intraspecific aggressiveness.[22] As for 'maternal' aggression, which the pregnant and above all lactating female (rat, mouse) displays towards a male intruder, the hormonal control is still little understood: prolactin does not appear to be an essential factor; testosterone exerts an inhibiting influence rather than a facilitating one.[23]

Alan Leshner's research (1981) into cortico-suprarenal hormones showed the major role played by corticosterone in the development of 'submissive' behaviour in mice. If an animal has undergone one or more 'defeats', it tends increasingly to display a submissive attitude when attacked by a congener. Faced with an attack, the animal's tendency to adopt this attitude does not depend on the level of plasma corticosterone present at that moment, but to what degree this level has decreased previously during a defeat. In other words, the reduction in the level of plasma corticosterone, provoked by defeat by a congener, determines — by its extent — the way in which this defeat is 'experienced' and memorized, and hence, the way in which it affects — in the sense of submission — the animal's subsequent behaviour when confronted with the same situation. Moreover, as we have already indicated,[24] the level of plasma testosterone also decreases in the 'defeated' animal at the end of a conflict, which has the effect of reducing its susceptibility to frustration, threat, and provocation. It is therefore clear that steroid hormones (suprarenal and gonadic) play an important role in processes whereby the perception and interpretation of a situation are modulated by previous experiences in the same or a similar situation. Furthermore, the researchers, who endeavoured to show from the experimental data a more general idea of the role played by hormonal factors in the determinism of different types of aggressive behaviour, reached the conclusion that these factors are in no way involved in the genesis of 'aggressiveness', but that they modulate various means of processing the information concerning a situation, and thus have secondary repercussions on the choice and implementation of the strategy judged to be appropriate.[25]

Research has also been carried out on groups of human subjects to

[21] R. J. Barfield 1984; K. E. Hood 1984.
[22] R. J. Barfield 1984.   [23] R. J. Barfield 1984.
[24] See Chapter 5: Role played by humoral factors, p. 112.
[25] R. Gandelman 1981, 1983; D. B. Adams 1983; R. J. Barfield 1984.

ascertain whether there was a correlation between levels of sexual hormones in circulation and the data supplied by certain tests aimed at assessing the subjects' 'aggressive' or 'hostile' tendencies.[26] As regards possible early hormonal influences, very slight aggressive tendencies were noted in children whose mothers had been treated during pregnancy—for one reason or another—with progesterone or a combination of oestrogen and progesterone. But as this hormonal treatment also had repercussions on the somatic development of the fetus and thus indirectly on certain parental attitudes towards the infant and young child, it is difficult to know exactly what meaning to attribute to the correlation observed between hormonal influence and less aggressive behaviour. The results obtained with adolescents or adult men are, at first glance, not very coherent. Whilst a few researchers have noted a correlation between plasma levels of testosterone and aggressive tendencies or behaviour, most studies have indicated the absence of any such correlation, even in groups of incarcerated subjects, including those incarcerated for sexual crimes. On the other hand, the overall results are more coherent if instead of considering open aggression or hostile tendencies as assessed by various psychological tests, an attempt is made to assess individual susceptibility to frustration or threat, or else the existence of certain attitudes of anxiety and suspicion; in these cases, on a more general scale, positive correlations are found with plasma testosterone levels. This leads to believe that circulating sexual hormones exert a certain influence not only on the way in which—at a given time in the ontogenesis—a situation is perceived and interpreted, but also on the way in which previous experiences have been undergone and memorized, and hence, on the repercussions they may have on subsequent perceptions and interpretations.

Primordial importance of emotions of an aversive nature

The part played by *aversive emotions* is only envisaged lastly, whereas we have emphasized its primordial character from the start; this is both because these emotions occur in a more immediate, more 'proximal' fashion in the determinism of aggression and because their genesis is modulated by genetic and hormonal factors, as we have just seen. In man just as in animals, an aversive emotion not only includes an unpleasant affective element, but also the overall activation of the brain, expressed by physiological changes (affecting the vegetative functions, secretion by the endocrine glands, muscular tone) which prepare the organism for action. This emotion, generated by any situation perceived as a threat to the individual's physical and/or psychic integrity, provokes behaviour which aims to put an end to the unpleasant affective experi-

---

[26] See D. Olweus *et al.* 1980; D. Benton 1981; H. F. L. Meyer-Bahlburg 1981*b*; P. F. Brain 1984.

ence and reduce the level of brain activation. For this activation undergoes self-regulation which maintains it at, or restores it to an optimum
level.[27] This dual objective is achieved by using aggressive behaviour
(likely to avert the threat) or by means of escape (which distances the
individual from the threat).

It thus appears that, far from being a mere epiphenomenon, an aversive emotion provides a mediating function that is vital for living beings.
This function developed when, during the course of evolution, links
between the brain's 'input' and 'output' gradually lost their rigidity, and
enhanced 'plasticity' made it possible to diversify adaptive types of
behaviour. An emotion then became the necessary mediator between
the changing circumstances of the environment and the adaptive responses which reflect the interpretation of these circumstances, based
on the traces left—within the brain—by individual past experience. It
would be far better to reflect on the role thus played by the emotions
before using—and abusing—the psychotropic substances that are supposed to help us get rid of them!

When establishing homologies among the animal species studied,
both the ethologist and the neurobiologist are tempted to list the phenomena observed and the mechanisms analysed in as small a number of
categories as possible. In the area we are concerned with here, Blanchard and Blanchard (1984) proposed for all mammals a clear dichotomy
between 'defensive' aggression underlain by 'fear', on the one hand,
and on the other, 'offensive' aggression underlain by 'anger'. This
dichotomy is based on concordant observations carried out in an experimental situation in which a 'resident' rat is confronted in its own cage
with an 'intruder' rat. In this situation the 'offensive' behaviour of the
resident and the 'defensive' behaviour of the intruder can be clearly
distinguished, not only by the respective postures adopted by each
animal, but also by the parts of the body at which their bites are aimed.
There is no doubt that the two types of behaviour reflect two different
ways of grasping a situation, two different ways of processing the
information arising from it; and it is likely that the emotion aroused,
while aversive on both sides, is not the same in both animals. We have
already reported[28] on experiments carried out in our laboratory where
manipulation of GABAergic neurotransmissions within the periaqueductal grey substance made it possible to modify the processing
of sensory information in such a way that the rat displayed a general
attitude of 'withdrawal' and 'avoidance' towards all stimuli. If the
rat thus treated is confronted with a congener, it regularly displays a
defensive type of behaviour, which would appear to correspond to a
reaction of fear (which can best be induced by manipulation of the
brain).

But can it be said that on the other hand, the so-called offensive

---

[27] H. Ursin 1985.   [28] In Chapter 5.

behaviour normally displayed by the resident towards the intruder corresponds to a reaction of anger? There is no way of ascertaining this directly. Moreover, what is the exact meaning of the concepts of offensive behaviour and defensive behaviour? The intruder is certainly obliged, in the face of the resident's attacks, to provide its own 'defence'. But doesn't the resident endeavour to 'defend' its familiar environment and to 'defend itself' against an assault on its tranquillity? Even if by doing so it uses a type of behaviour which differs in form from that used by the intruder, are the objectives of these two types of behaviour sufficiently different to justify the 'offensive–defensive' dichotomy, with the 'anger–fear' dichotomy we feel obliged to associate it with? It is not easy to answer these questions, for several reasons. On the one hand, as Scott (1984) emphasizes, a single type of behaviour (identical in form) may be used in an offensive or a defensive way: in a pair of goats the subordinate animal may use behaviour that is characteristic of an offensive attitude (butting with the horn), but it only does so in response to an attack by the dominant animal. It is therefore difficult to deduce, from what can be observed of the behaviour, the motivation and in particular, the affective state, which is supposedly responsible for it. More importantly, the affective state underlying a given type of behaviour develops with time. This is the case, for example, of the interspecific rat–mouse aggressive behaviour. If a rat is confronted for the first time with an unusual situation such as the intrusion of a mouse in its cage, any aggressive behaviour towards this animal of a different species is underlain by a state of motivation of an aversive kind. Subsequently, when on several occasions the rat assaults and kills mice introduced into its cage, this behaviour is itself reinforced in a positive way, since clearly 'appetitive' elements change the—initially aversive— significance of the situation. As we will see further on,[29] the aggressive behaviour of a rat which kills for the first time and that of the 'killer' rat with experience of this behaviour are not affected in the same way by a given manipulation of the brain; this clearly shows that information linked to the mouse's intrusion is not processed by the rat's brain in the same way in both cases.

Rather than stress the more or less arbitrary classifications of aggressive behaviour (based on criteria which are far from being unanimously accepted), it would seem preferable to make a fundamental distinction with regard to the targeted objective: on the one hand, to put an end to a transitory 'painful' emotion or more enduring mental suffering, by acting on the situation which gives rise to it; on the other hand, to appropriate a coveted object which one means to own, in anticipation of the pleasant emotion that will presumably result. This distinction is independent of the form taken by the aggression and it enables more than mere analogies to be established between animals and humans. In

[29] In Chapter 8.

the case of putting an end to an aversive emotion (and to the more or less marked brain activation that is an integral part of it), the latter may predominantly, but not necessarily exclusively, correspond to one modality (uncertainty, anxiety, fear) or the other (dissatisfaction, annoyance, anger). Whilst animals share the experience of fear with humans, it is reasonable to believe that dissatisfaction and anger are felt more often by humans, due to the—possibly 'hypertrophic'—development of the self and its social status. We will now consider more concretely the aversive emotions linked to 'neophobia', those created by physical pain and those caused by frustration.

Neophobia and aggressiveness

The 'novelty' of a stimulus, its unfamiliar nature, is an important factor in determining both approach responses and avoidance responses.[30] Rats, like many other animal species, display 'neophobia' (fear of anything new). This natural tendency to avoid unfamiliar objects in a familiar environment may be conveyed, depending on the circumstances, by 'intolerant' behaviour towards them. In a study covering aggressive behaviour displayed by wild rats towards human subjects, mice, or congeners, Galef (1970) noted that the novelty of the stimuli was always a necessary condition for triggering aggression, and that familiarity with a certain type of stimuli clearly decreased the probability of an aggressive reaction towards them, without affecting the probability of aggression in response to the other two categories of stimuli which had retained a novelty status. The neophobia of the wild rat is far more marked than that of the laboratory rat, and this difference can explain, at least in part, the fact that the proportion of 'killer' rats (of mice placed in their cage) is far higher in wild animals (around 85–90 per cent) than in laboratory animals (around 10–15 per cent). In fact, certain laboratory rats kill wild mice but not laboratory mice (towards which there is probably less neophobia) placed in their cage. Furthermore, early social contacts with mice result in a clearly lower incidence of interspecific aggressive reaction in the adult rat.

Very generally speaking, strangeness is a very efficient stimulus for triggering aggression, and Marler (1976) rightly emphasizes that it may well be familiarity that actually constitutes the most important factor in reducing the probability of aggression. It is above all familiarity gained in inter-individual communication, with the consequent ability to recognize easily the meaning of signals emitted by others and to respond appropriately, that reduces the probability of aggression; in many animal species, on the contrary, it is most particularly congeners 'strangers' to the group which provoke the most marked hostility.[31] Humans also

---

[30] D. T. Corey 1978.
[31] S. Green and P. Marler 1979; M. Bekoff 1981.

feel more at ease if every gesture and word of the 'other' are familiar and easy to interpret; and when this is not the case, lack of familiarity and uncertainty are likely to be conveyed by a rather hostile attitude. Although, as we will see, xenophobia, ethnocentrism and racism are rooted in many other sources, it would be fruitless, even regrettable, to deny the existence of this truly biological factor which it is possible to overcome by making conscious efforts. For if we merely consider the monstrous (that is, 'ab-normal') nature of these attitudes, we will hardly be moved (we who, of course, are 'normal' people!) to become aware of the problem and make the necessary efforts.

From pain to aggression

Many surveys have been devoted to aggression provoked by painful or merely unpleasant stimulation, the parameters of which can be more easily controlled than those of a situation considered to be unfamiliar.[32] In animals, the two following experimental methods have been widely used: a rat, with the major part of its body tightly enclosed in a cylinder, receives electric shocks to its tail, and an analysis is made of the way in which it bites an object placed within reach; or else, the behaviour is studied of two rats which are together subjected in a small cage to electric shocks on their feet. In both cases, the painful electric stimulation triggers aggressive reactions towards the inanimate object or the congener, and the frequency and intensity of these reactions can be monitored by manipulating the parameters of the stimulation used. Since this behavioural effect is provoked in a quasi-automatic and stereotyped way, the idea of 'reflex' aggression has often been used. But it should be stressed that in no way, in the case of the two animals subjected together to painful stimulation, can this be considered a reflexive response (for which the mediating intervention of an aversive emotion would not be necessary at all). In reality the two rats fight less if the size of the cage is increased or if they are allowed to devise beforehand a strategy likely to reduce the number of shocks they receive. Above all, they prefer escape to aggression, if they are given a choice of response. To recall the experimental data mentioned above:[33] in a 'non-killer' rat which has never shown the slightest hostility towards a mouse which has lived in its cage for weeks, an aversive emotion induced by electrically stimulating the neuronal aversion system is conveyed by marked aggressiveness towards this same mouse; but once the rat has learned how to stop the electrical stimulation by pressing a lever, it is hardly possible any longer to incite it to kill a mouse by thus activating its system of aversion (here too, the rat chooses the escape response in preference to the aggression response).

[32] See R. Ulrich and B. Symannek 1969; R. J. Rodgers 1981; R. R. Hutchinson 1983; L. Berkowitz 1984; R. J. Blanchard 1984.
[33] In Chapter 5.

Given that under the effect of the electric shocks applied to them, the behaviour of each of the two rats is very similar in form to that displayed by an 'intruder' rat confronted with an attack by a 'resident' rat, Blanchard (1984) believes that aggression provoked by pain should be considered as behaviour of a defensive type and that it is therefore caused by fear. But it is not sure whether this interpretation can be applied to all mammals, since it seems that in certain species, painful stimulation also gives rise to behaviour considered to be of an offensive type.[34] In any case, researchers investigating this problem in humans allude more to dissatisfaction and anger than to fear.[35]

In humans, painful or merely unpleasant stimulation can be induced, for example, by asking the subject to plunge and keep one hand in icy water. The repercussions of such stimulation on the subject's behaviour are then analysed in the 'examiner–candidate' situation. This situation, which has been widely used (and which has raised certain objections of an ethical nature), is as follows: the subject whose behaviour is being analysed assumes the role of an examiner who asks a candidate questions and who 'punishes' the latter's mistakes (by inflicting from a distance electric shocks to the skin, or by emitting unpleasant noises through headphones, with an intensity that he controls and selects); the candidate, who has been informed of the experiment and who does not really receive the shocks or hear the sounds, only simulates the pain or the displeasure he is supposed to feel, but the examiner does not know this. Under these conditions, it can be seen that painful stimulation has the effect of inducing—or accentuating—an aggressive attitude: the 'examiner' applies stronger 'punishment'. Other unpleasant stimulations (offensive odours, excessive room temperature . . .) have a similar effect. The aggression-provoking influence of the painful stimulation itself depends on a whole range of factors which can be mentioned briefly, even if this means going beyond the purely biological aspects of the question. Berkowitz (1984) showed, with his collaborators, that the tendency to punish was particularly accentuated: if the word 'pain' was pronounced by the researcher and the subject acting as the examiner fully expected to feel pain when plunging his hand in the water; if the subject, during the five minutes in which he kept his hand in the icy water, was asked to write a brief dissertation on the educational value of punishment (whereas the pilot group, in its discourse, discussed the pleasures of life on the Californian coast); if the questioned 'candidate' had an unattractive face and stammered (did Leonard Berkowitz possibly hear of the 'ugly mug offence'?).

The part played by frustration in triggering aggressive behaviour

Frustration is produced whenever there is an obstacle to the pursuit of

---

[34] R. J. Rodgers 1981.   [35] L. Berkowitz 1984.

an objective, or to the fulfilment of an expectation. It arouses an aversive emotion, anger, which often leads to aggressive behaviour. Whilst it has sometimes been lent excessive or too exclusive importance, there is nonetheless little doubt that frustration is a major incitement to aggression. In animals, the most frequently used method is that of suspending the positive reinforcement (the 'reward') which it obtained regularly by carrying out a certain task, or reducing the likelihood of it by only giving it intermittently.[36] Many studies have been carried out on this subject, principally with pigeons, but also with rats and monkeys. When a pigeon has learned a response which is reinforced by obtaining food and this reinforcement is suddenly stopped, it pecks at the head, and especially the eyes, of a close congener, whether the latter is alive or stuffed. In rats and in monkeys, stopping a food reward or suddenly being deprived of morphine triggers aggressive behaviour, with bites and vocalizations. In all cases frustration has both aversive and activating attributes.

In humans, frustration, the effects of which may be cumulative, leads to anger and aggression, especially if another conducive factor is superadded.[37] The additional factor may be an insulting remark connected with the frustration, or prior physiological activation which may have very diverse origins (the violent or erotic content of a film, for example). The effects of frustration can also be accentuated by internal conditions (high alcoholemia) or external ones (noise, crowds . . .). Using the experimental method of the 'examiner–candidate' relationship described above, Bell and Baron (1981) showed that displeasure due to a high ambient temperature led the examiner to apply harsher punishments. Berkowitz (1981) demonstrated the major role played by the 'legitimate' nature, or otherwise, of vexation or anger linked to frustration. He subjected pairs of students to an instrumental task, promising each of them a reward (a sum of money) if they succeeded. When he showed them that they had not succeeded, he attributed this, in certain cases, to the machine (in this case anger towards the partner would be 'illegitimate') and in other cases, to the partner (in this case, the subject could 'legitimately' feel angry with the latter). In both cases the same degree of physiological activation could be observed. But if these pairs of students were then placed in the 'examiner' and 'candidate' situation, it became clear that only the one whose frustration gave rise to 'legitimate' anger punished more severely than those subjects who did not undergo any frustration ('control' group).

(B) Factors linked to the personality of the subject

The question that arises right from the start is whether there exists a prominent feature or special constitution which would make it possible

---

[36] See T. A. Looney and P. S. Cohen 1982.   [37] See J. R. Averill 1982.

to identify unequivocally the 'aggressive personality'. There was a time when this question received a frankly positive answer. Lombroso, to mention but one, spent a long time on the 'born criminal', considering that this was a subject whose development had stopped at the stage normally reached by 'savages', a stage where primitive instincts still prevail. The antisocial behaviour of the born criminal thus becomes the inescapable expression of his 'moral madness', that is the absence of any predispositions that would have rendered him amenable to moral feeling.[38] The 'crime chromosome' mentioned above is the latest avatar to date in this school of thought. Very generally speaking, the concept of the 'aggressive personality'[39] is currently still considered to have descriptive merits but hardly any progress can be made in understanding aggressive behaviour by trying to 'explain' it by the existence of such a personality.

What is certain is that in the same environment not all individuals display the same frequency and regularity of aggressive attitudes and behaviour. Longitudinal studies undertaken over a period of many years have clearly demonstrated the remarkable stability of 'aggressive' personalities as well as of 'non-aggressive' personalities.[40] A psychological test such as the 'Rosenzweig frustration test' enables this individual character to be discerned by assessing the verbal aggression caused by frustration.[41] Since there are established individual differences regarding the probability of the appearance of aggressive behaviour, the source of these differences should be sought. We will analyse in more concrete terms the conditions which throughout a person's development are liable to encourage, or on the contrary, to prevent, the development of aggressive tendencies. In other words, rather than emphasize a specific and 'finished' structure of the personality, interest should focus on the shaping of this personality and the conditions which may orientate it towards a lesser or greater 'aggressiveness'. This approach gains true importance through the many studies carried out both on animals and on humans, which clearly show the major role played by experience acquired in a given environment. We thus return to the basic idea already stressed at the beginning of this chapter: aggression 'reveals' a personality shaped by actual experience, an individual and 'historically' established way of apprehending situations and events and facing up to them.

We will deal separately in our presentation with 'factors linked to the subject's personality' and subsequently, with 'factors linked to the situation'. Now, we have just seen that the personality should be apprehended in its development, which necessarily leads one to wonder about the influences exerted by the family circle, and generally, by the

---

[38] For a review of criminological ideas, see J. Léauté 1972. See also P. Tort 1985.
[39] Just as the concept of 'aggressiveness': see Chapter 2.
[40] L. Pitkänen-Pulkkinen 1981; D. Olweus 1984.
[41] S. Rosenzweig 1981.

sociocultural environment. The potentially 'aggression-provoking' situation will, when the time comes, be perceived and interpreted on the basis of references supplied by this same environment. In these circumstances any distinction between the social influences which contribute towards orientating the formation of the personality and those which contribute towards determining the way in which a situation will be perceived and interpreted at a given time in the ontogenesis, will obviously be somewhat arbitrary. A certain amount of redundance is therefore inevitable.

The role played by the level of emotional reactivity

Before examining a certain number of concrete facts showing to what extent and how certain experiences affect the development of behaviour in the sense of more or less marked 'aggressiveness', one aspect of the personality should be indicated which, in animals as in humans, contributes—all other things being equal—towards determining the probability that aggression will appear. This is the *level of reactivity*, which differs from one individual to another, and which develops during the early stages of the ontogenesis. In rats, the existence has been demonstrated of a close correlation between the intensity of the 'startle reaction', provoked by a tactile or auditory stimulation, and the probability that intraspecific aggression will be triggered in response to painful[42] stimulations. If two groups of rats are formed, based on their level of reactivity towards a tactile stimulation (high for some, lower for others) and each of these animals is then placed as an 'intruder' in the cage of a 'resident', it can be noted that the behaviour in response to the attacks by the resident, is not the same in both groups: the intruders whose reactivity is high inflict four times more injuries on the residents than those whose reactivity is lower; and the residents spend five times longer defending themselves from the former than from the latter.[43] Because social isolation has the effect in both rats and mice of greatly increasing the probability of aggression, it is worth indicating that this isolation also provokes a noticeable increase in reactivity towards tactile stimulations, in particular to those emanating from congeners.[44] As for interspecific rat–mouse aggressive behaviour, the probability of a rat assaulting a mouse placed in its cage is correlated with its own emotional reactivity, whether this is assessed by means of the 'open field'[45] test or by means of a 'conditioned emotional response'.[46] Moreover, as we will see in the next chapter, all hyperreactivity—especially towards stimulations of an aversive nature—induced by experimental manipulation, clearly increases the probability of triggering of both intraspecific

---

[42] M. Davis 1980.
[43] R. J. Viken and J. F. Knutson 1983; J. F. Knutson and R. J. Viken 1984.
[44] R. B. Cairns 1979.   [45] M. Vergnes *et al.* 1974.   [46] D. C. Bowers 1979.

aggression provoked by painful stimulations and interspecific rat–mouse aggression. On the contrary, manipulations which lower the level of emotional reactivity (early manipulation of the animals by the experimenter; an 'enriched' environment) have the effect of reducing the probability of interspecific aggression.[47]

Irritability and emotional susceptibility in humans

Caprara and his collaborators (1983) analysed, by means of the method which places the subject as the 'examiner' in front of a 'candidate', the repercussions of two features of the personality (which can be likened to the 'reactivity' taken into account in animals): on the one hand, 'irritability', defined as being the regular tendency of an individual to react in an offensive way to even a slight provocation; on the other hand, 'emotional susceptibility', defined as being the tendency to experience feelings of inadequacy or distress. Very 'irritable' subjects administered to the 'candidate' more severe punishments than the less 'irritable' subjects; likewise, very 'susceptible' subjects did more severely punish than less 'susceptible' subjects. In research of a psychophysiological nature carried out on imprisoned subjects, the degree of 'impulsiveness' and the likelihood of violent behaviour were also related to a 'reactivity' index, that of electrodermal activity (caused by a change in the resistivity of the skin): in subjects displaying a propensity of violence, particularly prolonged electrodermal activity was noted in response to intense or unusual sound stimulation.[48]

Structuring influences exerted by the social environment

In most mammals the socio-affective behaviour of adults depends to a large degree on the way in which the individual learned during the early stages of his life how to communicate with his congeners, how to develop his faculties of 'social perception' (which allow him to anticipate the behaviour of others and act in consequence), to build up social relationships and adapt his own behaviour to the dynamics governing them.[49] The essential role played in this socializing process by the progressive establishment of a set of socio-affective links becomes clearly apparent if one considers the repercussions of special rearing conditions on the development of the behaviour of monkeys.[50] If a macaque is reared in total isolation, it withdraws into a corner of its cage when fear responses are fully mature at around the age of 60 to 80 days, and it tries to 'erase' the surrounding reality. A little later, when aggressive behaviour has also developed, it will act very aggressively towards any

---

[47] J. A. Garbanati *et al.* 1983.   [48] J. W. Hinton 1981.
[49] See M. Bekoff 1981. See also Chapter 4, under Socio-affective behaviour, p. 67.
[50] See W. P. Meehan and J. P. Henry 1981.

animal as soon as it is placed in the presence of a group of congeners. It is incapable of interacting properly with the others, since it cannot recognize the prevailing hierarchy within the group and does not make the appeasing gestures expected after a defeat. If a macaque has grown up without any contacts except with its mother, it will be clearly less marked by fear. But neither has it learned to behave in an appropriate fashion in situations of social interactions; it is also exaggeratedly aggressive when placed within a group, and its 'irrelevant' behaviour condemns it to a subordinate position in the hierarchy. Lastly, if the young monkey has only had contacts with its peers, it will be better able to integrate into a group and cooperate with the others in the group; but it will not have benefited from the feeling of security normally provided by attachment to the mother, and is quite timid.

The gradual establishment of social links thus enables the individual to confront unusual situations and to face up to them effectively, without excessive fear. The nature and extent of these links obviously depend on the structure and internal dynamics of the group, which in turn depend on the ecological conditions in which the group develops. In the case of Australian rodents and certain Canidae, different species have a more or less elaborate social organization, with or without the intervention of social 'play' which facilitates the formation of links between individuals; in the least 'social' species, earlier and more marked development of intraspecific aggression[51] can be noted. As has already been mentioned above, the familiarization of congeners with one another plays an essential 'preventive' role in this respect. When mice are reared in environmental conditions that prevent any organization within the group, the development of marked aggressiveness can be observed, and all the males bear multiple scars. What is more, these conditions create very intense 'social stress' which is expressed by noticeable hypertension and both cardiovascular and renal lesions; moreover, after 9 to 10 months half the males are dead.[52]

It is in small laboratory rodents that it is easiest to manipulate the experiences which mark out the ontogenesis of an individual, and it is these which have produced the largest amount of experimental data. In rats, one of the research methods used is that of rearing an animal, whose behaviour is later studied, under conditions where it is exposed to frequent—or less frequent—aggressive interactions.[53] If the subject is reared, for example, with three castrated congeners, its experience of aggressive interactions will be reduced. This experience can also be reduced by rearing the subject, deprived of its sight, with three unimpaired congeners, under conditions of continuous illumination, 24 hours a day; it was in fact demonstrated that continuous lighting attenuated intraspecific aggressiveness in rats, whereas deprivation of sight

[51] M. Bekoff 1981.   [52] W. P. Meehan and J. P. Henry 1981.
[53] J. F. Knutson and R. J. Viken 1984.

did not change it at all. If the behavioural reactions triggered in these animals by painful stimulations are then studied, it can be seen that subjects which were less exposed to aggressive interactions during their development, prove to be less aggressive than those which had gained more experience of this kind of interaction. Though at first glance it may seem paradoxical, subjects reared together with less aggressive congeners turn out to be more 'offensive' than the others when placed as 'intruders' in the cage of a 'resident'. The most likely explanation of this fact is a dual one: on the one hand, these animals have seldom had the opportunity—since faced with less aggressive congeners—of behaving in a 'defensive' way; on the other hand, their own tendency to attack was not tempered by the 'punishments' they would have received from more aggressive congeners. Other experiments have shown the repercussions of prior contacts with a female on the behaviour of the male rat, when the latter is confronted, alone, in the capacity of a 'resident', with a male 'intruder': the resident which had a female in its cage was clearly more aggressive than the one which lived together with another male; and these repercussions caused by the previous presence of a female are more marked if the latter is unimpaired than if she is castrated.[54] By introducing a strange congener in a group of rats and by using as 'intruders' and members of the receptor group rats which had been, or which on the contrary had not been, isolated after weaning, Luciano and Lore (1975) noted that the intruder is only severely attacked if it has itself been deprived of social interactions after weaning and is placed in a group made up of animals which have always lived in a group.

In mice, research has mainly concentrated on the role played by interactions with the mother as well as on the effects of social isolation.[55] Male mice of the A/J stock, which are normally not very aggressive, are far more likely to display aggressive behaviour if they have been reared by females of the CFW stock, where the males are far more aggressive. Mice reared by a female rat, on the contrary, turn out to be less aggressive than mice reared by their own mother. Male mice, selected according to the weight of their brain (low, medium, or high), differ in adulthood in the frequency and intensity of their aggressive interactions, provided that the animals have been kept in isolation from the age of 21 days; for they need only be allowed to interact for a further ten days, up to the age of 31 days, for these interindividual differences to disappear. The many studies carried out on the repercussions of social isolation, which clearly show that one of these repercussions consists of an increase in intraspecific aggressiveness, also show that the precise nature and extent of the effects of isolation depend both on its duration and on the time at which it occurred during the ontogenesis. All the observations carried out (not only with regard to mice, but also with regard to

[54] G. A. Barr 1981.   [55] See N. G. Simon 1979; R. Lore and L. Takahashi 1984.

rats and monkeys) have indicated that a few brief confrontations with congeners of the same age, during the period following weaning, have the effect of attenuating, or even eliminating the damaging influence of subsequent long-term social isolation. If animals reared in a group are less aggressive than those reared in isolation, this is largely due to the 'defeats' experienced in confrontations with congeners. If a mouse reared in isolation and highly aggressive is placed within a group of congeners, its aggressiveness declines all the more rapidly depending on whether the receiving group is more aggressive. If this animal is isolated once again, but in a small wire cage placed inside the large cage of the group, its aggressiveness does not reach the high level observed previously, since it continues to perceive the signals which it learned to associate with the experience of defeat.[56] It is therefore evident that already in rodents and to a far greater extent in infra-human primates, the development of socio-affective behaviour occurs with many variations and is sensitive to the multiple influences exerted by the social environment.

The major determinants of an aggressive personality

Research carried out on children and adolescents has emphasized the great stability of the 'aggressive' or 'non-aggressive' character of developing personalities and the essential role played by certain factors during the first years of life. This does not mean that other factors are not liable to intervene later, but it is generally stressed that the latter only take full effect if the ground has been prepared by the former.[57] In reviewing the different studies concerning the stability of aggressive behaviour, Olweus (1984) made the following observations: individual differences observed at the age of 3 years are found again 2 years later; aggressive behaviour observed in boys of 8 to 9 years can often be related to similar behaviour observed 10 to 14 years later; lastly, the same aggressive attitudes and behaviour were found in subjects around their mid-thirties that had been observed 15 to 18 years earlier, when these subjects were adolescents. In a personal study undertaken in Sweden concerning two groups of boys (some aged 13 years, the others aged 16 years), Olweus endeavoured to define the factors which can be held to be the major determinants of marked interpersonal aggressiveness, as it thus emerged from assessments by classmates and teachers. It appeared quite clearly that the two principal determinants were formed by the mother's attitude towards the child: on the one hand, her 'negative' attitude of coldness and indifference, or even conveyed by straightforward hostility and rejection of the child; on the other hand, her 'permissive' attitude, by letting pass all aggression committed by the

---

[56] K. M. J. Lagerspetz and K. Sandnabba 1982.
[57] R. B. Cairns 1979; H.-J. Kornadt 1984; D. Olweus 1984.

child, by not trying to control it and teaching the child to control himself. A third factor plays a less important, though not negligible role: the violent behaviour of the parents towards the child, with repeated threats and corporal punishment. These same major determinants of an 'aggressive' personality can be found in the conclusions of several works (particularly American ones) which Olweus compared with his own conclusions. The study carried out in Sweden lends only minor importance to genetic factors; furthermore, no relation could be established between the 'aggressive' or 'non-aggressive' personality of adolescents and the socio-economic situation of their family.

An efficient antidote: acquiring 'prosocial' conduct

In a longitudinal study carried out in Finland (the same children were seen at 8 years, 14 years, and 19 years), Pitkänen-Pulkkinen (1981) concentrated mainly on the development of the traits of the 'non-aggressive' personality. She was led to distinguish, in children and adolescents with definite self-control, between a 'docile' attitude (the control concerned the genesis of emotions of an aversive nature, the genesis of impulses) and a 'constructive' attitude (this control mainly concerned the expression of impulses, with a search for strategies other than aggression). These conclusions are to some extent the negative image of those formulated by Olweus: the 'negative' attitude of the mother prevents the child from developing control over the genesis of emotions of an aversive nature; her 'permissive' attitude does not encourage recourse to means of action other than aggression, on the contrary. If the child learns to solve his problems by means of 'prosocial' behaviour, he tends not to adopt aggressive strategies. In an American longitudinal study spanning 22 years (the subjects having been seen first at 8 years, then seen again at 19 years and at 30 years of age), Eron and Huesmann (1984*a* ,*b*) also noted great stability in individual ways of behaving, and they showed that aggressive attitudes and prosocial (or 'altruistic') attitudes were correlated negatively, that is, they were to a great extent mutually exclusive. Like aggressive behaviour, prosocial behaviour is acquired very early in life. And again it appears that the quality of the parent–child relationship plays a very important role. The more a child is demeaned by his parents (who systematically criticize his behaviour and his achievements, who humiliate and punish him in public) the greater the chances that he will become an aggressive adult. If the child identifies well with his parents, on the contrary, this can be the best encouragement for him to adopt prosocial conduct. Patterson (1984) emphasized the fact that in our 'western' culture children are not asked to take care of siblings and are given no special responsibility, whereas in more 'primitive' cultures the emphasis placed on the care to be given to the younger ones and the need to participate in other household tasks contributes to a large degree to giving children a sense of responsibility towards others. The observations made by Ekblad (1984), which have already been mentioned, may be noted here:[58] Chinese children reared in a 'restrictive' environment know how to control their emotions and adapt their behaviour to collective standards, whereas Swedish children, reared in a more 'permissive' environment, display clearly more individualistic attitudes.

The essential role of the family milieu

Parental attitudes and interactions within the family milieu play an essential role, for at least three closely complementary reasons: they orientate the affective development of the child; they may or may not help the child to acquire the self-control necessary to develop his autonomy; they provide landmarks and models. Attitudes of coldness and even systematic disparagement have little chance of giving the child a warm and optimistic view of the world and of others. Furthermore, a child has a vital need of affection and tenderness, and lack of affection is very often expressed in long-term effects such as anxious agitation and impulsive reactivity.[59] In this respect it should be indicated that this lack of affection is regularly found in the most serious cases, where an adolescent was led to kill: in a study of 18 adolescent murderers, McCarthy (1974) emphasized the fact that all these young people had been rejected by their parent(s) and had grown up in a milieu where no consideration was taken of other people's feelings. Even if there is no actual lack of affection, the development of an aggressive personality can be encouraged by insufficient or improper control exerted over a child's behaviour. Hamburg and Van Lawick-Goodall (1974) believe that in general, tantrums brought about by the most diverse frustrations are a 'forerunner' of subsequent aggressive behaviour. If the parents, out of concern for their own peace and quiet, positively reinforce these tantrums by giving in more or less regularly to the demands of the child, they teach him that aggression is a type of behaviour that 'pays'. Later on, aversive interactions with the other members of the family may start off a veritable escalation of violence, with aggressive behaviour learned and reinforced, if the parents have imposed an incoherent, confused, and unsuitable kind of discipline which does not clearly spell out for each child the limits which may not be overstepped.[60]

As for landmarks and models, many of these are provided nowadays by television; and violence is omnipresent on the screen. Although much controversy has taken place on this subject, there seems to be no doubt nowadays that a child learns from his television screen, that aggression is a type of behaviour that is tolerated after all, or even

[58] In Chapter 4, under Structuring influence of the environment, p. 69.
[59] See R. Ebtinger and A. Bolzinger 1982.
[60] G. R. Patterson 1984; G. R. Patterson *et al.* 1984.

approved, and more often than not, very efficient.[61] Eron and Huesmann consider that children are particularly receptive between 8 and 12
years, but that the effects of violence seen on the television are cumulative over a long period. Moreover, they found a highly positive correlation between the extent of contacts with violence through the media at
the age of 8 years and the level of aggressiveness noted in the children of
these very same subjects, 22 years later. Parents are certainly not directly
responsible for the content of television programmes. But it is up to
them to control the type of pictures which contribute towards shaping
cognitive structures and feeding the imagination of their children.

(C) Factors linked to the situation and to the sociocultural context

As for the most extreme form of human aggression, murder, Picat (1982)
observed that a jurist will see it as a serious offence and a moralist as a
sin, but that in the eyes of a criminal psychiatrist, it is first and foremost
a situation of encounter and confrontation. It is hardly necessary to
stress that in this case as in many others, the 'situation' cannot be
dissociated from the personality of whoever 'took action' (the encounter
is not always fortuitous: it may be sought or, on the contrary, avoided),
nor from the multiple facets of social influence. Personality and social
influence, for their part, are inseparable, if one considers that the latter
'implies the linking up of all past experiences, that is, the history of
the individual in his physical and social environment, and all the
present conditions, in which affectivity, knowledge, and reason play a
part'.[62]

Multiple and ambivalent repercussions of the sociocultural context

The role played by the social structure of the group and by the social
relationships interwoven in it is both complex and ambivalent. On the
one hand, life within a group is beneficial for the individual from at
least two standpoints (without this adaptive quality, it would not have
developed as it has). Cooperation between individuals within a group of
monkeys gives them access to a greater number of resources and enables
them to provide treatment for those animals which need it.[63] As for the
human species, it is a well known fact that the community is better able
than the isolated individual to fulfil all kinds of needs. Interindividual
attachments which develop within a group also have the effect of attenuating the pathogenic repercussions of stress. We mentioned at the
beginning of this chapter the experiment which showed that two pigs
coming from the same group did not display, in a stressful situation, the
peak of plasma corticosterone observed in an isolated animal or in two

[61] J.-P. Leyens 1979; G. Comstock 1983; L. D. Eron and L. R. Huesmann 1984a.
[62] G. de Montmollin 1977.   [63] See R. A. Hinde 1983.

animals coming from two different groups. The Seymour Levine team at Stanford showed that the physiological repercussions observed in a young monkey, when it is separated from its mother, were far less marked in its familiar social environment than in an environment where the young animal did not enjoy this same 'social support'.[64] Meehan and Henry (1981) report on the observations carried out during the Vietnam war on doctors transported by helicopter to the front, under enemy fire: when they departed on their mission, these doctors showed no increase in their levels of plasma corticosterone, by virtue of the 'social support' they received in the form of the respect and admiration which everyone felt for them. In this case too, attachment to the group protected individuals against the pathogenic effects of particularly stressful situations.

Yet on the other hand, the development of diversified and enriching interindividual attachments and a sense of belonging, together— correlatively—with a clearer and more differentiated consciousness of self and a certain self-esteem, gives rise in humans, aside from the obvious benefits, to many grounds for frustration which may generate aggression: frustration linked to any threat or attack on his sense of personal integrity and dignity, to an advantageous interpersonal relationship, to his sense of identity and of belonging to a group. Moreover (and this 'other side of the coin' indirectly confirms the beneficial effects of socialization), it was amply demonstrated that social disintegration or destructuring was particularly 'aggression-provoking'. Scott (1975) reviewed many examples, taken from three categories of vertebrates (fish, birds, mammals), where social disintegration, for whatever cause, was always expressed by a clear increase in frequency of confrontation and aggression between individuals. The same applies in human communities. In the Ammassalimiut society (on the east coast of Greenland), the conjunction of accelerated acculturation, social destructuring, and the disappearance of certain traditional forms of regulation, brought about the clear development of aggressive and violent behaviour.[65] The Yanomami society of Venezuela is acephalous, with very little hierarchy and without authoritative institutions; and in this society, violence is an 'institutionalised means of social control for the purpose of ensuring respect for the law'. In the absence of established systems of authority, the practice of violence 'produces order and makes it possible to respond to the numerous tensions within the society.'[66]

Since we emphasized from the outset the close relationship between the personality and its own sociocultural context, we should also indicate two further influences which this context is liable to exert. Anxious agitation and impulsive reactivity, the result of a lack of affection and of incoherent discipline, are usually expressed by lack of interest in the future, as opposed to the present, with an inability to become involved

---

[64] C. L. Coe *et al.* 1985.   [65] J. Robert-Lamblin 1984.   [66] C. Alès 1984.

in any project which would generate a meaning and ensure a future. The sociocultural context may, however, at least to a certain extent, mitigate these deficiencies of a socio-affective nature or, on the contrary, facilitate their development and amplify them. In this respect, what about our own context now, at the end of the century? If one refers to a text written as an introduction to a series of 'interviews' devoted to the 'individual', the following ideas can be seen to follow in close succession: 'subjectivity crisis and identity crisis; dissemination of meaning and values; life, a meaningless trajectory, stretching from one void to another; uprooting and uselessness . . .'.[67] Even if this picture may appear exaggeratedly depressing, the context it depicts is certainly not of a nature to attenuate the deficiencies which may be generated by 'negative' and 'permissive' parental attitudes.

Secondly, social influence affects the control of the emotions and the use of aggression as a means of expression and action. Within the Yanomami society, mentioned earlier, there exists a 'permanent climate of persecution and aggression expressed sociologically by mutual suspicion, rancour and hostility by parties who were once in opposition'. And within the framework of this vindicatory system, 'the spirit of revenge is inculcated in the child from a very early age as being a totally positive and primordial value, and the revenge, usually carried out in the form of destruction of objects and people, is legitimate and legitimised'.[68] On the opposite scale, in certain rural Tahitian communities a veritable 'conspiracy' against aggression has been observed: the ability to control anger and the need to acquire a measured and discreet autonomy indicate 'a kind of unconscious consensus that rejects aggression as something unthinkable, a nightmare'; these communities appear to expend 'considerable energy in circumventing aggression'[69] in educating their children. When presenting a series of studies on 'the Ethnography of Violence', from which we have just taken two examples, Elisabeth Claverie emphasizes 'the culturally coded character' of violence and the fact that the works presented oppose 'the theories of violence of an ethological or biological nature, which show violence as a primordial impulse and a demonstration of "aggressive tendencies".'[70] It is to be hoped that if she reads the present work it may persuade her that certain biologists (and their number is growing) share her views to a large extent!

The wide diversity of 'aggression-provoking' situations

If we now turn to more precise and more clearly outlined 'situations', it should firstly be realized that these act in a more or less direct fashion, that is to say, the objective of an act of aggression is more or less closely

[67] R. Jaccard 1985.   [68] C. Alès 1984.   [69] J. F. Baré 1984.
[70] G. Lenclud *et al.* 1984.

linked to the very situation which gave rise to this behaviour. If the aggression is caused by a particularly irritating, stressful, or frustrating situation, the objective is directly linked to that situation, since it is the latter, the way in which it is perceived and the aversive emotion it arouses, that must be put an end to. It may also happen that a specific situation does not trigger aggression unless it was preceded by a different, longer lasting situation. Israël (1984) rightly stresses that to a workman who is bothered by the noise in his workshop, children shouting in the street can be 'the last straw'; and intolerance to noise will be expressed, possibly through aggression (which, isolated from its context, may appear exaggerated) as soon as the noise can be 'personalized'. A different kind of example is given by Cusson (1983*b*) who reports (quoting the autobiographies of Claude Brown who grew up in Harlem and James Carr who grew up in the black ghetto of Los Angeles) that in certain milieux where violence prevails, fathers teach their sons to make use of situations that enable them to show their fighting spirit in order to prevent them from becoming the whipping boy of their comrades. Lastly, in many cases a specific situation may merely appear to be a good opportunity to fulfil a certain desire, mainly in the case of subjects who are 'locked in the present',[71] especially if the aggression is very likely to be successful and does not put the aggressor at risk.

As for concrete situations that are liable to give rise to aggression, they are so many and so diverse, in the case of humans far more than of animals, that it would be impossible to give a detailed account of them. In the case of animals, five main categories of circumstances where aggression fulfils a specific function can be distinguished.[72] First of all, animals often compete for resources such as food, water, or shelter. The likelihood of fights between individuals therefore becomes greater if these resources are less abundant. Animals also compete for sexual partners, and in the case of many types of monkeys conflicts are particularly intense during reproductive periods. Here too, the presence of a limited number of females may give rise to savage fighting as has been seen with male baboons at London Zoo. Thirdly, there are the attacks on animals which are strangers to the group, and 'xenophobia' has often been observed in monkeys, whether the latter are at liberty or in captivity. It also goes without saying that all animals protect their own lives, whether they are defending themselves against a congener or against a predator. Lastly, animals defend their offspring, and a mother macaque will come to the aid of her young even if they have reached juvenile age.

The role of social systems and their internal dynamics

Since individuals of a single species compete with one another, two 'strategies' have developed which limit the frequency and severity of

---

[71] M. Cusson 1983*b*.   [72] See R. E. Passingham 1982.

conflictual interaction. In a territorial system, individuals or groups share available space; and in general the occupant of a given territory is the victor in a fight against an intruder. In a system of dominance, a hierarchy is established among individuals living in the same area, and it is the animal's faculties which determine its position within this hierarchy. To repeat the terms used by Passingham (1982): 'in a territorial system, the animal's chances depend on *where* it is, not *who* it is; in a hierarchical system, its chances depend on *who* it is, not *where* it is.' A dominant monkey will not give up his status of his own accord, and he will defend his rank within the hierarchy. In acquiring and maintaining a dominant position, the individual's combativeness is not the only determining factor; on the contrary, efficient support from third parties plays a major role. This position of the dominant individual often depends on the power of his allies, which in turn depends on the power struggle prevailing between rival sub-groups.[73] Intervention by other animals in dyadic conflicts is in fact very frequent, and has several purposes, from protecting a relation attacked by a dominant animal to preserving alliances and maintaining the statu quo. In his report on the detailed observations made on chimpanzees at Arnhem zoo, de Waal (1982) also stresses the role played by coalitions, and he considers these primates to be 'intelligent manipulators', especially when using others as 'social instruments'. In his foreword to this work, Desmond Morris in fact considers that 'there is hardly anything that occurs in the corridors of power of the human world that cannot be found in embryo in the social life of a chimpanzee colony'.

Marked differences can also be found between social systems and the social relationships formed within them, even when comparing closely related species. In his study on three types of macaques (the rhesus monkey, Java macaque, and Tonkean macaque), Thierry (1984) found interesting correlations between the degree of 'openness' of the social system, the degree of 'maternal permissiveness', the development of appeasing behaviour, and the intensity of aggressive interactions. Let us take the two extremes, placing the third species (Java macaque) in the intermediate position. In the case of the rhesus monkey, the social system is fairly 'closed': the mother is very protective and in a general way affiliative interactions take place mainly within the matrilinear clan, which encourages inter-clan conflicts; appeasing behaviour is little used and aggression is intense (biting in over 20 per cent of interactions where there is contact between the adversaries). In the case of the Tonkean macaque, on the other hand, the social system is more 'open': the mother is more 'permissive' and she allows the youngster to circulate freely and interact with all the members of the group; appeasing behaviour is widely used and aggression is much less intense (no bites were observed throughout the duration of the study). The nature of

[73] See R. A. Hinde 1983; in particular the contributions by S. B. Datta.

social interactions therefore clearly reflects the internal dynamics of the social system, and not a degree of 'aggressiveness' characterizing one species in particular.

Diversification of projects in humans

The number and diversity of potentially conflictual situations increases considerably if we turn from infra-human primates to man (whilst it is true that we have inherited from our 'lower brothers' competitiveness linked to the satisfaction of elementary biological needs, it must also be acknowledged that we remarkably succeeded in making this inheritance 'bear fruit'!). There is also the fact that conflict 'is not the objective outcome of a situation, even if the circumstances do carry considerable weight, but the result of the subjective will of persons, groups or communities who are trying to break down the resistance others are putting up against their intentions or their project'.[74] It is precisely the intentions and projects that developed and diversified themselves in relation to the development of language and the world of ideas. Language, which has indeed been and still is, a major factor in helping man to open out and progress, has also greatly contributed towards multiplying both the reasons for and the means of action in the domain of aggression and violence. On the one hand, it is through verbal communication and inner discourse that numerous concepts and symbols are gradually interiorized and incorporated into the representations borne by the individual brain. This enables the development of a clearer sense of self and the search for a better asserted social identity, at the same time as the need for true social recognition, particularly the desire to have one's own level of aspirations and expectations acknowledged by others. Language also plays a major role in the differentiation of behaviour and in the genesis of conflicts, both of which are connected with social status and people's roles in society. It should be noted in this respect that for every individual 'his status consists of the behaviour he may legitimately expect from others, and his role the behaviour which others legitimately expect from him'.[75] Obviously, language also plays a part in the process of categorizing and creating stereotypes, with all the ensuing prejudices, which lead to a 'selective perception' of the outer world, itself a source of conflict.

But language has contributed, too, to greatly enriching the variety of means of expression and action in the area we are dealing with. Aside from calumny, slander, insults, or threats, means of symbolic aggression which are curbed by law, there are—both in public or in private life—those 'murderous little words' that can demolish a person more surely than physical aggression, with complete impunity. Threats uttered during a conflict between spouses or in the context of a professional

[74] J. Freund 1983.   [75] J. Stoetzel 1978.

relationship, or else quarrels with neighbours, are often serious by virtue of the nature of the threat: in almost 80 per cent of cases they are death threats.[76] The author–victim link in these threats is usually no more than 'an instant in a pre-existent relationship' and 'the potential violence of the remarks is intensified by the words and posture accompanying them'.

Confronting otherness and 'fear of others'

When speaking of aversive emotions and their role in generating aggressive behaviour, we emphasized the fact that in animals and men alike, unfamiliarity and uncertainty often find expression in a somewhat hostile attitude. There is always a certain fear of others, and confrontation with otherness has to overcome real resistance. It is also a fact that, faced with a new experience, 'the prevailing tendency in humans is to identify resemblances and identical constituent elements, leaving aside differences and elements of otherness.'[77] Stoetzel (1978) notes that in his study on inter-personal relations, the sociologist Bogardus showed that it was possible 'to give an objective meaning to the expression of "social distance": the experiment showed a hierarchic order in the individual propensity to associate with people, ranging from parents and allies to strangers, going through friends and relations, neighbours, colleagues and fellow citizens'. Under the effect of anguish generated by confrontation with otherness, a subject may show aggressive behaviour, not because of his 'aggressiveness' but because he is trying to compensate for his shyness or a feeling of inferiority. Aside from this more or less deep fear of the other person, which forms an integral part of everybody's psychology and which can be overcome, there is quite a different element in attitudes qualified as xenophobic, ethnocentric, and racist. Billig (1984) stresses the major role played by social factors and social pressures in the genesis of the prejudice and the resulting discriminations. We have already indicated previously[78] that in the specific case of a subject with an 'authoritarian character' who turns the relationship with his entourage into one of subordination and exclusion, there takes place a veritable mutual reinforcement between the person and the group which forms through opposition—usually full of hatred—to an 'outside'.[79] On whatever scale it holds sway, any 'fundamentalist' ideology generates intolerance and hence potentially—or even effectively—violence. Furthermore, reference to a certain integrity or a certain orthodoxy is often a pretext to conceal the desire to preserve a position of power and dominance. This was the case, for example, in two news items which appeared next to each other in the same issue of

---

[76] C. Ballé 1976.   [77] J.-C. Benoît and M. Berta 1973.
[78] In Chapter 4, under Social and personal identity, p. 72.
[79] See H. Tajfel 1978; M. Billig 1984; L. Israël 1984.

Le Monde (31st July 1984), but apparently had nothing to do with one another: on the one hand, the South African authorities stated that it was forbidden for a white to have 'guilty relations' with a black ('Immorality Act'); on the other, Pravda severely censured the GDR, suspecting it of trying to establish 'guilty relations' with the FRG. Lastly, it is a well-known fact that when the concern of our leaders to hide their failures is combined with our own wish to find an outlet for our difficulties and our disappointments, a 'stranger' provides a convenient 'scapegoat'; and depending on the period, it will be 'perfidious Albion', the 'jerries', or the 'immigrants'. In other words, these hostile attitudes towards the 'other' are multidimensional, and it is important to be fully aware of the latter fact if they are to be acted upon effectually.

Aggression with a defensive aim

Defensive aggression is used by an individual whose life, freedom, dignity, property, or reputation is threatened; and there are many stressful or frustrating situations which may provoke aggressive behaviour, especially if the individual has no chance of escaping from the situation that oppresses or threatens him.[80] In his analysis of 'Violence, the answer to frustration', the Study Committee on Violence, Criminality and Delinquency emphasizes the fact that it is not so much the 'objective' situation which causes aggression, but rather the way in which an individual or a group perceives and interprets it; and the report notes that 'what matters is not that the individual or group analysed is mistaken regarding a value or even regarding an objective truth; what does matter is that following a real or even imaginary frustration, violence may occur'.[81] Furthermore, in the face of the constraints, tensions, and conflicts created by a number of everyday conditions, the search for an 'objective truth' is usually quite illusory. The Study Committee in fact analyses in a very subtle way the feelings of humiliation, alienation, or rejection which may arise at work; the solitude and silence threatening the family, which today plays a lesser role as the 'school of exchange'; the uncontrolled urbanization which has given rise to crowded, segregated, and anonymous populations. In these circumstances violence also becomes a 'substitute for dialogue', a 'cry': a cry of the 'mute', since effectively there has been a weakening of dialogue at all levels; a cry of the 'lost' because of the growing unintelligibility of the rules of social intercourse. To Mucchielli (1981), two fundamental reasons explain the increase in aggressive forms of social defence responses: on the one hand, many populations suffer from an enhanced sense of devalorization because they are better informed about other social groups and there is a lower threshold of tolerance towards inequality; on the other hand, the various aggressive behaviour models are well known to all

---

[80] See M. Argyle *et al.* 1981; M. Cusson 1983*b*.   [81] See A. Peyrefitte 1977.

since the mass media 'give priority to murders, war and violence . . . in order to "interest" their public'.

Aggression as a means of satisfying a desire

Other situations are not, in themselves, generators of violence: they merely furnish the opportunity for 'action' which will overcome boredom and create the excitement of adventure, or one of 'appropriating' property for a wide variety of reasons.[82] Here aggression is no longer a means of action intended to put an end to an intolerable situation, but an instrument for the purpose of immediately satisfying a desire. In its report, the Study Committee referred to above gives a very good analysis of our 'covetous society', which glorifies the act of consumption. Advertising, which endeavours to convince us that we only live to consume, urges us to spend; and of course it is not concerned with where the money comes from and how it was acquired. For his part, René Girard has given much thought to 'mimetic desire': we desire things because somebody else, simply because he desires or appears to desire them, makes them seem desirable. This mimetic desire gives rise to mimetic rivalry, which generates violence. Given the fact that people's desires are increasingly similar due to the gradual dedifferentiation of our societies, there are innumerable opportunities for rivalry, competition, and conflict.[83] It may come as a surprise that we discuss in a single paragraph the 'appropriation' behaviour studied by criminologists and the mimetic violence analysed by René Girard. In fact, since both cases allude to the origin of behaviour which threatens the integrity of others, any distinction between those which fall foul of the law and those which do not, is not of primary importance here.

Aggravating circumstances

Apart from 'situations' which in one way or another give rise to aggressive behaviour, there are circumstances, whether transient or more lasting, which encourage the onset of aggression and may aggravate the form it assumes. In the first place there is the increasingly widespread ownership of firearms (in one of his notes to *Le Monde* Bruno Frappat once mentioned that 'people own weapons today as though they were fishing rods'). The increase in weapons kept at home is doubly dangerous. On the one hand, Berkowitz (1974) showed that in the case of certain men, the very presence of a weapon had the effect of increasing the intensity of the 'punishment' they administered when they were angry (weapon effect). On the other hand, and above all, a 'shot' is

---

[82] See M. Cusson 1983*b*.
[83] See the texts compiled by M. Deguy and J.-P. Dupuy 1982; especially those of Ch. Orsini and L. Scubla.

obviously likely to do more damage than a 'punch'. Goldstein and Keller (1983) note that in 1981 the number of people killed by firearms was as follows: 48 in Japan, 52 in Canada, 8 in Great Britain, 42 in the Federal Republic of Germany, and more than 11 500 in the United States. What is more, in its annual report on 'Crime in the United States', the FBI indicated that 18 692 murders were reported in 1984 (and the vast majority of them were certainly perpetrated with a firearm). Does this mean that Americans are so much more 'aggressive' than other people? Or are these murders not due to a great extent to the proliferation of firearms and to the fact that 'directions for their use' are freely demonstrated on the television? Alcohol consumption is the second aggravating 'circumstance', since it affects the control a subject is able to exert over himself, and thus increases the likelihood of an impulsive, hasty, poorly controlled response. So long as it does not alter the subject's motor capacities, alcohol consumption increases the frequency of aggressive behaviour in different situations, both in animals[84] and in humans.[85] On the road, alcoholemia is often combined with speeding, because of deficient self-control and an artificial feeling of euphoria, which are behind much of the damage caused by 'car violence'. Whether in the case of owning firearms, of alcohol consumption, or certain types of quasi-criminal behaviour at the wheel, a certain degree of indulgence is all too often shown (under the pretext of course of defending individual freedom) in a wide-ranging interpretation of the slogan 'it is forbidden to forbid', whatever the price may be in terms of human life and serious and permanent disabilities. The prevailing amorality is unlikely to improve matters; we will return to this subject when considering the factors which are likely in principle to hold back a potential aggressor.

Confirmation of the instrumental value of aggression

Once a situation has been interpreted in such a way that aggressive behaviour appears to be the appropriate strategy, the effective use of this behaviour is all the more likely since experience will have confirmed that it is an efficient means of achieving the desired aim. In animals as well as in humans, the instrumental value of aggression is subject to a learning process: the brain records the results obtained through the use of this instrument and takes it into consideration when subsequently analysing the same or a similar situation. There is hardly need to emphasize that from a purely biological point of view, this 'plasticity' in using aggression enables an individual's behaviour to adapt to changing environmental conditions. We have already mentioned the fact that 'all-round aggression' would be complete nonsense in biological terms.

[84] K. A. Miczek and M. Krsiak 1981.
[85] R. J. Sbordone *et al.* 1981; D. Goldstein 1983; C. M. Jeavons and S. P. Taylor 1985.

In both mice and rats, the experience of 'victory' in a conflict interaction increases the likelihood of the victorious animal showing an 'offensive'[86] attitude in identical or similar circumstances. It has generally been considered that this attitude was strengthened in a positive way by the perception of submissive behaviour shown by the vanquished animal. But this positive reinforcement is even more marked if the animal which is losing the fight is removed, thereby 'frustrating' the winning animal of its 'victory'. In actual fact the aim of the aggression (which aim, if achieved, positively reinforces the behaviour used) is not the submission of the opponent as such but the disappearance, or at least the 'neutralizing' of a strange congener which has penetrated a familiar environment. If 'intruder' rats are repeatedly placed in the presence of a 'resident' rat or a colony of several animals, the attacks launched against them become increasingly frequent and increasingly severe. This is due both to the fact of having acquired efficient methods of attack and to the gradual lessening of the fear which the intruder provoked initially. In the case of a colony, it is the dominant male which is the instigator of 80 to 90 per cent of the attacks launched against the intruder. It is interesting to note that if this dominant male is removed from the colony, a previously subordinate animal quickly takes its place in order to attack the intruder.

The experimental activation of the neuronal system of positive reinforcement[87] makes it possible to create a contrived aggressiveness towards a congener where this attitude did not exist at all to begin with. Two rats are chosen which are perfectly placid towards each other. In one of them an electrode is implanted in the positive reinforcement (reward, pleasure) system; in order to check that the electrode has been implanted in the right place, the animal is given the opportunity to practise self-stimulation. Then, every time this animal shows the slightest inclination to be aggressive, the slightest sign of aggressive behaviour towards its congener (the latter nibbles a biscuit, which is just the one it wants!), it is stimulated by the implanted electrode. In other words, any inclination towards aggressiveness is rewarded, and an agreeable affective experience is regularly associated with any sign of aggressive behaviour. This rat can thus be seen to develop increasingly marked and stable aggressiveness[88]. This aggressiveness towards the congener was not innate; it was developed because we made it 'pay off', because it enabled the animal to sustain again a pleasant affective experience which was first associated repeatedly by intracerebral stimulation whenever there was an inclination to aggression.

Every time an aggressive action makes it possible to obtain the expected and desired result, there is an increased likelihood of the subsequent use of a strategy whose effectiveness has been confirmed. The

---

[86] See K. J. Flannelly *et al.* 1984; R. Lore and L. Takahashi 1984.
[87] See Chapter 5.   [88] T. J. Stachnick *et al.* 1966.

anticipated result may consist not only of a pleasant affective experience, but also of an end to an unpleasant experience. It may be a matter
of putting an end to a situation that is stressful because it is unfamiliar,
as we saw earlier on, or to painful stimulation. If two rats are exposed
together to electric shocks, one of them can be taught that it will be able
to put an end to these shocks if it attacks its congener 'properly'. Thus
it is possible to develop a veritable 'escalation' of the intraspecific
aggressiveness that this animal shows as a result of undergoing
the conditioning. Furthermore, the use of more and more intense
aggression occurs not only if the rat is again placed in the same situation
in which it learned the instrumental value of aggression, but also in
circumstances of social interaction of the resident-intruder type[89]

Effects of 'models' on aggressiveness

At any age humans learn, and are taught, that aggression is a course of
action which, especially if wielded unscrupulously and competently, is
liable to be profitable. Unlike animals, which in this respect have only
their own experience to go by, humans 'benefit' from the examples they
are given by 'models' freely dispensed to them. Patterson (1984) analysed the way in which a child acquires the ability, in a family where the
upbringing is inconsistent, to coerce others and become a real 'little
monster'; the escalation takes place due to the reinforcement unconsciously given by the parents themselves and the other members of the
family. As adults, all individuals have their personal way of resolving
problems, and it can be seen that certain individuals have learned to use
unpleasant, even openly aggressive behaviour to control their entourage.
This attitude turns into a veritable 'style of behaviour': if subjects who
have mentioned violent episodes in an account of their experiences, are
asked to simulate (in a role-playing session) how they would resolve this
or the other problem in everyday life, they usually use 'negative'[90]
behaviour. Those who during their childhood were exposed to violence
(whether they observed it, were victims of it, or themselves participated
in it) tend more than others to approve the use of violence for personal
or political purposes.[91]

The role played by 'models' is particularly clear in the case of the
criminogenic influence that young delinquents may exert on an adolescent. Cusson (1983b) presents several arguments which confirm this
influence to be real: there is a very close correlation between the existence of delinquent comrades and delinquency; when a young man is
questioned, he will declare of his own accord that it is associating with
delinquents that has led him to delinquency; if habitual criminals one
day decide to commit no further crimes, breaking off relations with their

[89] R. J. Viken and J. F. Knutson 1982; J. F. Knutson and R. J. Viken 1984.
[90] K. J. Gully and H. A. Dengerink 1983.
[91] D. J. Owens and M. A. Straus 1975.

delinquent companions is a major factor in this move. Adolescents are particularly vulnerable and receptive to this criminogenic influence if frequent failures during their school life and at work have convinced them that there is no way to satisfy their desires by legitimate means. Associating with delinquents not only has a contagious and stimulating effect, but later also makes it easy to backslide. For crime committed with others has every chance of being more profitable, less risky, and more exciting; therefore it is also much more likely to be repeated.

In every society there are rules defining circumstances where aggressive actions are legitimate. Thus a survey carried out in Finland showed that in the opinion of the adults questioned, aggressive acts committed in cases of self-defence, or in order to defend personal property, or to defend a third party, were all justified. But generally speaking, it was considered that human life was worth more than material goods; the subjects questioned felt that killing the aggressor was less justified in the case of aggression against personal property than in the case of self-defence.[92] Whatever the prevailing rules, it is clear that the understanding, or even the approval of the community makes it easier to use aggression as a course of action. Are these standards which transpired from the Finnish survey the same everywhere, and above all, in what direction are they developing? When asking this question one cannot help going back to the role played by television, which has already been discussed in the context of personality development. Repeated exposure to scenes of violence accentuates, in certain viewers anyway, a tendency to behave in an aggressive manner, like the models shown, for at least three complementary reasons: these scenes render violence banal, whilst diluting one's natural compassion for the victims; they often make a laughing-stock of social and/or moral constraints, conveying the impression that 'a real man' does not stoop to this kind of nonsense; they provide the opportunity to learn new ways of attacking their fellow men. Under these circumstances, is it really necessary to choose systematically the most violent pictures to announce the next detective film at the cinema or the next broadcast of a 'crime thriller' on television? And what is there to say about the proliferation of video cassettes showing violence in its pure state, which children buy at their parents' bidding before themselves becoming imbued with these degrading pictures?

What might restrain a potential aggressor?

Since aggression is a mode of action used for the purpose of achieving an aim considered beneficial by the individual, it is obvious that in the forward-looking simulation that precedes the planning of an aggressive act, assessing the respective importance of the anticipated benefits and

---

[92] K. M. J. Lagerspetz and M. Westman 1980.

the risk incurred plays an essential role. If the anticipated 'cost' of the action is high, if the difficulties likely to arise from it are major ones, there is less likelihood that aggressive behaviour will effectively be used. Far more so than animals, human beings behave in this respect as 'intuitive statisticians'[93], and the factors they may have to take into account are many and various.

Assessment of the risk incurred: in animals . . .

In both rats and mice, painful stimulation triggers aggressive behaviour, especially if the animal has no chance of escaping[94]; but the same painful stimulation may have the opposite effect, that is, to bring about the suppression of the aggressive behaviour, if it is applied immediately following the occurrence of this behaviour.[95] As we have already seen, the gradual disappearance of the fear caused initially by the intrusion of a strange rat and by the uncertainty as to the chances of being able to master the situation, increases the frequency and intensity of the attacks launched on the intruder. Conversely, the fear provoked by the presence of a cat leads to the suppression of the aggressive behaviour normally shown by the dominant male in a colony towards an 'intruder' congener; and a clear decrease in the probability of these attacks can still be observed when the cat is no longer physically present.[96] If mice rendered aggressive by prolonged social isolation are placed in a colony of congeners, their aggressiveness gradually declines; and this decline is all the more rapid if the animals in the colony counter-attack more vigorously.[97] Numerous studies have shown that the experience of 'defeat' brings about the rapid extinction of offensive behaviour, whilst at the same time tending to develop defensive attitudes; and this effect can be observed not only when the animal is again placed in the situation in which he underwent his defeat, but is generalized by being extended to other situations.[98] As far as interspecific rat–mouse aggressive behaviour is concerned, Baenninger (1970) showed that if a 'killer' rat was 'punished' for its aggressive behaviour towards a mouse placed in its cage by the administration of a painful electric shock, the rat quickly learned to avoid these painful shocks by abstaining from attacking the mouse. Furthermore, an affective experience of an aversive nature induced by intracerebral stimulation (it can be seen that the animal learns how to put an end to this stimulation if given the chance) has the effect of blocking an attack being carried out by a 'killer' rat which has already killed repeatedly and whose aggressive behaviour is therefore caused by motivation of an

[93] G. de Montmollin 1977.
[94] See above under From pain to aggression, p. 152.
[95] R. Ulrich and B. Symannek 1969; M. J. Follick and J. F. Knutson 1978.
[96] R. J. Blanchard *et al.* 1984.
[97] K. M. J. Lagerspetz and K. Sandnabba 1982.
[98] See K. J. Flannelly *et al.* 1984.

appetitive nature.[99] In more natural conditions, it can be observed that if two 'killer' rats are placed together in the same cage and a mouse is placed in their presence, only the dominant animal attacks and kills the mouse, whereas the other one, affected by the fear provoked by the congener, shows only a few sporadic signs of aggression. In wild rats, the time taken to kill the mouse is considerably lengthened if the 'killer' rat is merely transferred from his familiar environment to another which, due to its unfamiliar nature, renders the animal more fearful.[100]

... and in humans

In humans, an assessment of the risk incurred and the anticipated difficulties is different from that in animals, from a dual standpoint. On the one hand, it extends over a longer period of time, since it not only takes into account those difficulties which are likely to arise immediately from the aggressive episode itself; it also contemplates the reprisals, revenge, or penal sanctions which may occur subsequently. On the other hand, it encompasses, in addition to the physical dangers incurred, the risk of possible reprobation by the community. One criminologist emphasized this, referring to his own specific field, by writing: 'Everything leads to believe that crime occurs rarely where it is censured and flourishes where it is looked upon indulgently'.[101] Aside from actual criminal acts, behaviour used for the purpose of facing up to difficulties and solving problems is deeply influenced by the general attitude of reprobation or indulgence prevailing within the community with regard to aggressive behaviour. Considering sociocultural differences on a worldwide scale, Reynolds (1980) contemplates the historic development of three major regions: in the first (comprising Africa and South America), oral traditions and clear social rules determine unequivocally both how individuals should behave and how they should not behave; in the second (corresponding to Asia), 'models of man' have evolved in which 'he has been mostly construed as a spiritual entity or as an entity which can enter into some relationship with spiritual forces in the inanimate world'; in the third (which includes Europe, the Middle East, and regions to which European societies have flocked), the individual forms part of 'enormous bureaucratic structures' which consider the use of force legitimate in order to increase their resources and spread their ideas. This collective attitude has influenced the behaviour of individuals all the more strongly since 'western culture tends to create exterocentred individuals, that is, individuals whose personality is formed by reference to the reactions and expectations of others'.[102]

On a completely different scale, we have seen above that certain rural Tahitian communities reject aggression as something unthinkable, a

[99] P. Karli *et al.* 1974.   [100] P. Karli 1956.   [101] M. Cusson 1983a.
[102] Riesman, quoted by H. Touzard 1979.

nightmare, and that a veritable 'conspiracy' against aggression can be observed here. John Paddock became very interested in this phenomenon, which he qualified as 'anti-violence'; a consistent set of values, attitudes, and behaviour by which certain communities are distinguished from neighbouring communities, conveyed through particularly low rates of interpersonal violence. In his report on the symposium he organized[103] and in his comments on a book by Ashley Montagu,[104] he speaks not only of the observations made by others, but also of those he made himself in the Oaxaca region, in Mexico: in several villages and towns located in the same section of the Oaxaca valley, the rate of aggression, especially that of murders, is 'abnormally' low; this fact conveys 'a general attitude of rejection of all interpersonal violence'. The community takes every measure to anticipate and prevent aggressive behaviour, and there is a very strong consensus that considers interpersonal violence quite simply as a form of expression that is incompatible with belonging to the community; the results obtained 'demonstrate beyond argument that man is not inescapably a violent animal'.[105]

Wherever aggressive behaviour is tolerated or encouraged, it is matched by a systematic devalorization of the adversary, which lifts a restraint and thus encourages aggression (according to the well-known proverb: 'he who wants to drown his dog, accuses it of having rabies'). The Yanomami in Venezuela, whom we have already mentioned and who live in a permanent climate of persecution and aggression, show complete contempt for those who do not form part of their family or the community they belong to; the terminology of aggression is in fact derived from a term covering 'a category which has the semantic value of infra-humanity'.[106] Generally speaking, it is a well known fact that all persecutors are convinced, or endeavour to convince themselves, that their victims are 'less than nothing' and that they should therefore be 'crushed like bugs'. Check (1985) analysed the influence exerted by pornographic films with regard to aggressive attitudes towards women, depending on whether these films showed violence inflicted on women or whether, without showing scenes of violence, they gave a degrading, dehumanizing image of women; in both the latter and the former case, the pornography had the effect of accentuating the subjects' approval of violence perpetrated against women, diminishing their compassion for a woman who expresses her anguish at having been raped, and accentuating their own verbally expressed tendencies to commit rape. In these circumstances, it is particularly distressing that when visiting the Comic Strip Salon (in Angoulême), certain commentators could not help noting the 'obsessive omnipresence of blood, sperm and death', and the fact that 'today's comic strip is driven by hatred and contempt' and that 'we have entered the era of the most brutal and bestial pornosoliciting comic strip' (*Le Monde*, 27–28th January 1985).

---

[103] J. Paddock 1979.   [104] J. Paddock 1980.   [105] J. Paddock 1975.   [106] C. Alès 1984.

A powerful restraint: respect for the dignity of others

The very fact that the devalorization of others clearly facilitates aggression shows — as if it were necessary — that a 'positive' attitude towards others constitutes the best restraint against the use of aggression as a means of expression and action. Especially since recognition of and respect for the dignity of others usually go hand in hand with people's own sense of dignity, which they are anxious to preserve. The roots of this attitude go back to early childhood, and the ability to enter into affective relationships ('attachment') here plays an essential role. If a young child experiences encouraging and supportive parental attitudes imbued with an obvious spirit of solidarity, he will acquire a sense of his own worth and dignity, as well as a certain 'belief in the helpfulness of others' and 'a favourable model on which to build future relationships'.[107] Attachment is a process of 'psychobiological attunement' which enables an individual to place himself, together with others, 'on the same wavelength'.[108] It is largely on this ability to form attachments that the development of a feeling of belonging, a sense of commitment and responsibility towards others, and adhesion to common values are based. Criminologists agree that attachment and commitment towards a group are necessary conditions, though by no means sufficient, to prevent 'deviance', even though they then differ as to whether to emphasize the responsibility of the individual[109] or that of society.[110]

Since the processes of attachment, adhesion, and commitment do not take place in a vacuum, it is clear that the sociocultural context may facilitate them or, on the contrary, render them more difficult. If these processes are to be facilitated, the context should be such as to promote authentic affectivity, a certain 'standing back' from the inducements of the moment with a view to the longer term, and other values than purely material ones. What about our own prevailing context? Firstly, as far as affectivity is concerned, it must be admitted that people increasingly play around with feelings and express sentiments, that are false because they do not really exist. We receive personalized promotional letters which seek to give the impression that someone cares about our well-being, our personal happiness; they are trying to make us believe that somebody is 'thinking' of us, whereas their sole concern is to encourage us to 'not think'! In certain television broadcasts the presenter simulates genuine sympathy and shows the public — like in a circus — the feelings (this time real, but rendered ludicrous) of people whom they have generously showered with gifts. But one cannot use with impunity the affective feelings of others for purposes which have nothing to do with affectivity; falseness distorts feelings, and derision renders them ridiculous. As for 'standing back' from the present and the power to project oneself into the future, this implies the ability to control

[107] J. Bowlby 1984.   [108] T. Field 1985.   [109] M. Cusson 1983a.   [110] S. Box 1981.

oneself and acceptance of sustained effort. One has only to observe the behaviour of motorists (including one's own at times) to notice that self-control is certainly not the most common of characteristics. Furthermore, a veritable proliferation of 'games' of all kinds is taking place, leading people to 'invest' and 'invest their energies' in activities which in fact require neither a sustained effort, nor commitment, nor responsibility. In this situation, it is comforting to see that young people are concerned about this state of affairs, calling it an 'emergency' and showing their desire to become committed, 'not to pastimes, but to projects requiring (their) real responsibility' (*Le Monde*, 11th August 1983).

As for the common values to be adhered to, those most widely encouraged are such as to promote—or not—a spirit of generosity and solidarity. It would be trite to say that the watchword is above all 'to possess in order to show off', which has the effect of encouraging 'ostentatious' consumption. Attachment and commitment apply to objects and their intangible counterparts, ready-made ideas. This goes hand in hand with the surrounding amorality: not so much the immorality of those who deliberately infringe an ethic and in so doing acknowledge its existence; but the gradual obliteration of the limits between 'Good' and 'Evil', with the refusal to assume any personal responsibility. Jean-Pierre Dupuy (1982) emphasized the fact that modernity has shaken off all transcendency, and that 'the secondary resacralizations (which it has) produced are unable to stabilise social differentiation.' He considers that the lack of differentiation, which implies a 'trivializing of relations' and 'death of meaning', has the effect of 'precipitating people against one another'.

8 The neurobiology of aggressive behaviour

In their criticisms of those who believe in the existence of a rigid biological determinism and attempt to define and measure the 'qualities' thought to underlie this or that category of behaviour, Rose and his colleagues (1984) take issue with biologists who claim to explain human aggression by considering it as the expression of one and the same 'aggressiveness' as has supposedly been defined and measured in the rat that attacks a mouse introduced into its cage. And they add that this rat behaviour is sometimes termed 'muricidal', 'which presumably makes the experimenters happier that they are measuring something really scientific'. Setting aside the somewhat caricatural fashion in which Rose and his colleagues present positions that fewer and fewer biologists continue to defend any longer, such criticisms furnish an excellent starting point for one who 'discovered' rat–mouse interspecific aggressive behaviour and who, with his colleagues, has been studying it for thirty years. Not that we wish here to present a plea *pro domo*, which, in itself, would have only limited interest for the reader. But it is important to see, in the light of a specific example, what can actually be learned by the experimental analysis of aggressive behaviour in the rat, as long as the working hypotheses and the interpretation of the data obtained should be framed by an appropriate conceptual framework. We shall thus see that Rose and his colleagues are simultaneously right and wrong: they are right to criticize concepts on which many an experimenter bases the interpretation of his results; they are wrong if they believe that the neurobiological data furnished by the analysis of an aggressive behaviour displayed by a rat cannot in some way inform us about the brain mechanisms involved in human aggression. Since the author has thus been led to sketch the progress of his own thinking and to expound a position involving him alone, the use of the word 'I' becomes mandatory.

Some stages in personal development

It was completely by chance that I came to be interested in aggressive behaviour, on the occasion of a one-year stay (1954–1955) at Curt Richter's Psychobiological Laboratory in Baltimore. Not having initially a sufficient number of cages, I introduced some mice into cages already occupied by rats and noticed that some of the latter killed the mice thus introduced into their company, while others did not. Curt and I quickly agreed that this rat–mouse interspecific aggressive behaviour could provide an excellent 'model' for an experimental analysis of the brain mechanisms underlying (in general, as we thought) the triggering of aggressive behaviour. And we decided that my initial aim should be to find a means of abolishing by a brain lesion this aggressive behaviour in spontaneously 'killer' rats and, conversely, of eliciting by another brain lesion the same behaviour in spontaneously 'non-killer' rats. Thus it was that I was able to show that a bilateral lesion of the amygdaloid nuclear complex was capable of suppressing all aggression towards mice in previously 'killer' rats, while a bilateral ablation of the frontal poles of the brain caused this interspecific aggression to appear in subjects that had never shown it spontaneously.[1] The results obtained meant, in the perspective of those days, that the amygdala, in the capacity of a 'facilitating' structure, was an integral part of the 'neural substrate' of aggression and that, in 'non-killer' rats, the frontal pole of the brain exercised an 'inhibiting' influence on the activity of this same substrate.

These first experimental approaches quite naturally came to take their place in the context of the views prevailing at that time, which saw the relations between the brain and behaviour in terms of 'neural substrates', thought to underlie in a narrowly specific way this or that category of behaviour (for example, that of aggressive behaviour), these categories being seen as so many natural entities. In this way of looking at things, the brain was a sort of mosaic made up of 'motivational systems', each of whose activities (e.g. that of 'aggressiveness') generated a tendency or natural propensity of the individual organism to 'emit' the corresponding behaviour ('aggression', in our example). This did logically lead to research into the brain structures capable of 'facilitating' or, on the contrary, 'inhibiting', the activity of a given motivational system in a specific way, and, thus, of accentuating or attenuating the tendency to emit the corresponding behaviour in response to an appropriate triggering stimulus. It was within this general context that we carried out a long series of experiments with lesions and stimulations of various brain structures, and the results obtained in those experiments became the material for a synthesis published in the form of a diagram representing the 'neuroanatomy' of interspecific rat–mouse aggressive behaviour, i.e. an integrated set of facilitating and inhibiting structures.[2]

But it was just when this synthesis was being published that my way of seeing things changed progressively and profoundly, under the double influence of certain specific experimental facts and the questioning of some fundamental ideas. Three specific elements played a determining role in this connection. When a bilateral ablation of the frontal poles of the brain is carried out, causing the appearance of aggressive behaviour in rats which were previously 'non-killers', the ablation actually affects a whole group of structures in the forebrain (olfactory structures of the base, septum, caudate nucleus, and cortex). By more selective destruction of one or other of these structures, we noted that only the bilateral

---

[1] P. Karli 1956.   [2] P. Karli 1971; P. Karli *et al.* 1972.

ablation of the olfactory bulbs reproduced the behavioural effects induced by that of the frontal poles, namely, emotional hyperreactivity and the appearance of interspecific aggression. Destruction of the septum, which also causes marked hyperreactivity, did not bring about the appearance of 'muricidal' behaviour. It was thus difficult to establish any causal relation whatever between experimentally induced heightening of the emotional reactivity level and changes in the rat's behaviour towards the mouse. In 1972, two American teams[3] published results which apparently contradicted ours, since they had caused the appearance of interspecific aggressive behaviour by carrying out a destruction of the septum, while showing the existence of a correlation between the degree of post-operative hyperreactivity and the probability of the appearance of aggression. Were not these contradictory results due to the fact that I had carried out (in 1960) lesions of the septum in 'non-killer' rats which had previously become accustomed to the presence of a mouse in their cage, while my American colleagues destroyed the septum before confronting the rats operated on, for the first time, with a mouse? This hypothesis was fully confirmed. In fact, when the study of this question was resumed, it became clear that the behaviour of the rats after destruction of the septum differed profoundly depending on whether they were confronted for the first time with the intrusion of a mouse into their cage or, instead, they had had prolonged contacts with mice before the operation: in the former case, 60 per cent of the rats operated on killed a mouse introduced into their familiar environment; in the latter, only 8 per cent killed the mouse put back into their cage again.[4] But even before this clear confirmation it had become obvious that two factors played an important role in the determination of the probability of the appearance of interspecific aggressive behaviour, namely, the level of emotional reactivity and the degree of familiarity with the mouse species.[5] The third element was supplied by experiments in intracerebral stimulation which showed the existence of close correlations between two distinct effects of one single stimulation: on the one hand, the affective experience (appetitive or aversive) induced by the stimulation (the nature of this experience being verified by the fact that the animal quickly learned to trigger this stimulation by itself or to switch it off); on the other hand, its effect on the behaviour of the rat towards the mouse (either triggering aggression or else halting aggression already in progress). It can easily be imagined that all these experimental data carried us far from the conception according to which the display of aggression was supposed to be due to the activation of a specific motivational system, this activation itself resulting from the algebraic sum—now positive—of a set of 'facilitating' and 'inhibiting' influences exercised by brain structures that were to be identified.

[3] K. A. Miczek and S. P. Grossman 1972; W. M. Miley and R. Baenninger 1972.
[4] C. Penot and M. Vergnes 1976.   [5] See P. Karli *et al.* 1974.

At the same time as (and partly because) interesting experimental data were thus brought to light, the conceptual framework too was changing. I gradually became aware of an important idea which is not at first 'obvious' to a physiologist: the ways in which the brain relates to behaviour are not of a linear, unidirectional character; quite the opposite, brain and behaviour interact in a complex fashion and it is the continuous dialogue between the individual and his environment which gives these interactions their full significance. Indeed, it is the constraints imposed by, and the performances required for, an effective dialogue with the environment which define the specifications held in the 'specification sheet' of cerebral functioning. It is not, then, merely a question of analysing brain mechanisms and seeing how their operation 'explains' the behaviour; it is just as important to understand the modes whereby the brain functions, starting with a study of the vital functions that behaviour must provide for the living creature. From this point of view, behaviour constitutes a means of expression and action that allows the individual to gain mastery over the relationships he establishes with his environment. What triggers and directs behaviour is not, therefore, the external event as such (considered as a 'triggering stimulus' that is capable of activating a given motivational system), but, much more, its significance, which is endogenous; for it is born of the confrontation between current sensory information and the traces left by individual experience, or, more precisely, those internal representations formed and updated by and for dialogue with the environment.

If we are willing to accept that the essential function of any behaviour is not to 'respond' to this or that sensory information but rather to modify some information originating either internally or externally, in order to 'correct' it with respect to a set point or an internal representation, it is important to emphasize the complex processing operations undergone by sensory information, rather than concentrate our research on specific motivational systems which are mainly convenient semantic bridges allowing us to link the 'inputs' and 'outputs' of the brain. It was, then, necessary to recast our statement of the problem from a different point of view and to bring the experimental analysis to bear on the factors determining the way in which the individual brain perceives and interprets a given situation (and the mechanisms through which these factors act) and, therefore, the choice of an appropriate behavioural strategy. This perspective on things had in my view the further advantage of laying stress on processes and mechanisms not so closely bound up with a given species, so that this research therefore yielded data with more general validity, capable of being extrapolated — with all necessary prudence — to the analysis of the biological foundations of human behaviour.

Problems of a conceptual and methodological nature . . .

It is not easy to present a consistent synthesis of the experimental facts available to us at the present time, since these facts—both very numerous and necessarily of a fragmentary character—were not discovered or interpreted in the context of the general point of view that has just been outlined. Certainly, everyone (or nearly everyone) has abandoned the idea that all observable aggressive behaviour could be an expression of one and the same 'aggressiveness', and the search is no longer going on for the 'centre' or system that might be the generator of this unitary propensity for displaying aggressive behaviour. But the way of looking at things has not changed fundamentally if the quest for a single motivational system has given way to one for several more narrowly specialized systems. David Adams must be credited with having laid the theoretical foundations for a clearer distinction between 'offensive' and 'defensive' aggression, a distinction which has since undergone many developments.[6] Nevertheless, stress is still placed on observable behaviour (on its form rather than its function) and on the stimuli which determine the degree of activation of the corresponding motivational system and on those which can modulate the way in which this activation is externally expressed. In other words, the motivational systems thought to be responsible for 'offensive' or 'defensive' behaviour are theoretical constructions which depend on the distinction made, from a behavioural point of view, between these two categories of behaviour. This is to say that the heuristic value of these theoretical constructions depends closely on the validity of the distinction made on the plane of observable behaviour. Now, as we have seen above,[7] the 'offence–defence' dichotomy, as it has been defined on the basis of some particular experimental situations, cannot easily be generalized to all situations and to all mammals. And, above all, when we try to extrapolate this dichotomy so as better to understand the nature and origin of human aggression, we find ourselves constrained to appeal to the mediation of the affective states thought to underlie offensive and defensive behaviour by postulating the existence and the very general character of the double relationship of 'fear–defence' and 'anger–offence'.[8] Now here, too, there are reasons for believing that, in some ways, affirming the general validity of these relationships is rather an arbitrary step to take.

. . . illustrated by the analysis of 'muricidal' behaviour

The example of 'muricidal' behaviour will allow us to see in a more specific way the errors of interpretation which have been made when the first concern has been to place a manifestation of behaviour into a certain category, mainly on the basis of its observable characteristics, instead of analysing the situation with which a living creature is con-

[6] See D. B. Adams 1980.
[7] In Chapter 7, under Primordial importance of aversive emotions, p. 148.
[8] See D. C. and R. J. Blanchard 1984.

fronted, the way in which it perceives and interprets that situation, and the way in which it attempts to gain mastery over it by putting into operation a behavioural strategy that it thinks appropriate.

The aggressive behaviour of the rat towards the mouse is most often considered 'predatory' behaviour, i.e. behaviour aimed at obtaining nourishment; this has led some to wonder whether 'aggression' is actually involved. Baenninger (1978), who examines in detail the various aspects of this question, considers that even if the rat which attacks and kills a mouse ends by eating it wholly or partly, this does not necessarily mean that the attack was of a predatory nature from the beginning. And, in fact, much data have clearly shown that in the 'muricidal' sequence, attack behaviour and feeding behaviour are not closely linked to each other. We must first recall that, in the stock of Wistar white rats bred by us, the percentage of those animals which spontaneously kill mice introduced into their cages does not exceed 10 to 15 per cent. And, usually, rats do not eat the first mouse (or mice) they kill. Moreover, 'non-killer' rats permit themselves to die of hunger in the close proximity of a mouse, even after having eaten—quite rapidly—a mouse killed by the experimenter.[9] It is difficult to believe that the mouse represents a natural prey for the rat, since the latter dies of hunger in the presence of this potential source of nourishment. On the other hand, various experimental operations have allowed us to dissociate aggression from the quest for and ingestion of food. If a conditioned aversion is created in a rat by making it 'sick' (by administering lithium chloride) as soon as it eats the mouse it has just killed, the animal continues to attack and kill mice, although henceforth it no longer eats them.[10] When a bilateral lesion of the posterior part of the lateral hypothalamus is carried out in a 'killer' rat, temporarily abolishing both attack behaviour and feeding behaviour, the animal starts killing mice again several days, if not several weeks, before it resumes spontaneous feeding; and, during this period, it eats none of the mice it kills.[11] If we destroy the ventro-medial hypothalamus and so cause the appearance of interspecific aggressive behaviour in certain rats and an exaggeration of food intake (hyperphagia) in others, we note that there is no correlation between these two lesion-induced behavioural effects.[12] Experiments involving electrical stimulation have shown that, in the mesencephalon, a stimulation applied to the dorsolateral region triggered the reaction of inter-specific aggression but not the ingestion of food while, on the other hand, an activation of the ventrolateral region caused the animal to ingest food but not to attack mice.[13] A similar dissociation was observed under the effect of local injections of *d*-amphetamine: a bilateral injection in the central nucleus of the amygdala abolishes muricidal behaviour without affecting the intake of food and water; on the other hand,

[9] P. Karli 1956.
[11] P. Karli and M. Vergnes 1964.
[13] R. J. Waldbillig 1979.

[10] D. Berg and R. Baenninger 1974.
[12] F. Eclancher and P. Karli 1971.

injections carried out in the substantia nigra or in the ventral region of the caudate nucleus cause a reduction in this intake without altering muricidal behaviour.[14] This being so, we may certainly conclude that acquired experience progressively creates a linkage of the two forms of behaviour, but not that both of them result from the activation of one and the same motivational system, namely, that of 'predation'.

The attempt to classify muricidal behaviour under this or that label gives it too rigid a meaning and causes us to lose sight of the fact that, as a result of its interactions with the mouse, the rat's interpretation of the stimuli emanating from it evolve over time, as do the motivational processes underlying its behaviour in response to the stimuli. When a rat is confronted for the first time—or times—with the intrusion of a mouse into its cage, its fear of the new, its natural tendency to avoid unfamiliar objects in a familiar environment, will give rise to an affective experience of an aversive nature; and if, because of marked emotional reactivity, this aversive emotion reaches a certain degree of intensity, there is a high probability that the rat will display 'intolerant' behaviour aimed at ridding itself of the intruder and putting an end to the emotion it has aroused. But the significance and the motivating properties of the stimuli emanating from the mouse are probably quite different in the 'killer' rat which has acquired a certain amount of experience of interspecific aggression. Indeed, as this behaviour is repeated, the motivation underlying it develops to become more and more clearly appetitive in nature. Observations show that the rat kills more quickly and more 'coldly'; it rarely displays the emotional reaction, of whatever degree of intensity, observed during its first confrontations with mice. The fact that intolerant and defensive behaviour thus rapidly changes into appetitive behaviour is attested to by a body of specific observations.[15] On the one hand, it is observed that an experienced 'killer' rat will seek out and kill the mouse, even if its level of reactivity has been considerably lowered (by doses of chlorpromazine or reserpine as high as 15–20 mg/kg) or if it has been deprived of its olfactory, visual, and auditory afferences. On the other hand, the availability of a mouse for interspecific aggression may play the role of 'reward' in situations of instrumental learning. Interspecific aggressive behaviour thus positively reinforces itself. Such positive reinforcement is probably due to the conjunction of at least three factors:[16]

(1) the execution of a behavioural sequence characteristic of the species, with the proprioceptive feedback effects it contains, has its own reinforcing qualities;

(2) the rat has learned that such aggression is effective in terminating— and preventing—an affective experience of aversive nature;

[14] H. Yoshimura and K. A. Miczek 1983.   [15] See P. Karli *et al.* 1974.
[16] P. Karli *et al.* 1974; P. Karli 1982.

(3) the rat eats part of the mouse it has killed, often showing a particular
appetence for its brain.

This being so, one may easily suppose that, in a rat which has killed
repeatedly, the stimuli emanating from the mouse induce an 'expectation', i.e. an anticipation of the consequences to be derived from the
aggressive reaction, quite different from that induced in a rat confronted
for the first time with a creature of this alien species.

Quite obviously, the significance and the motivating properties of
these stimuli are by no means the same in a rat which has had the
opportunity over a long period of becoming so familiar with the presence of a mouse in its cage that this animal constitutes an integral part
of its familiar environment. We have already seen, and we shall see
again later, that familiarization has the effect of preventing the display of
muricidal behaviour which, in the absence of prior social contacts,
would certainly not fail to be provoked by some operation on the rat's
brain. We can also point out, in this regard, the following observation:
the state of hunger, which raises the organism's level of reactivity,
clearly facilitates the initial triggering of interspecific aggression; but this
facilitation induced by hypoglycaemia is much more marked in rats that
have had little or no previous contact with mice than in rats which have
lived in contact with mice over prolonged periods.[17]

When a rat is confronted with a mouse, the information-processing
operations set in motion thus differ, at least in part, depending on
whether the rat is a 'naive' one, confronted for the first time with the
intrusion of a mouse into its cage, or a confirmed 'non-killer' rat which
has been made familiar with the presence of a mouse in its environment,
or, again, a 'killer' rat having a long experience of interspecific aggressive behaviour and its consequences. This being so, it is not surprising
that certain experimental manipulations have not at all the same effect in
the three possible cases. Independently of the fact, already pointed out,
that some operation may cause the appearance of muricidal behaviour in
a 'naive' rat while being without effect in a confirmed 'non-killer' rat,
there exist experimental data whose apparently paradoxical nature results simply from the fact that the mouse does not have the same
significance for the 'naive' rat and for the confirmed 'killer' rat, and that
it does not arouse in them an 'expectation' of the same nature. This is
why ablation of the olfactory bulbs very markedly increases the probability of an initial use of interspecific aggression in a 'non-killer' rat,[18]
while it attenuates the tendency of a 'killer' rat to attack the mouse.[19] An
activation of GABAergic neurotransmissions (by intraperitoneal administration of certain drugs) has similar consequences: in a group of confirmed 'killer' rats, a decrease is observed in the proportion of them that

[17] L. Paul *et al.* 1973.   [18] M. Vergnes and P. Karli 1963.
[19] R. Bandler and C. C. Chi 1972; M. E. Thompson and B. M. Thorne 1975.

kill, while in 'naive' rats the probability of the appearance of muricidal behaviour increases.[20] 'Paradoxical' results are also obtained when electrical stimulation is applied to the neuronal aversion system in 'non-killer' and 'killer' rats, thus experimentally inducing an unpleasant affective experience: in the 'non-killers', it is easy to trigger aggressive behaviour which these animals never display spontaneously; in the 'killers', on the other hand, the same stimulation halts aggressive behaviour in progress, for it interferes with the appetitive motivation which had governed its initiation.[21]

It is clear that if, in our zeal to categorize, we look upon muricidal behaviour as something monolithic and unambiguous, underlain by immutable motivational processes and giving expression to the activation of a neural substrate that is quite distinct and capable of being individualized, we inevitably commit errors in interpreting the experimental data and we always fall short of a true comprehension of the behavioural processes and brain mechanisms involved. Let us take two specific examples. Kemble and his collaborators (1985) note that the fact of giving rats a repeated experience of 'offensive' behaviour towards fellow-creatures does not in any way increase the probability that later the same rats will display muricidal behaviour; and those which attack and kill the mouse do it with a 'form' of behaviour which is not the same as the offensive behaviour manifested towards fellow-creatures. The authors conclude from this that muricidal behaviour 'does not correspond' with an intraspecific attack because, they consider, the form of the two kinds of behaviour should be the same, and the experience acquired with the one should be transferred to the other, if the brain mechanisms involved were the same ones in both cases. What is surprising here is that the identity of these mechanisms could be hypothesized, merely because a similar behavioural strategy (a similar 'output' from the brain) might be employed in both cases. For it is quite obvious that the way in which the two situations are perceived and interpreted can hardly be the same: in one case, the rat is confronted twenty times with the intrusion of a fellow-creature, after having been brought up with it and having had ample leisure for interacting with it; in the other, it is confronted for the first time with an animal of an alien species. How could the nature of the information processed, and its interpretation with reference to the traces left by experience, be the same in these two situations? In reality, making an inventory and an analysis of the determinants of each of these instances of behaviour makes it clear that, among the factors involved, some are common to both situations while others contribute specifically to the way in which one or other of them is experienced by the rat and the way in which it will respond.

As for the 'neural substrate' of muricidal behaviour, a conception

---

[20] A. Depaulis and M. Vergnes 1984.   [21] See P. Karli *et al.* 1974.

which seeks to individualize it within the brain and envisages its deployment in a basically mechanistic way, is an equally serious misunderstanding of some important aspects of the biological determinism of such behaviour. Albert and Walsh (1984) make an inventory of the brain structures exerting an 'inhibiting' influence on muricidal behaviour, by considering that the appearance of this behaviour following the destruction of one or other of these structures is the consequence of a 'disinhibition'. Now, the brain structures involved participate, in reality, in very various processes, such as the modulation of emotional reactivity, the genesis of affective experiences, the establishment and use of the traces left by experience, and positive and negative reinforcement. It is thus highly probable that the various 'disinhibitions' capable of bringing about the appearance of interspecific aggressive behaviour are not of the same nature and that they do not correspond in any way to one unambiguous process. In any case, the workers cited have themselves recently observed that the behavioural effects of the various 'disinhibitory' lesions were not the same whether the rats had or had not been brought up in the company of mice,[22] and whether they had been brought up as a group or in social isolation.[23] A different argument may be added to show further that we are not always doing 'the same thing' when, by a brain operation, we evoke muricidal behaviour in rats which do not present it spontaneously: in some cases but not in others, in fact, aggressive behaviour induced experimentally in this way can be eliminated by carrying out a lesion of the medial part of the amygdala.[24]

It goes without saying that the foregoing exposition, underlining the conceptual and methodological difficulties strewn along the researcher's path, is not in any way intended to discourage the reader! The objective aimed at is of course quite different: to bring out the reasons for which I feel I must present the 'neurobiology of aggressive behaviour' in a markedly different way from the usual one. In fact we are not about to review the various brain structures and the various neurotransmitters participating in the 'neural control of aggression' by 'facilitating' or 'inhibiting' the observable behaviour. But we shall be looking at the factors analysed in the previous chapter which contribute to determining the probability that an individual will employ a form of aggressive behaviour, as a means of expression and/or action, in order to gain mastery over a given situation. And in the light of the ideas developed in Chapter 5 we shall attempt to see what are the brain mechanisms brought into play when one or other of these factors comes to take its place in the 'causal field' underlying the observable behaviour.

[22] D. J. Albert *et al.* 1984.   [23] D. J. Albert *et al.* 1985.
[24] S. Shibata *et al.* 1982*a*.

Perception and interpretation of a situation evoking aggression

Processing of chemical signals in macrosmatic species

It is clear that in mammals the information capable of giving rise to the employment of aggressive behaviour is of an extremely diverse nature and that, especially in the most evolved species, it may undergo highly complex processing. In the case of the so-called macrosmatic species, information of a chemical nature that is perceived by means of the olfactory function plays an essential role in individual recognition and thereby in the perception of any change occurring in the social environment such as to disturb the current relational balance. It is therefore important to point out that the significance and motivating properties of a chemical signal depend on a whole range of factors. This fact may be illustrated by data obtained in the mouse. First of all, it appears that experience—i.e. early interactions with fellow-creatures—is essential if chemical signals are truly to acquire the significance normally conferred on them. In fact, a mouse ceases to display aggressive behaviour towards its fellow-creatures if it is brought up from birth to weaning by a she-rat.[25] Furthermore, the motivating properties of a given signal depend on the *social context* in which it is transmitted. For instance, a strange female introduced within a small group of three to five adult females is often attacked by the group's members and the aggressive response of the female members of the group is so much more vigorous if the intruding female is in milk; on the other hand, a single female never uses aggression against one of her fellow-creatures. Moreover, the *endocrine state* of the animals plays a large role: if the milk-producing female is introduced into a group of males, only very exceptionally will they attack her; by contrast, when the group is made up of castrated males, not only is the intruding female severely attacked: the members of the group fight violently with each other.[26] We may add that in this case the effect of the low level of circulating male sexual hormones is precisely the inverse of its well known one in the determination of aggressive interactions in male mice, for castration suppresses any interaction of this type.

In the guinea pig it is also chemical signals that trigger aggressive reactions and the presence of these signals in urine is just as much controlled by circulating male sexual hormones: if a castrated male, whose anal-genital region has first been painted with urine derived from a female, a castrated male or an intact male, is introduced into the cage of a male guinea pig (i.e. into its familiar environment), we note that it is the 'intruder' bearing urine emitted by an intact male that excites the

[25] V. H. Denenberg *et al.* 1964.   [26] See P. Ropartz 1978; M. Haug *et al.* 1981.

most intense attacks on the part of the 'resident'.[27] More recent observations have shown that these odoriferous substances, whose production is controlled by androgenic hormones, play a more complex role than simply inciting aggression. Indeed, the behaviour of a male guinea pig confronted by a fellow-creature depends on both its own and the other's androgen level, which allows each of the animals to compare itself with or measure itself against the other; and the interaction is, in general, all the more violent as the hormonal levels are closer to each other.[28]

We have seen that, for a rat, the significance of the signals emanating from a mouse is very different depending on whether the rat is 'naive' regarding interactions with the mouse species, or has 'killer' or 'non-killer' experience. With reference to olfactory signals, it is proper to state that their significance is simultaneously a result of the processing of this information — particularly in the limbic structures of the temporal lobe — and an important determinant which, in turn, affects certain of the initial stages of its processing in the olfactory bulbs. It was mentioned previously that, in the sphere of feeding behaviour, the response of the olfactory bulb to food odours was subject to selective facilitation through the involvement of a 'centrifugal' influence probably originating in the hypothalamus, whenever the organism was hypoglycaemic due to having been deprived of nourishment. As far as the rat's socio-affective behaviour is concerned, recorded bioelectrical activity of the mitral cells of the olfactory bulb has shown that this activity was modulated depending on the 'frightening' (e.g. the odour of a predator, such as a fox) or reassuring nature of the information being processed and also depended on the level of vigilance, itself determined, at least in part, by the significance of the message emitted by the olfactory receptors.[29] It thus appears that, in such apparently 'simple' behaviour as rat–mouse interspecific aggression, and at such a peripheral level as that of the olfactory bulbs, there are already complex interactions between the processing of the objective data of sensory information, the organism's level of vigilance and reactivity, and the significance that the information being processed bears for that organism.

Complex role of the olfactory bulbs

In macrosmatic species, such as the mouse or the rat, the olfactory bulbs play a very important role not only because they are necessary to the processing of the olfactory signals whose perception is essential for the normal course of social interactions, but also because of the more global influence which they have in permanently moderating the level of reactivity of the organism. In fact, a rat deprived of its olfactory bulbs (a 'bulbectomized' rat) is characterized by increased irritability, and this

[27] A. P. Payne 1974.   [28] N. Sachser and E. Pröve 1984.
[29] M. Cattarelli *et al.* 1977; M. Cattarelli and J. Chanel 1979.

marked hyperreactivity is expressed in more frequent and more intense aggressive interactions with its fellow-creatures. Since this hyper-reactivity is accompanied by anosmia (absence of all olfactory percep-tion), it is not surprising that an ablation of the olfactory bulbs causes the appearance of interspecific aggressive behaviour even in a 'non-killer' rat that was previously given an opportunity to become accustomed to the mouse species. The preventing effect of prior familiarization, observed in other circumstances, is not distinguishable when the rat is no longer able to recognize a mouse on the basis of olfactory informa-tion. The role played by this information is just as evident in the be-haviour of the rat towards the young of its own species: if a 'killer' rat is presented successively with two or three mice, then with a 25-day-old rat (which has essentially the same size as an adult mouse), it rapidly kills each of the mice and then hurls itself at the young rat, and it is only at the last moment that the attack behaviour suddenly gives way to quasi-maternal behaviour; but when the rat is deprived of its olfactory bulbs it kills not only the mice but also the young of its own species (which it does not recognize quickly enough to be able to stop in time the aggression in progress).

Enhanced receptivity regarding certain information

The triggering of aggressive behaviour is accompanied by a selective accentuation of receptivity to certain information. This fact is clearly shown by observations made in the cat, in which visual and tactile information play a preponderant role. In 1962, Wasman and Flynn showed that electrical stimulation applied to the hypothalamus permitted the triggering of two different types of aggressive reaction: on the one hand, attack reactions without any apparent emotional display (the animal attacks a rat and kills it 'coldly') are provoked by stimulation of the lateral hypothalamus; on the other hand, by operating on a more medial region it is possible to trigger interspecific aggressive reactions of an 'affective' nature, less well directed and accompanied by intense emotional displays. Now, lateral hypothalamic stimulation triggering attack behaviour also has the effect of raising markedly the sensitivity of the skin in an area around the mouth, so that the least touch in this area provokes a movement of the head towards the stimulation and a biting reaction; at the same time, a particularly sensitive skin area appears on the backs of the front paws, and touching this area provokes a rapid response in the form of a 'cuff with the paw'.[30] Moreover, these lateral hypothalamic stimulations seem also to influence the processing of visual information: the cat turns on the rat to attack it only if it is located in the half of the visual field whose information is being processed by the cerebral hemisphere located on the same side as the point of applica-

[30] J. P. Flynn 1976; R. Bandler 1982.

tion of the hypothalamic stimulation.[31] We may conclude that the 'bias' thus introduced into visual perceptions plays an important role in the initiation of the experimentally induced aggression and that the bias introduced into tactile perceptions facilitates the execution of the behavioural sequence by giving rise to appropriate feedback effects.

In spite of their interest, these data leave two basic questions entirely unanswered. First, it is important to know how, in natural conditions, the neural circuits are activated which are brought into play by electrical stimulation applied to the lateral hypothalamus. Next, we need to understand more clearly the nature of the processes which are triggered—or modulated—by hypothalamic stimulation. It happens that the lateral hypothalamic area is a place where numerous ascending and descending neural pathways pass near each other, so that many and various processes are capable of being influenced.[32] If we consider only the ascending dopaminergic fibres, it must be recalled that their activation facilitates both focusing of attention and motor initiative, while accentuating the motivating virtues of the signals and the tendency of the organism to 'move towards' these stimuli.[33] Adams (1971) has observed that lateral hypothalamic lesions provoked in the rat a marked reduction in 'offensive' aggressive behaviour (territorial fighting), while the 'defensive' aggressive behaviour caused by the administration of electric shocks remained unchanged. But is it necessary, in this case as in some others, to appeal to the 'offence–defence' dichotomy and to differentiate the 'neural substrates' considered to control respectively — and specifically — offensive and defensive aggressive behaviour? Should we not rather place the emphasis on the fact that motor initiative ('spontaneity') plays a much more important role in the use of 'offensive' behaviour than of a 'defensive' reaction? In this connection, a very interesting fact has been shown by Van den Bercken and Cools (1982): by manipulating the cholinergic transmissions within the caudate nucleus (on to which come to be projected the nigro-striatal dopaminergic fibres) we modify, in the programming of behaviour, the relative priority given to intra- and inter-individual constraints; in the case of the monkey's social interactions, this is expressed by a modification, in the animal so operated, of the balance between those forms of behaviour with regard to which it takes the initiative and those which are excited by its partners. This being so, trying to 'explain' the repercussions on behaviour of activation or interruption of the ascending dopaminergic pathways by invoking a modification of the level of 'aggressiveness' and the animal's 'propensity' to behave 'offensively' or 'defensively' does not make very much sense. Presenting things in this way cannot have more than a purely descriptive value (concerning the probability of appearance of this or that type of behaviour): it 'explains' nothing. The

[31] R. Bandler and S. Abeyewardene 1981.   [32] See P. Karli 1976, 1981.
[33] See Chapter 5, under Role of the cerebral catecholamines, p. 114.

explanation should rather be sought in the alterations which, in a much more general way, affect this or that information-processing operation leading in particular to the use of aggression by way of a behavioural strategy thought to be appropriate.

When we examine more closely certain aggressive behaviour triggered by electrical stimulation applied to the lateral hypothalamus, we note also that it is difficult to maintain that this stimulation activates the neural substrate of any kind of aggressiveness. Speaking of the behavioural effects of stimulation of the neuronal reward system in the macaque, we have indicated that such stimulation had the effect of attenuating the fear reactions provoked by a snake and accentuating the degree of dominance manifested towards a fellow-creature.[34] But there is no question here of 'push-button' triggering of stereotyped reactions; quite the contrary, the effects of the hypothalamic stimulation depend clearly on the behaviour of the fellow-creatures, and in particular on the dominance relations previously established. Furthermore, one and the same stimulation can trigger a behaviour of aggression or of submission, depending on the rank occupied by the partner in the group hierarchy.[35] In the rat, too, lateral hypothalamic stimulation can induce an attack reaction against a subordinate fellow-creature while not inducing the same effect against a dominant fellow-creature.[36] At certain hypothalamic sites, stimulation induces in the rat active transportation of biscuits and of a mouse (which the rat takes 'delicately' between its teeth to move it in the same direction as the biscuits), whenever the rat is confronted with both categories of object; but the same stimulation incites the rat to attack and kill the mouse if only the latter is present in its environment.[37] It is thus important to carry out more precise research into the way a given manipulation of the brain affects the way in which that brain apprehends and interprets the events occurring in the familiar environment and, thereby, the way in which it employs behaviour in an attempt to master them.

(A) Genesis of aversive emotions

In the previous chapter we emphasized the primary role played by the aversive emotions generated by any situation perceived as a threat to the physical and/or psychical integrity of the individual; emotions of this kind call forth behaviour aimed at terminating the displeasing affective experience and lowering—towards an optimal level—the degree of cerebral activation. It is by aggression (capable of removing the threat) or by flight (which removes the individual from the threat) that the living creature achieves this double objective. The neuronal aversion system

[34] P. E. Maxim 1972.   [35] R. Plotnik *et al.* 1971.   [36] J. M. Koolhaas 1978.
[37] P. Schmitt, unpublished observations.

plays a preponderant role in the genesis of the aversive emotion and in the development of the behaviour ensuing from it.

We saw earlier that Wasman and Flynn (1962) observed two types of aggressive reaction to electrical stimulation applied to the hypothalamus of the cat, the more medial stimulations evoking an 'affective' and 'defensive' type of reaction when confronted with a rat. In the rat, too, electrical stimulation of the hypothalamus, like that of the mesencephalic tegmentum, is able to trigger two types of aggressive behaviour towards a mouse: on the one hand, well orientated aggressive behaviour unaccompanied by any apparent emotional display (an immediate attack reaction provoked in 'killer' rats which, spontaneously, kill only several hours after the introduction of the mouse into their cage), triggered from stimulation sites located in the lateral hypothalamus and in the ventro-medial tegmentum of the mesencephalon, at the level of which the rat also presents self-stimulation responses; on the other hand, poorly orientated aggressive behaviour, interspersed with attempts at flight and accompanied by intense displays of emotion (a defence reaction induced in rats which, spontaneously, are 'non-killers'), triggered from other sites (situated within the periventricular aversion system: the periaqueductal grey matter and medial hypothalamus) at which the rat also presents responses switching off an imposed stimulation.[38] In the first case, stimulation has 'appetitive' effects which are shown by the fast acquisition of behaviour permitting the animal to stimulate itself; in the second, the 'aversive' nature of the effects induced by the stimulation is shown by the learning of switch-off responses. In the cat, the differing nature of the two types of aggression triggered by hypothalamic stimulation appears distinctly in the following observation: if an attempt is made to transfer to the intracerebral stimulation a learned avoidance response (the animal jumps on to a chair so as to avoid a painful electric shock), it is noted that only the hypothalamic stimulation provoking the aggressive reaction of 'affective' and 'defensive' type will prove capable of triggering the previously learned avoidance response.[39]

Role of the periaqueductal grey matter

The periaqueductal grey matter plays a very important role in the development of the defence reactions in response to any threatening — or simply stressful — situation. This fact is attested to by a variety of convergent data.[40] On the one hand, one can trigger the defence reactions characteristic of a given species (rat, guinea pig, cat, monkey) by stimulating electrically this periventricular area of the mesen-

---

[38] See P. Karli *et al.* 1974.
[39] D. Adams and J. P. Flynn 1966.
[40] See J. P. Chaurand *et al.* 1972; J. P. Flynn 1976; D. B. Adams 1980; C. L. J. Stokman and M. Glusman 1981; J. Mos *et al.* 1982; R. Bandler *et al.* 1985.

phalon; conversely, lesions carried out in this same region attenuate or even abolish defence reactions, whether they be provoked by a natural threat or by electrical stimulation of the hypothalamus. On the other hand, the triggering of such a defence behaviour is accompanied by a marked accentuation of the activity of certain neurons within the periaqueductal grey matter. Finally, data obtained from neuro-pharmacological manipulations supply additional confirmation while giving further detail of an aspect of the matter which is perhaps essential. One may in fact cause the appearance of defence reactions in a cat by means of microinjections of carbachol[41] into the ventrolateral part of the periaqueductal grey matter;[42] in the cat, as in the rat, reactions of this type may be triggered by local microinjections of excitatory amino acids which cause an activation of the neurons by acting directly on their cell bodies, and this behavioural effect may be reproduced in the rat by a local administration of bicuculline, which blocks GABAergic transmission.[43] Now, in all these neuropharmacological experiments, we observe a clear 'lateralization' of certain experimentally induced behavioural effects: the defence reaction is triggered in a more pronounced, if not exclusive, way if the animal is stimulated on the opposite side to that of the periaqueductal microinjection. Here, again, then, we find the correlation we have already mentioned between a certain kind of attitude displayed towards stimuli from the environment and the reactivity of the individual in its quantitative (intensity) and qualitative (appetitive or aversive) aspects, which itself reflects the way in which information about the environment is processed in the brain structures involved. But here again we still have to work out the details of how exactly these processes are linked and how they are put into operation under natural conditions.

It is not without interest to add some data concerning the periaqueductal grey matter which, at first sight, seem quite paradoxical in relation to what has just been expounded. As a matter of fact, in the 'killer' rat, a destruction of this peri-ventricular region *facilitates* the display of muricidal behaviour (a rat which habitually killed only after some delay now kills much more quickly); and electrical stimulation applied to this same region is capable of *inhibiting* an aggressive reaction which is going on.[44] But, if we recall that the interspecific aggressive behaviour of an experienced 'killer' rat is underlain by an appetitive type of motivation, it is easy to see that the display of this behaviour may be facilitated by the destruction of an essential component of the aversion and retreat system; it is also noted that this destruction provokes a facilitation of other appetitive behaviour (increase of food intake; accentuation of self-stimulation at the lateral hypothalamus). And it cannot be

---

[41] Carbachol is a substance which activates cholinergic transmissions.
[42] C. L. J. Stokman and M. Glusman 1981.
[43] See R. Bandler *et al.* 1985.
[44] J. P. Chaurand *et al.* 1972; P. Karli *et al.* 1974.

surprising that stimulation that induces an affective experience of aversive nature interferes with the course of appetitive behaviour, for thus a sort of 'motivational conflict' is being created. It is, in any case, sufficient to increase the intensity of the stimulation to provoke not only inhibition of muricidal behaviour but a true flight reaction (if there is no mouse — no inducement within the cage to 'move towards' — then the lowest intensities are enough to trigger flight). Electrical stimulation applied to the periaqueductal grey matter may also have a further effect deserving of mention: stimulation of this type can attenuate or even abolish the biting behaviour triggered by painful electric shocks applied to a rat's tail.[45] Given that this stimulation exerts an analgesic effect (it attenuates the painful sensations by the operation of a descending control which moderates, at the spinal level, the transmission of nociceptive messages), the reduction in the probability of appearance of biting behaviour certainly does not find its 'explanation' in the fact that the stimulation may have made the animal 'less aggressive'. In this case, as in the others, the experimental intervention has modified the way in which aggression-inciting information is processed by an individual brain and, thus, the way in which it responds.

Functional heterogeneity of the neuronal aversion system

The notion of a 'neuronal aversion system' incorporating the periaqueductal grey matter and the medial hypothalamus is essentially based on the fact that stimulation applied to one or the other of these two periventricular structures provokes a flight reaction, or 'defensive' aggression, or, again, instrumental behaviour by means of which the animal switches off the stimulation being imposed on it. But we must make it clear that the system so defined does not correspond in any respect to a homogeneous functional whole. For one thing, in certain respects, the relationships between the periaqueductal grey matter and the medial hypothalamus have a directional character. In fact, if hypothalamic stimulation no longer triggers a defence reaction once the periaqueductal grey matter has been destroyed, a lesion carried out on the medial hypothalamus does not in any way prevent such a reaction being triggered by stimulation of the periaqueductal grey matter.[46] In the same way, an injection of atropine (which blocks cholinergic transmissions) into the periaqueductal grey matter prevents the triggering of 'affective' aggression by an injection of carbachol (which activates cholinergic transmissions) into the medial hypothalamus, while an injection of carbachol into the periaqueductal grey matter remains effective even if atropine has previously been injected into the medial hypothalamus.[47] On the other hand, while a stimulation applied

[45] J. W. Renfrew and J. A. Leroy 1983.
[46] A. Fernandez de Molina and R. W. Hunsperger 1962.
[47] C. L. J. Stokman and M. Glusman 1981.

to one or the other of these two components of the aversion system produces analogous behavioural effects, the repercussions of a lesion are in no way the same in the two cases. As was indicated earlier, a lesion carried out in the periaqueductal region attenuates or even abolishes the defence reactions provoked by a natural threat or by electrical stimulation of the medial hypothalamus. By contrast, a lesion affecting the medial hypothalamus *raises* the animal's reactivity level and thereby augments the probability of appearance of a defence reaction in response to certain situations. It must be added that we do not know the precise moderating influences whose attenuation or elimination—as a result of the lesion to the medial hypothalamus—provokes a marked hyperreactivity: they are probably multiple, and capable of originating or relaying in this region, or even simply passing through it. However this may be, it is important now to see *whether* and *how* an increase in the reactivity level, produced by an experimental intervention within the brain, affects the probability of the triggering of aggressive behaviour.

Reactivity level and probability of triggering of aggression

Most of the experimental facts obtained in this field concern the following structures: olfactory bulbs, septum, ventro-medial hypothalamus, and raphe nuclei. An approach consisting of the partial or total destruction of one of these structures and an analysis of the repercussions of that intervention on certain of the animal's behavioural characteristics only looks simple. On the one hand, in fact, these various brain structures are linked with each other and, more generally, each interacts with the remainder of the brain in such a way that the observable behavioural repercussions do not express solely the absence of the structure that has been destroyed but also the anatomical and functional reorganization induced by the lesion in the remainder of the brain; in other words, the behaviour of the animal after the operation reflects, after a certain time, the capacities and performances of the brain so reorganized. On the other hand, when the 'reactivity' of an animal is assessed by subjecting it to various stimuli or situations, it is only rarely that a monolithic and unambiguous dimension is measured. One is often apprehending both quantitative aspects (a certain degree of excitability) and qualitative aspects (particular sensitivity to the affective attributes of sensory information and to the nature of these attributes) characterizing the sensitivity and reactivity of a living organism confronted with promptings emanating from the environment.

We must also underline the fact that increased emotional reactivity may have a very different—in fact a precisely contrary—effect, depending on the type of aggressive behaviour under consideration: more precisely, depending on whether the aggression is underlain by motivation of an *aversive* or, instead, of an *appetitive* nature. For instance, a lesion of the septum or of the ventromedial hypothalamus, leading to marked emotional hyperreactivity, facilitates the triggering of aggressive behaviour in response to a stimulus or situation having an aversive character, whereas it disturbs the behaviour through which a position of dominance is expressed, in rats[48] as in guinea pigs.[49] On the neuro-chemical plane, Daruna and Kent (1976) have noted that a fast serotonin turnover[50] (in the brainstem and in the telencephalon) characterized both 'less emotional' rats and rats proving to be very aggressive in a situation of competition for food, whereas a slower turnover was found both in 'very emotional' rats and in spontaneously 'killer' rats.

1. *Ablation of the olfactory bulbs.* In the rat, the bilateral ablation of the olfactory bulbs causes, in addition to anosmia (loss of the sense of smell), an enhanced 'irritability' which is shown by more frequent and more intense aggressive interactions with fellow-creatures. Moreover, following this intervention, one often observes the appearance of inter-specific aggressive behaviour in animals which were previously 'non-killers'.[51] Anosmia (when produced by destruction or removal of the nasal mucosa) does not, on its own, cause the appearance of muricidal behaviour in the 'non-killer' rat;[52] so it is undoubtedly the conjunction of anosmia and emotional hyperreactivity that causes this change of behaviour towards the mouse.

The probability that interspecific aggressive behaviour will appear after removal of the olfactory bulbs depends on both certain genetic characteristics and the endocrine state of the animals being studied.[53] By using only 'killer' animals (males and females) for reproduction over a two-year period we have not increased the proportion of 'killer' rats in our stocks; but the probability of inducing muricidal behaviour by a bulbectomy has shifted from approximately 50 per cent to nearly 100 per cent. By contrast, we have observed a reduction in this same probability in rats in which we had lowered the level of the circulating steroid hormones by carrying out—at the age of 1 month—a castration com-bined with a unilateral adrenalectomy. But we should stress the role played by another factor whose influence is largely determining, namely the post-operative environmental conditions. Indeed, the probability of the appearance of muricidal behaviour is high if the bulbectomized rat is kept in isolation after the operation, no matter what the age at which the latter takes place; as opposed to this, it is much lower if the rats operated on are kept in a group and are thus exposed to interactions with their fellow-creatures before being confronted by a mouse.[54] It is likely

[48] K. A. Miczek and S. P. Grossman 1972; S. P. Grossman 1972.
[49] D. M. Levinson *et al.* 1980.
[50] The rate of renewal or turnover rate of a neuroactive substance reflects the level of activity of the brain mechanism in the operation of which the substance is involved.
[51] M. Vergnes and P. Karli 1963.
[52] See P. Karli 1981.   [53] See P. Karli 1981.
[54] F. Didiergeorges and P. Karli 1966; M. Vergnes and P. Karli 1969.

that this preventing effect of intraspecific post-operative interactions is due to two complementary processes: first, the 'irritability' of the rats falls off progressively (interactions with their fellow-creatures have initially a particularly aggressive character, until the animals calm down more and more markedly); and secondly, because of the very fact of progressive attenuation of the aggressive interactions, a negative reinforcement may be produced which affects—more generally—the use of the aggressive components of the behavioural repertoire.

2. *Lesion of the septum.* Destruction of the septum causes the appearance of all the signs of a marked hyperreactivity in all species that have been studied (including the human species). In the rat, this hyperreactivity is accompanied by a noticeable facilitation of the triggering of both intraspecific aggression—whether spontaneous or induced by painful electric shocks—and interspecific aggression towards the mouse.[55] There is every reason to think that this increase in the probability of the triggering of aggressive behaviour is due to exacerbated sensitivity and exacerbated reactivity in the face of stimuli and situations of an aversive character. Indeed, the probability of aggression decreases in the measure that the hyperreactivity provoked by the septal lesion is attenuated.[56] If this lesion is carried out at an early age (7 days, for example), it induces a *lasting* hyperreactivity and it is then noted that, several months after the operation, an increased probability continues to exist that the intrusion of a mouse will trigger an aggressive reaction on the part of the rat.[57] Furthermore, painful electric shocks are capable of triggering interspecific aggression in a rat that has suffered a septal lesion,[58] whereas they do not have the same effect in an intact rat.[59] Observing intraspecific interactions in rats, Blanchard and his colleagues (1979) show clearly that the destruction of the septum induces a more marked defensive attitude (hyperdefensiveness) in response to signs of threat emanating from fellow-creatures, rather than an increased 'aggressiveness'.

The way in which hyperreactivity is specifically expressed in the behaviour of the animal operated on depends in large measure on the significance it attaches to the situation, with reference to its individual life-history. We have already mentioned that the destruction of the septum did not in any way cause the appearance of interspecific aggression in a rat that had been given the opportunity to familiarize itself with the presence of a mouse in its cage. As concerns intraspecific interactions, Lau and Miczek (1977) showed that the repercussions of a septal lesion were very different depending on whether the pre-operative experience of the rat was that of a dominant or a dominated animal. In

---

[55] See P. Karli 1976, 1981.
[56] K. A. Miczek and S. P. Grossman 1972; D. J. Albert and K. N. Brayley 1979.
[57] F. Eclancher and P. Karli 1979a.
[58] W. M. Miley and R. Baenninger 1972.   [59] P. Karli 1956.

the hamster, a lesion of the septum markedly accentuates aggressive behaviour towards a fellow-creature; but this more aggressive behaviour does not appear if, before the operation, the animal had suffered a defeat at the hands of that fellow-creature.[60] If it thus appears that the destruction of the septum in no way prevents the experience acquired before the operation from affecting the behaviour displayed during the post-operative period, the absence of the septum does prevent the intraspecific interactions from affecting, i.e. reducing, the probability of the appearance of muricidal behaviour. Indeed, if the probability is lower in intact rats brought up in groups than in animals brought up in social isolation, this difference (which probably expresses the fact that these social interactions have the effect of moderating the degree of reactivity, through the mediation of the septum) is no longer observed after both groups have undergone destruction of the septum at the age of seven days.[61]

The role played by the septum in the development of socio-affective behaviour certainly cannot be reduced to the contribution it makes in the adult brain to a certain moderation of emotional reactions. We must first recall that, in the course of the early stages of ontogenesis, septal functioning and social contacts interact with each other so as to give the individual a general level of reactivity which will be one of the dimensions of his 'personality'. Furthermore, it is partly in the competitions ('play fighting') characterizing social interactions between young animals, with the successes and failures that punctuate them, that certain attitudes (assurance, dominance, etc.) of the adult are formed. Now, neurobiological research into the 'social play' of young animals has clearly shown that in this regard the septum plays a non-negligible role: a lesion of the septum raises the frequency and duration of competitive interactions as well as the frequency with which the animal which has undergone the operation (the male more than the female) takes the initiative.[62]

It should not be forgotten, either, that the septum interacts with many other brain structures and that its proper functioning is part of a structured, dynamic, and evolving whole. We shall limit ourselves here to mentioning the probable existence of functional interactions of the septum with the raphe nuclei (which we shall discuss later). A lesion of the raphe, which brings about an increase in the level of reactivity as well as a facilitation of the triggering of aggressive reactions to painful electric shocks, also has the effect—at the level of the septum—of reducing the activity of choline-acetyltransferase (which is involved in the biosynthesis of acetylcholine). The hypothesis has therefore been formulated that the behavioural repercussions of the lesion of the raphe could well

---

[60] F. J. Sodetz and B. N. Bunnell 1970.
[61] F. Eclancher and P. Karli 1979a.
[62] See W. W. Beatty et al. 1982; D. H. Thor and W. R. Holloway 1984; J. Panksepp et al. 1984.

be due, at least in part, to a deficiency of certain cholinergic transmissions at the level of the septum. It has turned out that one may actually attenuate these behavioural effects of a lesion of the raphe by dosing the animal with physostigmine, an inhibitor of cholinesterase, i.e. by slowing down the degradation of acetylcholine.[63] Furthermore, the usual effects of a lesion of the raphe nuclei (heightening of the reactivity level, facilitation of the reactions of intraspecific and interspecific aggression) are not observed if, a month previously, the rats have suffered destruction of the septum.[64] Noting that, after injections of muscimol[65] into the septum, rats become more irritable and kill mice put in their presence more quickly, Potegal and his colleagues (1983) consider that this manipulation suppresses a moderating influence which the septum normally exercises by means of the raphe nuclei and their ascending serotonergic projections. It is appropriate to recall here the relaxation of certain inhibitions—shown by an increased 'impulsiveness'—which seems to be due to a defective action of serotonin on the substantia nigra.[66]

Studying the relations between the septum and 'impulsiveness' in the rat, Rawlins and his colleagues (1985) have made the following interesting observation: when intact and thirsty animals are placed in a T-maze, they prefer the arm in which the reward (water) is given only after a waiting period of 10 seconds, but regularly, to that in which the reward is immediate but of a random nature; if, by contrast, the animals have previously undergone a lesion of the septum, they prefer the immediate reward despite its random character.

3. *Lesion of the ventro-medial hypothalamus.* A lesion of the ventro-medial hypothalamus increases the organism's sensitivity to painful stimuli at the same time as it accentuates, more generally, its reactivity towards many sensory stimuli.[67] Since the first experiments carried out by Wheatley (1944) in the cat, numerous studies have shown that, in this species, the probability of the triggering of aggressive behaviour was markedly augmented following a lesion of the ventro-medial hypothalamus.[68] In the rat, too, it has been amply demonstrated that such a lesion clearly facilitates the triggering of intraspecific aggression between males—whether spontaneous or induced by painful electric shocks—and of interspecific aggression towards the mouse.[69] In the latter case, the facilitating effect of the lesion to the ventro-medial hypothalamus is particularly marked in rats which have previously undergone an ablation of the olfactory bulbs.[70]

When it proved that the effectiveness of this lesion—in augmenting

[63] M. Vergnes and C. Penot 1976*a*.   [64] M. Vergnes and C. Penot 1976*b*.
[65] In this case, muscimol activates inhibitory GABAergic transmissions.
[66] See Chapter 5, under Role of cerebral serotonin, p. 116.
[67] See P. Karli 1981; G. Sandner *et al.* 1985.
[68] See J. P. Flynn 1976.   [69] See P. Karli 1981.
[70] F. Eclancher and P. Karli 1971.

the probability of the appearance of muricidal behaviour—was markedly increased when carried out at an early age (7–8 days), the question arose whether the observed difference was not due, at least in part, to the fact that the rats that were operated on at an adult age were 'non-killer' rats that had become accustomed to the presence of a mouse in their cages, whereas the behaviour of those operated on at an early age was tested without their having had prior interactions with mice. The experiment carried out in an attempt to answer this question showed that it was sufficient to bring up the subject rat with a mouse (for one month, starting at the age of 30 days) for the early hypothalamic lesion to cease to cause the appearance of interspecific aggressive behaviour.[71] Two conclusions were thus indicated: the preventing effect of prior interactions with mice—regarding the triggering of muricidal behaviour by virtue of experimentally induced hyperreactivity—proves to be less marked following a lesion to the medial hypothalamus than following a lesion to the septum, when these lesions are carried out at an adult age (this fact was confirmed by Albert and his colleagues in 1984); if the lesion to the hypothalamus thus seems to attenuate a preventing effect already created by prior social interactions, it does not prevent the subsequent development of such an effect if it is carried out at the age of 7–8 days. We shall see later the essential role played by the amygdala in mediating the influence of acquired experience. Now, the amygdala has an important projection on to the ventro-medial hypothalamus, and it is very possible that a lesion to the latter interferes with a preventing influence originating in the amygdala (while the anatomic and functional reorganization brought about by a very early lesion could permit the amygdala to exercise this influence almost in a normal fashion). It is interesting to note, in this regard, that lesions to the ventro-medial hypothalamus carried out at the age of 3–4 weeks, when the young animals are meeting each other in their 'social play', not only increase the frequency and the duration of the interactions (as we have already seen for lesions to the septum) but also cause a qualitative 'deterioration' in these interactions (which become much more aggressive), as though the animals operated on were 'unable to correctly interpret the playful gestures'.[72]

4. *Lesion of the raphe.* The ascending serotonergic fibres originating from the (dorsal and medial) raphe nuclei have many projections.[73] For one thing, they project on to a set of structures (amygdala, septo-hippocampal system, and, in passing, the lateral and medial hypothalamus) which play an essential role in the association of a certain significance with the objective data of sensory information and in the genesis of the emotional reaction resulting from that. A destruction of the raphe nuclei causes a marked hyperreactivity, an intensified sensitivity of the

---

[71] F. Eclancher 1983.   [72] J. Panksepp *et al.* 1984.
[73] See Chapter 5, under Role of cerebral serotonin, p. 116.

organism to stimuli coming from the environment. Further, they project on to the nigro-striatal system and it seems that defective action of serotonin on the substantia nigra is expressed by a heightened 'impulsiveness', by the dissolution of certain inhibitions and by frequent 'acting out'. This being so, it cannot be surprising that destroying the raphe nuclei—or, more generally, depleting the brain serotonin—causes a marked increase in the probability of the triggering of aggressive behaviour which is underlain by a motivation of aversive nature, whether it be a case of intraspecific aggression induced by painful electric shocks or the *initial* triggering of interspecific rat-mouse aggression.[74] As in the case of the hyperreactivity caused by destruction of the septum, prior familiarization with mice has a marked preventing effect on the triggering of interspecific aggression. Given that prior removal of the olfactory bulbs accentuates the hyperreactivity caused by the lesion of the raphe at the same time as it interferes with recognition of the mouse, it is not difficult to understand that an association of the two lesions has the result of bringing about the appearance of interspecific aggressive behaviour in an increased percentage of the rats thus operated on.[75]

An increase of the probability of the appearance of aggression may reflect heightened sensitivity to certain environmental stimuli, marked 'impulsiveness' corresponding to a general facilitation of behavioural expression, and a reduction in fear of the partner's reactions. Concurring data lead us to believe that the last factor, too, may well contribute to the genesis of the behavioural changes observed in the case of an experimentally produced deficiency in brain serotonergic activities. If the spontaneous activity of the neurons of the dorsal raphe nucleus is studied, it is found that this activity is maximal, in the shrew, when the animal displays a fearful and defensive attitude, while it is noticeably reduced in the case of more frankly offensive behaviour; the activation of these neurons does seem to be linked with an appreciation of the context which is inspiring fear and a defence reaction.[76] In a situation of competition (for water), the dominant rat loses its status of dominance if by neuropharmacological methods its brain serotonergic transmissions are activated, while the subordinated rat becomes dominant when the same serotonergic transmissions are blocked; moreover, a rat that has suffered a lesion of the raphe nuclei always proves dominant when confronted with a partner that has not been operated on.[77]

In the case of social interaction of the 'resident–intruder' type, a general reduction in serotonergic activity, produced by the administration of parachlorophenylalanine, has the effect of facilitating in a specific

[74] See P. Karli 1981; M. Vergnes and E. Kempf 1981; O. Pucilowski and W. Kostowski 1983; P. Soubrié 1986.
[75] M. Vergnes *et al.* 1974.
[76] H. Walletscheck and A. Raab 1982.
[77] W. Kostowski *et al.* 1984.

way the triggering of offensive behaviour, whereas defensive behaviour remains unchanged, whether the animal thus treated be the resident or the intruder.[78] We may add that, in many investigations carried out in man, a correlation has been found between a marked tendency to 'act out' (aggressive actions or attempts at suicide) and reduced serotonergic activity, suggested in particular by the existence of a low level of metabolites of serotonin in the cerebrospinal fluid; but the discovery of such a correlation raises a certain number of problems that have not yet been resolved, and it would be premature to state any causal relation between this neurochemical characteristic and a marked propensity to 'act out'.[79]

(B)   Repercussions from past individual experience: the mediating role of the amygdala

The amygdala—in conjunction with other structures of the limbic system—is deeply involved in the processes through which significance, particularly affective significance, is associated with the sensory information by reference to the traces left by past experience, and in the processes by means of which this significance may be modulated under the effect of the consequences that flow from the behaviour. We have also seen that experiential factors are extensively involved in determining the probability of the triggering of aggressive behaviour.[80] The amygdala will therefore play an important role whenever, in the face of a situation having the potential of provoking aggression, reference is made to past experience, to the individual life-history.

In reality, the amygdala could have had its place in the foregoing pages which dealt with the level of emotional reactivity. For it is quite obvious that sensory information that does not acquire its cognitive and affective significance does not generate the emotion precisely corresponding to that significance. From the first experiments in which bilateral temporal lesions (which included the amygdala) were carried out on monkeys, Klüver and Bucy (1937) noted that the animals operated on, which proved to be unable to recognize the significance of the objects present in their visual environment ('psychic blindness'), were characterized by markedly attenuated emotional reactivity. And we have stressed that, very generally, bilateral lesions of the amygdala lead to a diminution of emotional reactivity, as shown by a certain indifference towards the environment and by markedly attenuated reactions to various social stimuli. It is not, therefore, surprising that these lesions of the amygdala markedly lower the probability of the triggering of intraspecific aggression in rodents as in man, and that they should abolish — at least transiently—the reactions of interspecific aggression in the rat

---

[78] M. Vergnes *et al.* 1986.   [79] See P. Soubrié 1986.   [80] See Chapter 7.

as in the cat.[81] Furthermore, our initial study had already shown that bilateral destruction of the amygdala was able to abolish muricidal behaviour, while causing a lowering of emotional reactivity that was particularly striking in wild rats.[82] This lesion disturbs the processes by means of which the signals emanating from the mouse acquire their motivating and reinforcing properties and, thus, the genesis of the appetitive motivation underlying this aggressive behaviour in the experienced 'killer rat'. Prior or simultaneous lesions of the amygdala prevent the appearance of aggressive behaviour following an operation which would normally have the effect of causing its appearance: no muricidal behaviour, if a removal of the olfactory bulbs is preceded by or accompanied with bilateral destruction of the medial region of the amygdala;[83] no marked accentuation of the defensive attitude, if destruction of the septum is accompanied by an amygdaloid lesion.[84] And if one first causes the appearance of muricidal behaviour by operating on the brain, a subsequent lesion of the amygdala is capable of abolishing it (after a lesion of the ventro-medial hypothalamus[85] or after removal of the olfactory bulbs[86]).

Prevention of aggression by prior familiarization

This functional disturbance produced by the bilateral lesion of the amygdala is only transitory, and the 'killer' rat whose aggressive reactions towards the mouse are abolished recovers its interspecific aggressive behaviour in one to three months. By contrast, it does seem that there is no possibility of replacement (so the deficiency produced by a lesion would seem irreversible) in the amygdala's involvement with the aggression-prevention that normally results from prior interactions with the alien species. As was emphasized earlier, the hyperreactivity induced by destruction of the septum markedly facilitates the triggering of aggressive behaviour in a rat confronted by a mouse for the first time, but not in an animal which has had the opportunity to familiarize itself with the presence of a mouse in its cage. Now, the preventing effects of this prior familiarization are never observed if the rat had suffered a bilateral lesion of the cortico-medial part of the amygdala before its prior contact with the mouse.[87] Deprivation of food and depletion of brain serotonin (by, for example, administration of parachlorophenylalanine) each have the effect of very markedly increasing the probability of an initial triggering of muricidal behaviour; but this triggering does not occur in 'non-killer' rats which have previously been familiarized with mice. In these cases, too, the aggression-preventing effect of prior social contacts is not displayed if the rats have been deprived of the medial

[81] See P. Karli 1976, 1981; M. Vergnes 1981.   [82] P. Karli 1956.
[83] S. Shibata *et al.* 1982*b*.   [84] D. C. Blanchard *et al.* 1979.
[85] P. Karli *et al.* 1972.   [86] S. Shibata *et al.* 1982*b*.
[87] C. Penot and M. Vergnes 1976.

region of their amygdala before these contacts.[88] In the course of ontogenesis, early social contacts with mice reduce practically to zero the
proportion of rats which prove to be 'killers' in adulthood. In contrast, in
a group of rats which had been amygdalectomized at the age of eight
days and then brought up for several weeks in the company of mice, 70
per cent presented muricidal behaviour in adulthood.[89] This high percentage of 'killer' rats is due to the fact that an early lesion of the
amygdala does not just have the effect of hindering the aggression-
prevention that normally results from interactions with mice; in addition, and at first sight paradoxically, the lesion causes a lasting hyperreactivity to which we shall have occasion to return. We may again point
out another observation demonstrating the vital role played by the
amygdala in mediating the behavioural effects of social contacts. If the
olfactory bulbs are removed at the age of 25 days, thus causing persist-
ing hyperreactivity, it is noted that 90 per cent of the rats kill at adult age
if they are brought up in isolation following bulbectomy, whereas only
10 per cent kill if they are brought up as a group and if therefore interactions between fellow-creatures had the effect of reducing the probability
of the triggering of aggressive behaviour. But this preventing effect of
interactions between fellow-creatures is noticeably weaker if, at the age
of 25 days, a bilateral lesion of the amygdala is combined with the
removal of the olfactory bulbs: under these conditions, it is not 10 per
cent but 60 per cent of the rats operated on and raised in groups which
turn out to be 'killers' in adulthood.[90]

Effects of the experience of defeat

In the case of aggression between fellow-creatures it is defeat which
constitutes a major experiential factor capable of influencing—i.e.
reducing—the probability of the triggering of such an aggression. And
in this case, too, the amygdala turns out to be deeply involved. Confronted with a fellow-creature which has emerged 'victoriously' from a
previous interaction, a rat adopts a rigid attitude ('freezing behaviour')
and keeps its distance. But these behavioural effects of defeat are very
strongly attenuated by a bilateral lesion of the amygdala.[91] It should be
added that the amygdala is involved from the time acquisition of social
learning begins: defeat has only a weak effect on the animal's subsequent behaviour if the amygdaloid lesion is carried out before defeat, or
if, during the experience of defeat, electrical stimulation is administered
which puts the amygdala temporarily out of action. But the amygdala is
quite as much involved in the process of retention of learning, and an
amygdaloid lesion is even particularly effective (in reducing the effect of
defeat on subsequent behaviour) when carried out 48 hours after the

---

[88] M. Vergnes 1981.   [89] F. Eclancher *et al.* 1975.
[90] F. Eclancher *et al.* 1975.   [91] J. M. Koolhaas 1984.

experience of defeat. Most particularly, it is the posterior part of the medial nucleus of the amygdala and the transition zone between the amygdala and the hippocampus that are involved in the mediation of the behavioural effects of this particular experience.

Noting thus that lesions of the amygdala are capable of both abolishing the interspecific aggressive behaviour of a 'killer' rat and facilitating the appearance of this very behaviour in a 'non-killer' rat (just as they can facilitate intraspecific aggression towards a victorious and dominant fellow-creature), we obviously do not 'explain' anything by saying that the amygdala is involved both in mechanisms which facilitate aggression and mechanisms which inhibit it. In reality, in the one case as in the other, an amygdaloid lesion disturbs the behavioural expression of previous learning which, under normal conditions, confers full significance on the signals emanating from the mouse: in the 'killer' rat, learning accentuates the incentive and reinforcing properties of the mouse, thus positively reinforcing and maintaining muricidal behaviour; in the 'non-killer' rat, familiarization with mice (which amounts to learning) makes neophobic reactions, particularly muricidal behaviour, inapplicable. Given that both the information processed and the processes involved are different in the two cases, we easily understand that experimental interventions on the amygdala or on the main efferent pathways permit their dissociation. Indeed, a bilateral interruption of the stria terminalis interferes with the aggression-preventing effect of prior social interactions with mice, while it does not in any way modify the 'killer' rat's aggressive behaviour; on the other hand, the latter behaviour may be abolished by a bilateral interruption of the diffuse ventral connections between the amygdala and the hypothalamus.[92] We may also add an experimental dissociation of a different type: a bilateral interruption of the stria terminalis which facilitates the appearance of muricidal behaviour does not modify the probability of the triggering of aggression in response to painful electric shocks; bilateral lesions of the lateral nucleus of the amygdala, which strongly reduce this probability, do not in any way modify the muricidal behaviour of a 'killer' rat.[93]

The following observation, made by Zagrodzka and her colleagues (1983) in the cat, shows that, among the information processed, it is important also to consider the global context and not just the single 'aggression-eliciting' object: when it has suffered bilateral destruction of the medial region of the amygdala, the cat no longer kills mice, whatever the situation in which these mice are presented to it; following bilateral lesion of the dorsal region of the amygdala, on the other hand, a cat loses its social status within a group and no longer kills mice while it remains within the group, whereas it continues to kill them if it finds itself in their presence alone. Correct interpretation of the data furnished by experimentation thus requires that one keep sight of the diversity of

---

[92] P. Karli *et al.* 1972; M. Vergnes 1976.   [93] M. Vergnes 1976.

the information being processed and the processing operations which they undergo, while bearing in mind the fact[94] that the various cortical unimodal association areas project on to distinct regions of the amygdaloid nuclear complex, and that each of these regions thus seems to be subject to the privileged influence of a given sensory modality.

Maturation of individual reactivity

In the course of the maturation undergone conjointly by the brain and behaviour, the amygdala is extensively involved in a double modulation, both quantitative and qualitative, of individual reactivity: on the one hand, a marked attenuation of the state of activation of the brain takes place following weaning, in consequence of the putting into place of moderating processes originating in the telencephalon; on the other hand, experience acquired in social interactions leads the individual, in a more selective way, not to respond to certain stimuli from the environment (because of familiarization, punishment, defeat, etc.).[95] Under these conditions, it is not surprising that early lesions of the amygdala (carried out in rats at the age of 7–8 days or 25 days) cause the development of a marked hyperreactivity which persists until adult age.[96] It is likely that such early amygdaloid lesions have behavioural effects which are all the more profound as they also reduce the frequency and duration of 'social play',[97] i.e. the social interactions which, under normal conditions, exert an important influence which structures and 'personalizes' behaviour. In the macaque, Thompson (1981) carried out early bilateral lesions of the amygdala (aspiration of the left amygdala towards the age of two months, and aspiration of the right amygdala three weeks later) and analysed the behaviour of the amygdalectomized animals throughout an eight-year period. Behavioural disturbances appeared progressively, worsening with age: the animals were hyperactive, persisted in behaviour that had become inappropriate, and proved to be incapable of compelling recognition within their group.

The amygdala interacts with other brain structures

Whenever the amygdala is involved in processes determining the way in which the individual brain perceives and interprets a given stimulus or situation, it interacts with numerous other brain structures. We have previously seen that, in the cat, a certain individual predisposition to display an 'offensive' or a more fearful and 'defensive' attitude when confronted with the most varied threats is correlated both with func-

---

[94] This fact was pointed out in Chapter 5, under The structures making up the limbic system, p. 95.
[95] See Chapter 5: Processes that moderate behavioural arousal, p. 80.
[96] F. Eclancher and P. Karli 1979*b*.
[97] J. Panksepp *et al.* 1984; D. H. Thor and W. R. Holloway 1984.

tional characteristics intrinsic to the amygdala and with certain characteristics of the transmission of neural messages between the amygdala, the ventral hippocampus and the ventro-medial hypothalamus.[98] On the other hand, an activation of the serotonergic neurons of the raphe seems also to be linked to an appreciation of the context arousing fear and a defensive reaction.[99]

We are thus led to think about possible interactions between the raphe nuclei and the amygdala in the genesis of the motivating properties of the environment and the attitudes adopted towards it. It is then interesting to note that a manipulation of the ascending serotonergic systems, which modifies the probability of the triggering of aggression, affects — among other things — cholinergic transmissions within the amygdala; and that manipulation of these transmissions is capable of affecting — among other things — the probability of the triggering of aggressive behaviour. Indeed, when the triggering of rat–mouse interspecific aggressive behaviour is facilitated by destruction of the raphe nuclei or by an administration of parachlorophenylalanine (an inhibitor of the biosynthesis of serotonin), an increase results in the choline-acetyltransferase activity (necessary to the biosynthesis of acetylcholine) in the amygdala. Conversely, when muricidal behaviour is blocked by augmenting the levels of available serotonin by various experimental manipulations, it is observed that choline-acetyltransferase activity is markedly depressed in the amygdala; and this blockage may be removed in a transitory way by the administration of physostigmine, an inhibitor of cholinesterase, i.e. of the enzyme which degrades acetylcholine.[100]

The probability of the triggering of aggression may also be modified by acting directly on the cholinergic transmissions within the amygdala: in the rat, blocking these transmissions in the basolateral amygdala (by locally injecting scopolamine) markedly reduces aggression in response to painful electric shocks;[101] in the macaque, cholinergic activation (by injecting carbachol into this same basolateral part of the amygdala) facilitates the appearance of aggressive behaviour towards a fellow-creature, especially if this neurochemical manipulation is effected in the dominant animal.[102]

Changes that are more or less profound, affecting the transmission and processing of neural messages within an interactive ensemble of brain structures, thus may modify certain 'personality traits'. Repeated electrical stimulation of the amygdala ('kindling') has the effect of transforming, in a lasting way, an initially offensive attitude on the part of a cat into a more and more fearful and defensive attitude. Now, this

[98] See Chapter 5: The amygdala does not operate in isolation, p. 106.
[99] See above under Lesion of the raphe, p. 204.
[100] G. Mack 1978; P. Mandel 1978; P. Mandel *et al.* 1979.
[101] R. J. Rodgers and K. Brown 1976.
[102] J. H. L. Van den Bercken and A. R. Cools 1982.

repeated stimulation causes a permanent change in certain neurophysiological characteristics of the amygdala and in modes of message transmission between the amygdala, the ventral hippocampus and the
ventro-medial hypothalamus.[103] Moreover, paroxysmal crises triggered
from certain sites in the amygdala modify the excitability characteristics
of structures with which the amygdala is connected: under the action of
these crises, stimulation of the hypothalamus more readily triggers
affective and defensive aggression in the cat.[104] These observations take
on their full interest if we remember that, in man, certain forms of
epilepsy originate in the limbic structures of the temporal lobe. Following a crisis, the sufferer may present a confused state with a disturbance
of affectivity, and he may react violently to certain stimuli, especially
tactile ones; between crises it is not uncommon to observe personalities
characterized by a peculiar emotionalism and impulsiveness, which may
be the reflection of changes induced by the crises in the dynamics of the
functional interactions of the limbic structures with themselves and with
more posterior structures, diencephalic and mesencephalic.[105]

Interactions of the amygdala with the temporal cortex and the frontal cortex

Since we are considering the functional interactions of the amygdala
with other brain structures, it is important to mention the temporal
cortex and the prefrontal cortex. In the rat, a bilateral section of the
connections between the amygdala and the temporal cortex which
covers it has behavioural effects similar to those produced by a bilateral
destruction of the basolateral amygdala (attenuation of neophobia,
defective acquisition of a conditioned aversion).[106] In this case, it is
difficult to decide whether interruption of the connections has really
disturbed an information-processing operation that requires interaction
between the two structures or whether we have not simply interrupted
the linear transmission of neural messages. The observations made by
Arthur Kling's team[107] in the monkey suggest more clearly the involvement of functional interactions between the amygdala and the temporal
cortex in the processes by means of which sensory information acquire
their full meaning. Following a lesion of the infero-temporal cortex, and
more clearly still following a lesion of the temporal pole, the bioelectrical
responses recorded in the medial part of the basal nucleus of the amygdala are strongly attenuated, and this attenuation is particularly noticeable for responses evoked by visual or auditory stimuli indicating a
threat (whereas the response to a pure sound or to music remains
perfectly unaltered). At the same time, it is noted that the behaviour of
animals which have undergone bilateral destruction of the temporal

---

[103] R. E. Adamec and C. Stark-Adamec 1984. [104] A. Siegel 1984.
[105] See J. Engel and S. Caldecott-Hazard 1984.
[106] R. E. Fitzgerald and M. J. Burton 1983. [107] A. Kling 1981; A. S. Kling *et al.* 1984.

pole is modified: they avoid social contacts, for they seem to have become incapable of correctly interpreting the social signals emanating from their fellows, and cannot themselves develop affective states that would be significant for the group. As far as the orbito-frontal subdivision of the prefrontal cortex is concerned, we have already pointed out the fact that, in the monkey, lesions to the orbito-frontal cortex, like those to the amygdala, cause profound changes to personality and social behaviour: deficient perception of the emotions expressed by others, impoverishment of spontaneous facial expressions, a tendency toward social isolation.[108] After becoming less capable of correctly processing sensory information emanating from its fellows and responding appropriately, a dominant animal loses its status, falls quickly to the lowest level of the hierarchy, and presents submissive behaviour to the other members of the group. In a species like the rat, in which the 'corticalization' of the cerebral functions is less highly developed, the effects of orbito-frontal lesions are more discrete: a slight increase in the frequency of aggressive interactions goes together with a certain accentuation of the reactivity of the animal operated on in regard to harmful stimuli emanating from its partner.[109] In order to bring to a close this brief glance at the interactions between the amygdala and the cerebral cortex, we may recall that opiate receptors seem to participate in a kind of 'affective filtering' of sensory messages at the cortical level, by intervening in the mechanisms by which the affective states induced by the mediation of the limbic structures (particularly the amygdala) influence the nature of the sensory messages on which attention is focused.

Confirmation of the instrumental value of aggression

Since behaviour is employed because of its expected instrumental value, it is clear that the probability of its effective use is modulated by any confirmation—or invalidation—of its real effectiveness. We have already seen that the limbic system plays a vital role in the comparison of results effectively obtained with those which are expected at the moment the behavioural strategy was evolved; in the genesis of the affective experience—pleasant or unpleasant—ensuing from it; and in the operation of a process of positive or negative reinforcement, which either increases or diminishes the probability of subsequent use of the instrument this behaviour provides for use in dialogue with the environment. But before carrying out, in this connection, a brief examination of the specific case of rat–mouse interspecific aggressive behaviour, it is appropriate to recall and to emphasize two more general aspects of behavioural biology: the living being attempts to master the relation-

---

[108] J. P. C. de Bruin 1981; B. Kolb 1984.   [109] J. P. C. de Bruin *et al.* 1983.

ships which it establishes with its environment, and the method it uses is largely determined by its relatively early experiences.

In many situations aggressive behaviour reinforces itself positively, from the moment that it allows the animal to master the situation and to dominate it; such positive reinforcement is particularly marked if the experimental conditions are such as to guarantee the success of the strategy employed and to minimize its cost.[110] An observation made in another field confirms the important role played in confronting a given situation by an ability to exert control and the cost inherent in that control: when subjected to a highly stressful situation, a rat develops few gastric ulcers if it is given the ability to control the situation; but this ulcer-preventing effect is correspondingly less marked if the task that has to be accomplished in order to exert the control is more complicated.[111]

If rats are observed in various social situations, it is noted that each individual generally and consistently presents one or the other of two very different 'behavioural styles'. Certain animals are characterized by their 'competitiveness': they seem always to be in control of their environment, whether this means facing up to an 'intruder' or responding to the threats of a dominant animal; and in these situations they present a substantially raised blood pressure. The others behave much more passively and their blood pressure does not rise so markedly.[112] Now, this 'behavioural style' is evolved very early, in the context of the 'social play' which develops after weaning. This play corresponds to a genuine need to interact with and measure oneself against others, for it is markedly amplified following a period of social isolation. And it does seem, in view of the data furnished by certain longitudinal studies, that there is an 'emotional continuity' between the successes recorded in this early social play and the processes underlying the subsequent development of an attitude of self-confidence and dominance.[113]

Experimental facts, some of which have already been quoted, show the relationship that exists between the activation of the neuronal reward system (by electrical stimulation or, under natural conditions, by the evocation and expectation of the 'gratifying' consequences of behaviour) and an increase in the probability of the employment of a strategy which has previously given the expected results. An experienced 'killer' rat kills a mouse coldly, without any very apparent emotional reaction. By dint of repetition, the interspecific aggression is itself positively reinforced and is ever more firmly underpinned by a motivation of appetitive nature. This appetitive behaviour can be abolished by carrying out a lesion (in the lateral hypothalamic area or in the ventro-medial region of the mesencephalic tegmentum) to disturb the normal functioning of the positive reinforcement system. This

[110] See M. Potegal 1979.

[111] A. Tsuda *et al.* 1983.

[112] D. S. Fokkema and J. M. Koolhaas 1985.

[113] See J. Panksepp *et al.* 1984.

is how mesencephalic lesions, which abolish (in a transitory way) muricidal behaviour, also abolish self-stimulation in the lateral hypothalamus.[114] In the other direction, a rat which habitually kills only after a delay of several hours can be caused to kill immediately by stimulation at certain sites in the lateral hypothalamus or in the medial tegmentum of the mesencephalon. Now, we note that all of these stimulation sites without exception are self-stimulation sites, sites in the positive reinforcement system.[115] And if the animal is given the opportunity to stimulate itself at one of these sites, it does so and by that means excites itself to attack and immediately kill a mouse which, otherwise, it would have attacked and killed only after several hours.

The fact that one and the same stimulation of the lateral hypothalamus induces both an effect of appetence (shown by self-stimulation) and a facilitation of muricidal behaviour cannot be considered a mere coincidence, for an interaction is observed between the genesis of the appetitive effect and the actual presence of a mouse. De Sisto and Zweig (1974) picked out, within the posterior region of the lateral hypothalamus, two categories of stimulation sites from which they triggered feeding or muricidal behaviour respectively. They then gave the rats the possibility of stimulating themselves and noted that the presence of food accentuated the self-stimulation performances at sites in the first category, while the presence of a mouse accentuated these performances at sites in the second category. The reinforcing effects due to the activation of the positive reinforcement system are thus additive with the reinforcing properties of the mouse, properties which are due precisely to the fact that muricidal behaviour previously furnished results that activated the positive reinforcement system. We may also recall here the experiment in which Stachnick and his colleagues (1966) created marked aggressiveness in the rat towards a fellow-creature (an aggressiveness which did not exist at all initially), by regularly associating a pleasant affective experience—induced by stimulation of the lateral hypothalamus—with the smallest hint of aggressiveness, and thus training the animal that aggressive behaviour on its part could be 'rewarding'.

Processes capable of restraining a potential aggressor

Taking account of the foreseeable cost of aggression

Man, and animals, behave as 'intuitive statisticians' when, faced with a given situation, aggression seems to be—to a first approximation—the appropriate strategy. For they assess the respective amounts of expected benefits and the risks to be run. And the probability of the actual employment of aggressive behaviour will be low if the foreseeable 'cost'

[114] J. P. Chaurand *et al.* 1973; P. Karli *et al.* 1974.   [115] P. Karli *et al.* 1974.

of this aggression is high. This is how the experience of 'defeat', like that of 'punishment', reduces the probability of subsequent use of aggression as a means of expression and action. In the interpretation of a situation, which conditions the final choice of strategy, this balance of the affective attributes (appetitive and aversive, positive and negative) plays an essential role. Given the number and diversity of the factors that may be taken into account, it is clear that a complex set of brain mechanisms is in operation. But it does seem that, within this set, the amygdala plays a central role. It is particularly involved in the attenuation of the hedonic and incentive virtues of a stimulus or a situation by negative reinforcements linked, for example, with an experience of defeat or punishment.

This modulation of the hedonic virtues of an object in the environment, like the modulation of the appetitive attitude that it excites, is more easily apprehended in the case of feeding behaviour. The amygdala is involved in reduction of appetence for a food, whether this reduction be progressive, as in natural satiation, or more brutal as in the case of conditioned aversion.[116] Given the fact that positive and negative reinforcements interact within the amygdala, it is understandable that experimental manipulations (lesions or activations) may have the effect of either increasing or diminishing the ingestion of food, depending on the exact location of the amygdaloid lesion[117] and whether the stimulation applied to the amygdala is adrenergic or cholinergic.[118] Even if the continuum from 'offensive' to 'defensive' attitudes, which characterizes social interactions in various situations, has to do with a more complex determinism than the continuum from 'appetite' to 'satiety', we must nevertheless recall that the probability of an offensive attitude may be markedly reduced by experimental manipulations of the amygdala: transiently, by acting on the cholinergic transmissions in the basolateral amygdala; permanently, by applying repeated electrical stimulation which modify the functioning of the amygdala and its connections.

Control over the emotions and impulses

When an instance of aggressive behaviour is one expressing, and aimed at terminating, an aversive type of emotion, its effective employment will be all the more probable if, because of a raised reactivity level, the aversive emotion is more intense and if, due to marked 'impulsiveness', the tendency to respond rapidly and actively is stronger. Conversely, the probability of aggression may be reduced if the individual is able to control the development of his emotions and if, being able to defer his reaction somewhat, he gives himself time to carry out a more complete evaluation of the situation and/or to choose a strategy other than that of aggression. It is thus appropriate to recall here that moderation of the

[116] See P. Karli 1976.   [117] E. Fonberg 1981.   [118] S. P. Grossman 1964.

emotional reactions implies the involvement of a normally functioning septum, and that the way in which the septum functions in the adult brain depends in large measure on the way it has or has not been brought into play in the early stages of ontogenesis (in a spontaneous way in social interactions, and in a more willed way in educative processes). As for impulsiveness, we have seen earlier that it can be increased as a result of deficient serotonin action on the substantia nigra, without our yet being able to state the existence of a causal relation between this neurochemical characteristic and a marked propensity to 'acting out'. But, however this may be, it must be emphasized that appropriate functioning of the ascending serotonergic projections and the absence of impulsiveness are not simple phenotypic expressions of a certain portion of an individual's genetic heritage, any more than are the appropriate functioning of the septum and a 'normal' level of reactivity. As in many other areas, this brain function and that behavioural trait do not simply make a contribution to fashioning the dialogue between the individual and his environment; they also, in turn, undergo the structuring influence of the specific modalities of that dialogue, above all in the course of the early stages of ontogenesis.

Neurobiology of 'affiliative' behaviour

A 'positive' attitude to others, which is based on the faculty of attachment and the feeling of belonging, constitutes the best brake on the use of aggression as a means of expression and action.[119] The neurobiology of the social emotions and 'affiliative' behaviour is still in its infancy, probably because of the 'prevailing scientific aversion to topics that deal with emotions, processes that are generally considered resistant to coherent inquiry'.[120] But the data already available are sufficient to show that the brain structures which play a vital role in the control of affiliative behaviour (namely the amygdala, the anterior temporal cortex, and the orbito-frontal cortex, in primates[121]) are also those involved in the processes determining aggressive behaviour. This is not surprising in any case, if we remember that these structures are involved in a primary way with the processing of information which has affective significance for the individual. We must recall in this regard that many affective significances are acquired very early in life, and that the 'social play' of the juvenile period contributes very extensively to fashioning the social relations and social roles of the adult individual. Now, as we indicated earlier, an early lesion of the amygdala reduces the frequency and duration of social play, which has the consequence of disturbing the structuring and 'personalizing' influence this play normally exerts over behaviour undergoing development. It is highly possible that a similar

---

[119] See Chapter 7: A powerful restraint: respect for the dignity of others, p. 179.
[120] J. Panksepp *et al.* 1985.   [121] H. D. Steklis and A. Kling 1985.

sequence of pathological processes is responsible for certain major disturbances of affectivity in man. We may, in fact, feel that a primary brain disorder has a more or less profoundly disturbing effect on interactions with parents and peers, and that this disturbance to early social experiences has a secondary effect on certain modalities of brain functioning.[122] A kind of 'vicious circle' is thus set up which might be broken by improved understanding of the interactions between the brain and behaviour in the course of their common maturation.

In the development of the affective state of 'social comfort' and the affiliative behaviour aimed at creating it, the opiate systems—which release endogenous morphines—play a very important role.[123] These systems are very active in the course of social play, and it may be considered that social attachments are formed, at least in part, by the fact that body contacts cause the release of opiate substances which contribute to creating the state of social comfort. The opiate systems seem also to be particularly concerned in the female during the periods of gestation and lactation; more generally, their involvement seems to be important during essential stages of socialization, such as puberty. In all cases of 'discomfort' or 'distress' due to a separation, the administration of an agonist of the opiate receptors reestablishes the affective state which is normally that of social comfort. If there is therefore no doubt that the opiate systems contribute to the development of social attachments and socio-affective exchanges, it is also important to recall that these attachments and these exchanges in turn affect the maturation of the receptors which bind the endogenous morphines.

[122] See G. W. Kraemer 1985.   [123] See J. Panksepp *et al.* 1985.

9   By way of conclusion: what can be done?

Since 'aggressiveness', supposedly an integral part of our biological heritage and of 'human nature', is often referred to in order to explain or justify certain types of behaviour, the main conclusion that has gradually become clear and confirmed throughout the preceding chapters should be emphasized from the start: this type of 'aggressiveness' explains nothing and justifies nothing. For if the idea of aggression is useful insofar as it enables what is taken as this means of expression and action to be described and defined by a generic term, it is abusive and misleading to reduce the extremely diverse objectives and underlying motivations to mere aggressiveness resulting from spontaneous endogenous generation. This also means that biological 'fate' should not be held responsible for human baseness, nor for the hatred that humans know so well how to sow and nurture before harvesting, some day or other, its bitter fruits. At the level of development reached by the human species, making use of aggression is no longer ineluctably determined in advance unless men refuse to assume, individually and collectively, the full extent of their responsibilities.

Thus whilst it appears that man is 'naked' in the face of an overall situation which gives cause for concern and falls within his own responsibilities, the question formulated above (what can be done?) must be asked with particular acuity. In order to draw up the outlines of an answer and include a few more concrete elements, we must go back to the subject matter given in Chapters 6 and 7 in particular: what are the main factors that contribute to determining the probability of aggression, and how can behaviour be changed (more exactly, how can the probability of aggression be reduced) by acting on one factor or another?

Several complementary analyses have been given in previous chapters for the purpose of informing the reader in an objective and critical way. 'Noticing' qualities of language were used to ascertain, describe, analyse, and compare phenomena and processes. Not that the 'performative' qualities of the discourse were completely forgotten, for presenting the facts with a view to frustrating certain myths which constantly recur in given areas, is certainly an intentional 'act of language'. But matters change significantly if after the analysis and for the purpose of answering the question it raises (what is to be done?), I am confronted with the necessary complementarity of the 'prediction' based on scientific positivity, and the 'prophetism' based on adherence to certain moral requirements. For 'prediction without prophetism remains empty and vain; it is a mere play of the mind. Prophetism without prediction is likely to be but illusion and utopia.'[1] Bourdieu (1985) warns us 'both against irresponsible voluntarism and fatalistic scientism', and advocates 'rational utopism' which according to him would be 'capable of making use of knowledge of the probable in order to make the possible happen'. Prophetism, which is not only a sign of a certain voluntarism, but 'is considered to inspire direction',[2] also designates the 'preferable' amongst the possibilities which could be made to happen. Given these circumstances, my discourse is not just expression of the knowledge which I share with many others within the scientific community, but the reflection of a personal commitment whose aims transcend this knowledge.

Since the concern to meet certain moral requirements and inspire a certain orientation is combined with a definition of possibilities based on an objective analysis of the facts, I must, for reasons both of clarity and honesty, 'state my case'. Obviously my approach will only make sense if I maintain my faith in human beings and in their freedom, and if I consider that 'humanity can only take possession of itself through the conscious enterprise of men of conscience freely pursuing their own aims'.[3] In its choice of aims and the means it employs, a human community must endeavour to become 'a people whose ideas, projects and actions are based on justice, a people which is constantly striving to achieve a more authentic community of persons, where each individual feels himself to be accepted, respected and valued'.[4] In the area which concerns us, respect for the dignity of others plays an essential part. But this attitude is based on our concern for our own dignity, stemming from the concept of the infinite within us. Emmanuel Levinas (1985) is therefore right in saying that this idea of the infinite in us 'takes shape in my relationship with other men, in the sociability that is my responsibility towards my fellow-men'.

I have purposely called on several very different 'witnesses' in order to draw up a general direction. For it appears that a diversity of messengers in no way prevents the message from being a cohesive one, and it is this convergence which gives it its full meaning and strength. But then why is it not universally understood and put into practice? Because adherence to such a message is far from always being sincere: there is no problem as long as the sense of responsibility and the efforts . . . of others are called upon; but when it is necessary to assume all one's own responsibilities and make all the required efforts oneself, there is often no one there! And it is this more than the very real complexity of the problems posed, which is likely to make a good number of objectively 'possible' solutions 'utopian'.

[1] G. Rocher 1968.
[2] G. Rocher 1968.
[3] A. Gorz 1985.
[4] John-Paul II, quoted in *Le Monde*, 27th December 1985.

In concrete terms, what should be done about aggressiveness?

If we now examine from a more concrete angle the means likely to be used to fight against the development of individual and collective aggression, it is worth recalling briefly that the likelihood that aggression will take place in a given situation can be changed in two ways: either by acting on the situation itself, in order to modify its objective characteristics; or by acting on the individual (or collective) relationship to an objectively unchanged situation.[5] In practice, the following courses of action can be pinpointed: carrying out social changes, making use of procedures to structure or restructure the personality, interventions on the brain.

Direct interventions on the brain

A breakdown of justificatory reasoning

It is always very tempting to give priority among these methods, to the last one, for two reasons. On the one hand, neurobiology is considered to provide a relatively simple and efficient course of action, which as we will see, is quite illusory. On the other hand, this school of thought follows an old and very widespread ideology with a pessimistic and conservative vision of human beings and human society. Social Darwinism, supported by sociobiology, states that our aggressive behaviour is the result of natural selection and that it ensures the survival and reproduction of the 'fittest'. If within human society 'brother often turns on brother', this must correspond to a 'law of nature'; and it would be futile, even dangerous for the survival of the species to try to change this 'natural order'. Emphasis is always placed on a kind of innate perversity in man. There is no question of affection, friendship, devotion . . ., for it would be difficult to explain what need would be satisfied by developing through the same natural selection two perfectly paradoxical attitudes. People may say that positive attitudes are not necessarily due to natural selection; but if that is the case, it is difficult to see why aggressive attitudes should necessarily be the result of such a selection and therefore, the reflection of a 'law of nature'.[6] In reality, it is important, once again, not to confuse behaviour as such, which all individuals may use as a means of expression and action, with a so-called 'instinct' that would necessarily lead to its effective use. Let us take the example of that marvellous instrument, the human hand. There is little doubt that it was the evolution of genetic inheritance, under the pressure of a number of factors connected with environmental interchange, which made possible the development of the most varied and subtle gestures. But it

[5] See Chapter 6.   [6] See S. A. Barnett 1981, 1983.

is certainly not our genes which determine directly and ineluctably that this hand should kill, wound, destroy, and curse, or on the contrary, fondle, heal, create, and bless!

The idea of the 'born criminal' is misleading

In the field of criminology, the innatist and pessimistic idea of social Darwinism meets its match in the Italian 'positivist' school of criminal anthropology, established by Lombroso.[7] It is clear that the essentially innatist approach of 'moral madness' said to afflict the 'born criminal', described by Lombroso 'excludes any serious re-educational undertaking, just as it considerably reduces the 'sociological' concerns which might prevail when determining the causes of criminal behaviour'.[8] The 'crime chromosome' (double Y sexual chromosome) is the latest development to date in this innatist concept; we have seen what should be concluded in this respect.[9]

Beware of genetic engineering!

In a way, 'genetic engineering' is likely to reinforce sociobiology and the related innatist concepts of social Darwinism and criminal anthropology. It is true that one may be 'alarmed by a political school of thought that treats a human being like a machine exclusively programmed by its DNA, completely rejecting the influence of the outside world on to the expression of the genome'.[10] Was this remark made by an ideologist eager to condemn the dangers of an 'assumption of power' by biologists? Certainly not, since it was lucidly formulated by the biologist in charge of a 'Research group on normal and pathological human genetics'. In the case of certain hereditary diseases ('errors' in the metabolism, immunitary deficiencies) which turn out to be monogenic, that is, due to the breakdown of a single gene, 'gene therapy' can certainly be envisaged, consisting of transferring a normal gene to the afflicted organism, even if the problems of inserting the grafted gene in the genome of the receiver and controlling its expression have by no means been solved.[11] But problems of a very different kind are raised by the suggestion that genetic engineering methods could be used to eradicate from 'human nature' its 'more harmful and dangerous' side, that is, 'the human instinct of aggression and self-destruction'.[12] We have seen in the previous chapters that the likelihood of aggression being used as a means of expression and action was in no way determined by the individual level

---

[7] See Chapter 7: Factors linked to the personality of the subject, p. 154.
[8] P. Tort 1985.
[9] See Chapter 7: In man: the polemic concerning the 'crime chromosome', p. 144.
[10] J.-C. Kaplan 1983.
[11] See W. F. Anderson 1986.
[12] J. Glover 1984.

of innate 'aggressiveness', by an impulse given by an inherited 'instinct', but that it depended on a multitude of factors and mechanisms. In such circumstances there is no question of identifying the gene (or genes) of an 'aggression instinct' in the hope of subsequently succeeding in 'preventing them from doing harm'. If one considers that the various factors and mechanisms involved in the genesis of aggression do not act independently of one another, and that they are themselves the result of complex interactions between genome and experience, it is inconceivable that any intervention on the genes could foreseeably change the way in which the brain assesses a situation and/or the way in which it responds. Should we therefore deplore the complex, indirect, and changing character of the relationship between genome and behaviour because it deprives us of the possibility of intervening by means of genetic engineering? Certainly not, for two reasons. On the one hand, such irreversible alteration of a human being would in any case be an unacceptable breach of his inalienable dignity. On the other hand, and more importantly, our freedom lies precisely in the number of factors and the complexity of their interactions, in the indirect and changing character of the determinations.

The unwarrantableness of psychosurgery

What about psychosurgery? We have already seen why it is inconceivable that it can change behaviour in a strictly defined and determined sense, and in a foreseeable and regular manner.[13] But in the case of aggressive behaviour, the problems raised are not only of a scientific and technical kind but also—and in most debates, principally—of an ideological and ethical nature. Is it justifiable to entrust the neurosurgeon, who is supposed to work in the clearly defined interests of the individual, with the task of solving difficulties of 'adaptation' which broadly speaking, are the problem of society? Is it admissible to breach the most intimate and valued aspect of the integrity of a human being, in the hope of 'curing' society of the violence afflicting it? The answer to these questions will be decidedly negative.

Even if psychosurgery, thanks to the highly selective destruction of a specialized 'centre', were able to rid a human being of any intention to act in an aggressive way or to deprive him of the ability to actually carry out such an intention, only a dictator or a totalitarian regime could dream of establishing a 'consensus' in such a way. But since the brain does not contain any actual aggressiveness-generating centre, the probability that aggressive behaviour will be used can only be reduced by impairing the individual's socio-affective relations in a far more general way. When the first experiments carried out on animals showed that bilateral lesions of the amygdala had the effect of 'taming' the wildest

---

[13] See Chapter 6: Historical outline of psychosurgery, p. 131.

animals, attention was inevitably focused on this spectacular attenuation of aggressiveness. When on Christmas Eve 1954 I carried out my first amygdalectomies on wild rats (which are normally handled with great circumspection, even if one is wearing thick gloves), I was able to keep the operated animals at home over the holiday and let my young sons play with them, run after them and pull their tails, without the slightest fear that they would be bitten. When studying the effects of a brain lesion on aggressiveness, we are not too concerned at first with the onset of 'side' effects. Only gradually did the experiments on animals show that lesions to the amygdala, as to the cingular cortex, caused a far more general alteration in the genesis and expression of emotions, producing in primates considerable changes in personality and social behaviour. This more overall alteration in affectivity in fact did not escape the attention of those in favour of psychosurgical operations on the human amygdala. In this respect I would like to bring up another recollection which remains imprinted in my memory. At a scientific meeting devoted to 'the neurobiology of the amygdala',[14] Narabayashi presented the results he had obtained in 'hyperkinetic' children,[15] showing on the projected slide the phrase 'more obedient' (the children had become more obedient). Interrupted by an American colleague who asked whether he believed that in the case of a child treated in this way the essential point was really whether he had become more obedient, Narabayashi answered that he had talked to the fathers of these children and that they were very pleased with the results achieved; but he added with great honesty that the mothers had also been to see him, often in secret, and that they had in effect said to him: 'Doctor, if only you could give me back my child as he was before!' These mothers were in fact suffering from the affective indifference which the child showed towards them, whereas they had very warm affective relationships with him before the operation.

In their criticism of the use of amygdaloid and cingular lesions in 'violent' human subjects, Carroll and O'Callaghan (1981) rightly emphasize that psychosurgery has experienced, and is continuing to experience two major weaknesses on a purely scientific and technical level. On the one hand, these operations were not based on solid and indisputable scientific facts. One has not always proven able to resist the temptation to extrapolate experimental data from animals to humans hastily and prematurely, without paying much attention as to whether actual homologies both as regards behavioural processes and brain functions existed or not. On the other hand, the methodological strictness required in assessing post-operative results (the precise nature of the observed 'improvement' and the exact extent of undesirable 'side' effects and possibly even more serious risks) seems to be absent from

---

[14] This meeting was held in 1971 at Bar Harbor, in the United States.
[15] See Chapter 6: Historical outline of psychosurgery, p. 131.

most of the reports published. Considering that moreover, the most stringent ethical reservations have been expressed from several quarters, it should come as no surprise that the 'militant' trends in psychosurgery belong to the past. I have already mentioned that Mark and Ervin, who advocated the rather widespread use of amygdaloid lesions with a view to preventing the repetition of violent acts, had gradually toned down their position, admitting that violent behaviour was not necessarily due to anomalies of a biological nature.[16] Delgado, who for his part has always taken a lively interest in the development of sophisticated techniques to make it possible to act on the human brain and ensure 'physical control over the mind', whom Rose and his collaborators (1984) consider to be the 'chief science fiction visionary' of the seventies, distanced himself somewhat from his quasi-prophetic activism when he admitted that 'human beings are born with the capacity to learn aggressive behaviour, but not with established patterns of violence' and that 'prevention of human hostilities is not related to the organisation of motor performance, but to the neurological traces of hates and ideological conflicts which are triggers for the harmful use of established patterns of behaviour'.[17]

'Anti-aggressive' effects of psychotropic substances

Making use of psychopharmacology involves acting on one or several modes of neurotransmission operating within the brain. As has been emphasized several times, many intracerebral processes involving the different neurotransmitters and neuromodulators help to determine the way in which the brain perceives and interprets situations, by association with pleasant or painful emotions, as well as the way in which it selects and uses appropriate behavioural strategies. Given these circumstances it is not surprising that the substances which help to trigger aggressive behaviour,[18] like those which reduce the probability of such behaviour,[19] are characterized by the great diversity of their 'targets' and modes of action. When barbiturates or neuroleptics are administered in a psychiatric hospital to patients who are—at times—particularly agitated and aggressive, the attenuation of their aggressiveness is due to the sedative properties of the substance administered, that is, an inhibiting action exerted on all the psychomotor functions. As for the various drugs used in cases of more 'chronic' aggressiveness, none acts directly and selectively on aggressive behaviour. Their 'anti-aggressive' effects, ascertained by rule-of-thumb, are only one facet of more general repercussions on the subject's affectivity or on his more or less impulsive way of responding to certain inducements in the environment. It should also be stressed that the behavioural effects brought on by a given

[16] See V. H. Mark and W. A. Carnahan 1980.   [17] J. M. R. Delgado 1981.
[18] See W. B. Essman 1981.   [19] See A. Cools 1981a; T. M. Itil 1981.

pharmacological agent can vary widely from one case to another, depending both on the subject's personality and his predominant type of behaviour, as well as on the more general context in which the psychopharmacological treatment takes place.

Manipulation of the level of circulating sex hormones may be considered as a special case of intervention by psychopharmacological means. This manipulation, which aims to reduce androgenic impregnation (by male sex hormones) of the organism, particularly of the brain, was used not only to try to prevent a relapse in the case of sexual aggression, but more generally, for the purpose of attenuating violent behaviour, especially in certain prisons.[20] The most radical method, which is obviously irreversible, is castration; certain studies published in the fifties report on several hundred castrated subjects. Newly invented 'anti-androgen' substances were subsequently used, which lower the level of androgen hormone secretion or prevent the latter from acting normally at receptor level. The administration of oestrogens (ovarian hormones) was abandoned due to the undesirable 'side effects' they were liable to cause (gynaecomasty, that is, development of breasts; water retention with headaches; thrombophlebitis; aggravation of certain types of epilepsy). Overall, these attempts to 'treat' violent subjects by manipulating their level of circulating sex hormones led to rather 'pessimistic' conclusions and comments: not only were the risks incurred by the subjects thus treated far from negligible, but the actual efficacy of the treatment applied is often left in doubt.

Before taking action, one sure criterion: whether it is reversible or irreversible

Since all these physical interventions on the human brain are the subject of lively debate and I have my own opinion on the question, it is not easy to take sides objectively and dispassionately. Nevertheless, a twofold distinction should be made. Reservations or even definite opposition are particularly justified in the case of an intervention that constitutes an *irreversible mutilation*, such as castration or a psychosurgical intervention (if the latter has only a transitory effect, it is useless anyway). On the other hand, the *reversible* effects of certain drugs (antidepressants, anxiolytics, sedatives) can usefully be combined with—and expedite—the beneficial effects of appropriate psychotherapy; of course they should not be abused, especially as we still know little about the possible irreversible effects which might be caused by repeated and prolonged absorption of these drugs. A second distinction should be made between aggressive fits which form part of a much more complex clinical picture (psychosis, brain tumour, epileptic state) and the far more frequent aggression perpetrated by individuals whom there is no

---

[20] See T. Whitehead 1981; P. F. Brain 1984.

reason to consider as ill. In the first case, a pathological process must be treated, the nature of which has been ascertained as appropriately as possible, and it can reasonably be hoped that the aggressive fits will disappear, together with other symptoms, if the treatment turns out to be effective. In the second case, the source of the aggressive behaviour and the problems it poses for society, are 'medicalized' to an abusive extent, and biology and medicine are expected to play a role that they are not suited for. In this respect I share the opinion of Paul Brain (1984), a biologist like myself, who believes that exaggerating the role played by biological determinants in the genesis of human aggression and the consequent medicalization of the measures to be taken 'distort the truth, raise false expectations in many quarters and create fears that are not easy to resolve'.

The individual and society: which should adapt to the other?

Although the nature of things is infinitely complex, two opposite but equally simplistic viewpoints are usually categorically supported: some people believe the individual should be helped to adapt to the 'natural' constraints of social organization; for others, this organization should be changed so as to respond fully to individual 'legitimate' needs and aspirations. The choice between these two viewpoints and the concrete proposals resulting from them is not of a scientific, but of an ideological nature. This became quite clear to me for the first time when I was invited to take part in the United States in discussions among 'experts' which led to the drawing up of a report in October 1973 at the request and on behalf of the American Government (Department of Health, Education and Welfare). We were asked to do two things: firstly, to make an inventory and a coherent synthesis of scientific knowledge concerning the 'biology of violence'; and then to formulate 'recommendations' (which the American Government naturally reserved the right to adopt or reject). Despite certain differences in interpretation, discussions on the state of affairs, in this particular area, took place in a calm and courteous atmosphere. But the air became charged with electricity when it became necessary to voice an opinion on the relevance of different 'treatments' to be applied to the brain of violent subjects. The widespread use of psychosurgery for really hardened criminals, even the systematic addition of an 'anti-aggressive' substance to the drinking water distributed to the country's citizens, were advocated by some. The others, including myself, believed that the integrity of the human brain had to be respected and therefore any intervention which might have 'depersonalizing' effects should be rejected, hence the need to reflect carefully on all other methods and means which would make it possible to work effectively towards reducing the frequency of violent behaviour. The final report, in its conclusion concerning the different medical procedures for treating violence, states that 'the scientific and medical literature available at this time is inconclusive in regard to the efficacy of these procedures'.

Honesty compels me to add that it is not sufficient to have a clear conscience and to award oneself a medal for 'right spirit' in order to solve these problems, if one is amongst those responsible for dealing with them and if possible, for solving them. Even if one advocates 'social change' (which is easier to proclaim than to actually carry out), it would be unrealistic to think that one could make things happen fast enough to make it possible to rule out in the very near future any measures likely to alter the human personality. In effect, depending on the level of 'responsibility' attributed to a violent criminal, article 64 of the Penal Code can send him to prison or to a psychiatric hospital, possibly for a long period. Now, as Ebling (1981) very rightly emphasizes in his 'Ethical considerations in the control of human aggression', doesn't such confinement also bring about 'depersonalization' to a greater or a lesser extent? Sensational declarations are always out of place here; on the contrary, a great deal of lucidity and humility are needed, and these, as is known, are the most difficult virtues to put into practice.

Restructuring the personality or encouraging social change?

Whether it is a question of analysing those factors liable to affect the likelihood of aggression or of intervening tangibly in any one of them in order to reduce such a probability, it would be artificial to envisage separately the determinants linked to the psychological structures of the personality and those linked to the family circle and the sociocultural context. It should be stressed yet again that using aggression as a means of expression and action 'reveals' a personality shaped by experience, an individual and historically established way of apprehending situations and events and facing up to them. And it is quite clear that the social environment provides both the crucible within which the personality is structured and numerous references on the basis of which the subject interprets the situations he encounters and chooses the strategies which appear appropriate to him. Before successively considering measures for structuring or restructuring the personality and those which endeavour to promote social change, it is therefore necessary to demonstrate their close interdependence by making a few brief preliminary remarks.

Me equals us

There are many reasons which lead us to believe that every 'me' is an 'us', that 'collective identity is immanent in rather than transcending the individual identity', and that 'the more he (the individual) learns about the single us or the several us to which he belongs, the more he appropriates the network of bonds which makes him unique'.[21] And social cohesion requires that there be a collective identity and conscious-ness whose values, standards, models, aspirations, and projects stem from a community's background and continue to evolve in time. As the personality develops the 'social Me' is sustained by the elements of the collective consciousness; the individual internalizes them before they get expressed in 'role playing' and in 'role expectation' (by others). But aside from absolutely positive values, models, and projects, the collec-tive consciousness also conveys non-values, false values, disputes, even certain lasting hatreds. In an era characterized by 'the passiveness with which an individual endures anonymity, levelling out and uniformity', in which he 'willingly exchanges his freedom for comfort' while taking part in 'a collective consensus for social regimentation',[22] there is the danger that his personality may be reduced to a function and that an individual sense of responsibility dissolves into a kind of collective irresponsibility. This is why it is more necessary than ever for an indi-vidual to proceed in a resolutely 'personalist' way and to distance him-self from the internalized contents of the collective consciousness in order to be able to view them from a critical standpoint and pass lucid and personal judgement on them. In other words, the personal identity and consciousness, while continuing to draw strength from them, should transcend the collective identity and consciousness, so that an autonomous 'I' develops and asserts itself towards the 'Social Me' and the 'roles' it assumes in existential reality. For this 'I', which is fully responsible for its choices and hence, truly free, alone can contribute usefully to creating a more humane situation, with greater dignity, responsibility, and freedom.

When studying the genesis of the motivations underlying our be-haviour, I emphasized the enriching 'interactionist' viewpoint according to which these motivations are 'customary types of interaction built up by the individual with his environment', and which considers that 'the subject taking the action and the world of action only exist by virtue of one another'.[23] The attitudes and behaviour of the individual therefore cannot be separated from the social environment and the sociocultural system within which they develop and are expressed. In cases of 'ordin-ary' violence which is very frequent and does not fall within the prov-ince of the law, studies show clearly the multideterminism of aggress-ive behaviour and the complexity of the interactions which bring into play the personality of the 'actors' (aggressors and victims) at the same time as a number of social representations, standards and models. This is why efficient prevention and control of violence requires that appropriate action be taken simultaneously at various levels and by very diverse means, whether in the case of violence by parents towards

---

[21] P. Fougeyrollas 1985.   [22] L. E. Pettiti 1986.
[23] See Chapter 4: What determines behaviour?, p. 57.

children,[24] violence affecting relationships in married couples,[25] or violence which disrupts school life.[26]

The theory of stigmatization: An illustration of the 'interactionist' viewpoint

In the field of criminology, interactionist views have likewise led to a shift in the interest traditionally shown in the 'delinquent' personality structure in favour of a series of dynamic processes of social interaction. In this connection we should mention the special form of interactionist view known as the theory of 'stigmatization' or 'labelling'.[27] This theory postulates that individuals who are rightly or wrongly suspected of having committed acts that violate the rules of social behaviour, are 'stigmatized' due to the reactions that these acts, real or imagined, draw forth from a set of people, groups, and institutions. As a result of these stigmatizing processes the individuals acquire the social status of 'deviants' and reorganize their personality around the social role thus attributed to them, which has the effect of establishing them even more firmly in their 'deviancy'. This being so, 'it is not deviancy that leads to social control but it is the social control itself that leads to deviancy'.[28] It should come as no surprise that in its most radical form, this theory advocates that repression should be widely suppressed. Neither is it surprising that such an extreme standpoint should be counterbalanced by a diametrically opposite position which states that 'the problem is not one of brushing off on society the responsibility for deviancy, but of protecting it against deviancy'.[29] Needless to say, by repudiating individual responsibility or by contesting collective responsibility, any real chance is lost of ever successfully remedying a situation in the genesis of which both kinds of responsibility are indissolubly linked.

Of course, aggression is only one means of expression or action amongst others, and we are usually free to resort to it or, on the contrary, to renounce its use. So if our personal responsibility is involved, society for its part creates situations and exerts influences that are far from being 'innocent' ones. Without discussing here potentially aggression-causing situations (to which I will return further on), suffice it to emphasize the twofold influence that a society exerts on the development and expression of the 'moral sense' of its members in this particular area: a society may or may not encourage the achievement of full cognitive and affective maturity, making possible personal moral judgements which go beyond opportunism and conventions and bring

---

[24] H. R. Keller and D. Erne 1983.
[25] D. Goldstein 1983.
[26] B. Harootunian and S. J. Apter 1983.
[27] See R. Gassin 1979; G. Levasseur 1979; M. Cusson 1983*a*.
[28] Lemert, quoted by R. Gassin 1979.
[29] Ch. Debbasch 1979.

about an awareness of and consideration for the dignity of others, thus lessening the likelihood of aggression; through the media, which are the reflection and one of the mainsprings of its cultural development, society may or may not render commonplace or even legitimize the use of aggression as an appropriate behavioural strategy. It is therefore not by chance but in a perfectly lucid and relevant way that the Study Committee for Violence, Criminality and Delinquency wrote in its general report: 'Should not prevention consist of first endeavouring to eliminate the roots of the evil? To a certain degree, the entire organization of society is therefore, or should be, preventive: the family, school, job, public equipments, legal regulations . . .'[30]

Further comment should be made concerning the idea of 'stigmatization', which is just one particular case in the 'categorization' process universally used to structure the social environment and which as I have already stressed, may in certain respects have a depersonalizing and dehumanizing effect.[31] In society generally, certain groups of deviants or delinquents are considered to be 'inferior' by virtue of the surrounding stereotypes; this devalorization is internalized, leading these groups to devalorize themselves and become rooted 'in a depreciatory social identity and generalised negativeness'.[32] Young people are particularly sensitive to any form of rejection, and it is therefore important to prevent them from suffering such reactions which are very detrimental to their social reintegration. This is why 'one of the primary concerns of children's judges, almost from their inception, has been not to stigmatise a delinquent or deviant minor who is at moral risk and for whom educational assistance is indicated.'[33] In other cases, the dangers of stigmatization are not necessarily clearly apparent at the outset. Keller and Erne (1983) report that with a view to preventing the development of aggressive parental behaviour, American teams have observed the behaviour of mothers during childbirth and during the first feeds in order to discern which ones appeared to have difficulty in bonding with their newborn baby. Only gradually did they realize that precautions should be taken to prevent these women from feeling stigmatized, even potentially, as 'bad mothers', with affective consequences that would not make a preventive procedure any more successful. When taking measures to prevent ill-treatment being inflicted on the child, it should also be remembered that 'prevention which respects the individual should avoid any risk of systematic controlling and categorizing, surrounding the parents with an atmosphere of suspicion which hinders any possibility of progress'.[34]

[30] See A. Peyrefitte 1977.
[31] See Chapter 4: Social and personal identity, p. 72.
[32] H. Touzard 1979.
[33] G. Levasseur 1979.
[34] M. Rouyer and M. Drouet 1986.

Should people or actions be condemned?

To close these preliminary reflections, a straight question should be asked which concerns us all: why the hurry to discredit a certain cat-egory of *people* (deviants, delinquents, criminals . . .) by stigmatizing them, rather than loudly censuring this or that type of *action*? Is it not because it is easy for us to persuade ourselves that we certainly do not belong to any of these categories (I am not one of 'them'), even though we may simply be clever enough not to get caught and thus avoid any stigmatization; whereas it might be more difficult for us to persuade ourselves that we are absolutely incapable of committing such acts and being really safe from the reprobation they would arouse? Furthermore, René Girard (1982) showed very clearly that in times of crisis (and especially cultural and moral crises) people are less likely to question their own mistakes. On the contrary, they need 'scapegoats', 'other individuals who seem to them particularly noxious for reasons that are easy to grasp'; and 'the further one goes from the commonest social status one way or the other, the greater the risks of persecution'. In light of the eclipse of values, the disintegration of the moral sense and human relations, is there really no more constructive attitude than this 'desper-ate wish to deny the evidence' which leaves the way open for the search for a 'scapegoat'?

The use of psychotherapy

We saw in Chapter 6 that different types of psychotherapy could be used, certain of which have a more or less direct effect on the malad-justed behaviour itself, whereas others aim to act on the personality of the subject, on his mental functioning as more globally dealt with. Even if this distinction is somewhat artificial, it should be maintained, since there are effectively differences of a theoretical nature, and therapists (or educators) regularly call on one or the other of these 'schools'.

Efficacy of behaviour therapies

'Behaviour therapies' endeavour to remedy interpersonal difficulties by leading subjects to acquire and use certain 'social skills'. Of course, as Boisvert and Beaudry rightly point out (1984), it is not easy to define the idea of 'social competence', or the amount of skills required to achieve this fully. But in more concrete terms, a set of strategies can be taught and learned in order to make interpersonal communication more efficient, prevent conflicts or help to solve them, thus rendering inter-personal relations more satisfying. To the extent that behaviour therapies use techniques based on principles drawn from theories of conditioning and learning (with the distribution of 'rewards' and 'punishments'), they may act directly on the aggressive behaviour itself. Arnold Goldstein describes in a detailed and critical way the different techniques used for this purpose.[35] He clearly shows preference for the deprivation of a reward (material or socio-affective gratification) following aggression, rather than the administration of punishment (which in his opinion should only be a verbal rebuke and not corporal punishment). He believes that punishment only puts an end to aggressive behaviour temporarily, may have negative 'side effects' and above all does not teach the subject other means of expression and action. In this respect he emphasizes strongly that any intervention aimed at reducing the likelihood of aggressive behaviour should be combined with the positive reinforcement of 'prosocial' behaviour liable to replace undesirable behaviour. He adds that positive reinforcement of the behaviour one would like to see develop is more efficient if a purely material reward is accompanied by gratification of a socio-affective kind (a word of approval or congratulations, a smile of thanks . . .). Rewards and punishments should be given immediately, systematically, and consistently so that the connection with the act in question is always clearly understood.

Instead of acting on the aggressive behaviour itself, behaviour therapy can also reduce the likelihood that such behaviour will take place by acting on a particular factor that helps to determine this probability.[36] Given the role which aversive emotions play (particularly through dissatisfaction, vexation, anger) it is important to learn how to control one's own emotions and to help others to recover their composure. In order to prevent conflicts, it is important to know how to communicate adequately and to be able to negotiate and enter into agreements. Learning how to master everyday situations brings greater self-confidence and a more optimistic and serene approach to relationships with others. Aggressive behaviour becomes less and less 'probable' as soon as other types of behaviour make it possible to enter into and develop totally satisfactory interpersonal relationships. Certain therapists or educators stress only the cognitive aspects of the social skills to be acquired. Others attach a great deal of importance to the 'empathy' which, in addition to cognitive ability, implies affective sensibility which enables the aspirations, viewpoints, and suffering of others to be better taken into consideration.[37]

Since one aims to remedy 'social maladjustment' and to develop skills leading to 'social competence', there arises the problem of whether these very 'directive' measures are legitimate. Charrier and Ellul (1985), committed to preventive action, are familiar with criticism to the effect that 'this preventive work merely consists of our present society recuperating those elements which are likely to challenge it.' But their aim is a

[35] A. P. Goldstein 1983*b*.
[36] See A. P. Goldstein 1983*b*; N. D. Feshbach 1984; I. G. Sarason and B. R. Sarason 1984.
[37] N. D. Feshbach 1984.

very different one, and they have made it clear that the purpose of prevention 'is not to adapt a young person to society, but to help him to form a strong enough personality to be able to find or make a meaning in life, and similarly, overcome his maladjustment'. Furthermore, most 'maladjusted young people' are unhappy, some of them live in a state of psychological misery and it is inconceivable that we should not help them on the grounds that an ideal society, where there are no more maladjusted people, will come about tomorrow. As for the methods to be used in 'prevention groups', Charrier and Ellul consider that while young people should be able to frequent these freely, non-directive pedagogy is completely wrong, because 'a completely non-directive attitude is extremely harmful and inefficient when dealing with maladjusted young people'. I myself have always been in favour of clearly directive education, and to those who express surprise I reply that a young person is no less free by being helped to acquire the means to attain complete freedom. With his newly gained backbone, having become an adult and a citizen in the fullest sense of the term, he will be able to challenge whatever seems to him questionable in a well-considered, responsible and free manner.

Even so, the problems raised, and the potential risks inherent in using techniques of 'control' and 'manipulation' of human behaviour should certainly not be played down. There may be the temptation, merely through the systematic use of rewards and punishments, to produce a kind of robot who would necessarily be 'well behaved', perfectly efficient, and totally submissive. As Nuttin (1980) said: 'One may be of the opinion that after all, it would be demeaning to a man to try to improve or even save him "without himself". Without himself means in this case: by using the mechanism of his elementary motivations which are spontaneously directed towards easy rewards.' This is why care should be taken to develop a moral judgement as soon as the cognitive functions allow, for 'human beings benefit from drawing up and assessing by themselves the pros and cons of the alternatives presented to them'. Rogers (1966) discusses, with understandable concern, the views of Skinner (who is among those who have studied in greatest depth the role played by the reinforcement process in conditioning) which, if they became real, would turn men and their behaviour into a perfectly planned product in a completely controlled society. Rogers quotes several very explicit passages from Skinner's writings, and it might be of interest to mention two of his quotes here: 'By means of a carefully thought out cultural plan, we do not control the final behaviour, but the *inclination* to behave: the reasons, the desires, the wishes. Strangely enough, in this case *the question of freedom never arises*.' Skinner states that because of the 'tremendous power of positive reinforcement' there would no longer be any constraint or revolt; what is more, 'the hypothesis that man is not free is essential for the application of a scientific method to the study of human behaviour. The inner man, who is free and held responsible for his behaviour [. . .] is only a pre-scientific substitute for the different sorts of causes that are discovered as the scientific analysis proceeds. All the different causes are *external* to the individual'. Skinner must be credited with the honesty of having clearly proclaimed, in the very title of one of his works (*Beyond freedom and dignity*) that his own perspective was aimed beyond the obsolete and useless ideas of freedom and dignity. It may thus be thought that he also accepts that people fundamentally disagree with him (without however underestimating in the least the 'tremendous power' of positive reinforcement!).

The education of the 'moral sense'

The acquisition of 'prosocial' behaviour through behaviour therapy does not necessarily guarantee that it will definitively replace aggressive behaviour which, in our society, is given repeated positive reinforcement. Moreover, we have just seen that it is not totally satisfactory to ensure that a subject behaves himself 'correctly' merely by using 'the mechanism of his elementary motivations'. Under these circumstances it might be better to act on the personality in greater depth by resorting to 'moral education', which should enable the subject to base his actions on a certain number of moral principles. Since the subject has to reach his own, autonomous moral judgement, and insofar as reaching this state of moral development requires sufficient cognitive and affective maturity, the educational process affects several aspects of the mental functioning and aims at avoiding the arrest of the process at an immature stage of development.[38] In the present context, that of social interaction, 'moral sense' basically corresponds to a 'sense of justice', that is an awareness and recognition of the equal dignity of all human beings, as well as the acceptance of true reciprocity in all human interactions, which implies full recognition of the rights of others and taking into consideration their aspirations and feelings. Is such moral education efficient? It certainly seems to be. Zimmerman (1983) quotes a study carried out by Kohlberg and Turiel showing the existence of a positive correlation between the maturity of moral judgement and the behaviour of subjects in an experimental situation which brought together an 'examiner' and a 'candidate'.[39] Amongst those who had achieved full maturity in moral judgement (as revealed in the 'Moral Judgement Interview'), 75 per cent refused, in their capacity as examiners, to administer electric shocks to the 'candidates', whereas only 13 per cent of those who were at less mature stages of development expressed the same refusal to inflict pain on their 'candidates'.

But the very idea of 'moral education' has given rise to twofold

[38] See D. Zimmerman 1983.
[39] See Chapter 7: From pain to aggression, p. 152.

criticism. Firstly, the *relative* nature of the values which such education endeavours to promote was put forward (values linked to a historical period, a place, a particular group . . .). In reality the moral principles mentioned briefly above have, or should have universal and timeless cogency. The same goes for the sincerity and loyalty that form the basis of confidence, facilitate mutual understanding and enable stable and gratifying interpersonal relationships to be established and flourish. Of course these are not absolute values that exist outside oneself and with which one could be endowed. It is our own responsibility to acquire, cultivate, and transmit them, and this is an essential aspect of our human dignity. Shotter (1980) notes in this respect the words of Hannah Arendt: 'Even if there is no truth, men can be truthful; even if there is no reliable certainty, men can be reliable.'

The second criticism concerns the risks of 'indoctrination'. But one cannot speak of indoctrination if the development of the powers of reasoning and judgement is combined with 'case studies' where problems of a moral nature are discussed frankly and openly.[40] Education should aim at encouraging personal self-development and achieving autonomous moral judgement. This is in fact the only way that 'corresponds to that risky mode of functioning known as the human being.'[41] What is the situation in our country? When the results of the '1986 vintage' of the *baccalauréat* were announced on the 'television news', a philosophy teacher expressed his feelings about an examination topic he had just corrected. The candidates had to deal with the following subject (I quote from memory): 'Can a moral problem be given a complete and definitive solution?' The teacher was surprised to note that all sorts of problems were discussed, none of which was strictly speaking a problem of a moral nature; and he wondered, without concealing his concern, whether the candidates were aware of the important role played by the 'moral sense'. There is no need to exaggerate or generalize the significance of this, but one may still wonder whether the clear absence of any 'indoctrination', as in this case, is really a matter for satisfaction. If one is opposed, and rightly so, to 'censure' and 'the moral order', doesn't such a standpoint often embody a certain hypocrisy? To what extent does this entail a search for facility, the refusal of efforts called for by a sincere acceptance of certain moral requirements, and perhaps also the confused notion that it would be difficult to give children or adolescents something that is sadly lacking in oneself?

Acquiring moral principles and adhering to them independently is one thing, putting them regularly into practice in everyday behaviour is another. For it can happen, in this area of aggressive behaviour[42] as well as in others,[43] that a more or less marked 'change of attitude' may take

[40] See D. Zimmerman 1983.   [41] J. Nuttin 1980.   [42] See D. Zimmerman 1983.
[43] See Chapter 4: Cognitive coherence of attitude and situational determinants of action, p. 71.

place when facing a concrete situation, compared to the basic attitude as expressed when forming a 'timeless' judgement. For the action of the moment, the sociocultural context may provide a number of 'situational' determinants; and the repercussion of the latter will be more or less marked depending on whether the subject has or has not learned to control himself, to take precautions against the inducements of the moment and to renounce immediate gratification. Here we find once again the complex interactions of the personality and the sociocultural context. Albert Memmi (1985) gives a fine analysis both of the 'fear of others' and the 'need for others', and he emphasizes the fact that our relationships with others may be 'imposed' (a thirst for power and dominance, the search for control over others in order to satisfy needs and desires) or on the contrary, 'negotiated' (negotiation for purveyance, for mutual satisfaction). He states that 'we will be released from barbarism when, having acknowledged our dependency, ceasing to be the predators and murderers of our fellow-men, we negotiate reciprocal purveyances'. But this is where, over and above the personality and the moral sense of both sides, the 'values' preached by the sociocultural context come into play. If, once the elementary needs of people have been satisfied, emphasis is placed on the joys of affection, the mind, art, nature, what could be easier than to negotiate ways of sharing which would only multiply them! But if society remains, or even becomes increasingly one of envy and ostentatious consumption (of goods which would only be reduced by sharing) there is little likelihood that solidarity will progress from discussion level to become a daily reality, and barbarism will continue to thrive!

The spectacle of violence: an illusory catharsis

The 'cartharsis' process has been considered by some as an efficient psychotherapeutic measure to free oneself in a harmless way from the 'aggressive energy' which according to them is inevitably accumulated. In an analysis of the concept of catharsis in relation to aggressive behaviour, Seymour Feshbach (1984) notes that A. A. Brill, the psychiatrist who introduced Freud's psychoanalytical method to the United States, recommended attending a boxing match once a month. Practising any 'harsh' sport was supposed to have the same effect. Those who today endeavour to justify showing scenes of violence, often to the point of nausea, in films and on television, also make out that these pictures enable the spectator to let off his 'excess aggressiveness'. Apart from the fact that there is no longer any basis for considering the idea of aggressive energy, a natural entity held to be the result of spontaneous endogenous generation, this catharsis often positively reinforces aggressive behaviour, that is, it increases the likelihood of its being set in motion. So a remedy for increasing violence should certainly not be sought in that direction.

Measures of social defence and social change

Turning now to interventions which seek to act on the determinants linked to the social environment, by carrying out changes of a social nature, it should be stressed straight away that these measures are of two kinds: on the one hand, the measures (of prevention and repression) which society takes for a precise purpose called 'social defence'; on the other, transformations of a more general nature which this society agrees or does not agree to impose on itself. The easy solution consists of entrusting the former to specialized institutions and considering that the problem of the latter . . . does not arise. In reality both kinds of measures depend closely on one another. I pointed out earlier on that the immediate social environment, itself an integral part of a wider sociocultural system, was the crucible within which the characteristic attitudes and behaviour of a personality were shaped, while at the same time providing situational determinants for the action of the moment. A further dimension should now be added to the scope and impact of the sociocultural context, since it is the latter which determines the nature, the legitimacy (as it is perceived), and the efficacy of concrete measures of social defence. Whether it is a question of social defence (against internal aggression) or national defence (against external aggression), it is important to be able to respond to the 'why' of this defence if the 'how' is to be worked out suitably and effectively. It is essential to know whether our society is 'worthy' of being defended and therefore the 'values' to be defended should be given careful thought. The criminologists who declared that 'if one attempts to build a criminal policy, agreement must be reached on the values to be protected and even on the hierarchy of these values'[44] or that 'these options of criminal policy thus depend on the development of the morals, ideas on and conditions of social life and of society's scale of values'[45] were evidently well aware of this.

Some basic values

Those values which should be defended because in everybody's view they deserve to be, constitute the indispensable basis of the legitimacy of the means to be used as well as the intention to use them. This means that values are required to which most of the members of the community will adhere and be committed (and defence of them will be perceived as legitimate) not only because they are deeply gratifying at the present time but also because they bring hope for the future (which creates the will to defend them). The problem that thus arises is that of the choice of these values, which should enable people to achieve

[44] G. Levasseur 1979.   [45] M. Ancel 1985.

fulfilment thanks to the simultaneous development of their own inner life and fraternal exchanges with others. Just as receptive as he is to the Evangelical message, I will willingly adopt the formula of Paul Ricoeur: 'Man certainly needs love, and justice even more, but above all he has need of a meaning'.[46] How I wish I could affirm that our 'developed' society, proud of its 'modernity', fully satisfies these essential needs! Insofar as love is concerned, there is certainly increasingly abundant literature and pictures which attempt to teach us how 'to make it'; but they usually encourage attitudes and behaviour that lead to others being used as sexual objects for consumption, which debases love instead of contributing towards its complete fulfilment. And does anybody teach young people — and the not so young — to 'experience' love by including sexuality in a relationship meant to be mutually enriching, with full respect for the sensibilities and dignity of the other? As far as justice is concerned, we are certainly lucky to live in a law-abiding country. But it is not enough to accumulate laws and regulations, they have to be adhered to and respected; not only through 'fear of the policeman', but because a fraternal and solidarity-showing community agrees to defend, over and above their differences of opinion which are inevitable but may become constructive, common values, shared projects, and prospects of a meaningful future. As for the search for a meaning, this is left to a few laggard dreamers! It is fashionable to declare with a knowing and vaguely disgusted air that nothing has any meaning, and that everything is 'rotten', that there is no future and so one might as well 'have a ball' here and now. Without realizing how painful it is for young people . . . and dangerous for their elders, to be deprived of a future and of hope. Under the section 'Young people write to us',[47] a 19-year-old girl from Avignon called upon us to: 'Give us back the strength to believe, not in a better world, but in one that is rich in solidarity and goodwill . . . For disappointed hopes will gradually give way to a devastating emotion which you can do nothing about: anger!' To paraphrase Hannah Arendt, quoted above, I would say that 'even if there is no meaning in the universe, man *can* give a meaning to his inner life and his commitment within a human community worthy of the name'.

People may object that it is easy (and 'elitist'?) to be concerned mainly with 'intangible' values such as love, justice, and meaning when one is oneself provided with whatever is necessary and even superfluous. In actual fact, I certainly do not fail to appreciate the importance of fair division of material goods, for there is no point in having a sense of justice that would be above and ignore these 'meanly material' considerations. But it is a well-known fact that in the absence of love, a moral sense and faith in the future, there is no chance of sharing fairly in a non-violent way. Now, we are dealing here with measures which, it is

---

[46] See P. Ricoeur 1986.   [47] *Le Monde*, 13th November 1984.

hoped, will contribute towards ridding ourselves of violence by finally rendering it perfectly useless.

It has become commonplace to say (but not yet to acknowledge in concrete terms and act in consequence) that the crisis our society is undergoing is not only of an economic nature but also, and principally of a cultural and moral nature. When, together with others, he was studying the 'issues of the end of the century', Augustin Girard (1986) stressed that 'mastering our future is a cultural matter' and that a common plan, without which there could be no future, 'is neither technical nor economic, but derives from the idea that man has of himself; of his human dignity'. And he is right to remind us of the warning given by Pierre Emmanuel: 'The worst disaster that could threaten a people is not military destruction, but indifference to the shape of its future.' This concern is shared by many people who question the future of our human communities, and Aurelio Peccei, founder of the Club of Rome, echoes this in a 'Cry of Alarm for the Twenty-first Century', expressing the opinion that 'man's best hope of having a happy future lies in an ethical revolution that will enable him to understand himself and to give a meaning to the world in which he lives'.[48]

The damaging effects of an ossified language

If there is agreement as to diagnosis and prognosis, surely the treatment should follow? Aside from the fact that this is a particularly arduous undertaking, since it is extremely complex, long, and exacting (and we hardly master any long-lasting action), two major obstacles, closely connected, block the way: language, a reflection of people's attitude of mind, and the way in which it is used, especially by those who, in positions of responsibility, ought to be able to help us make slow but sure progress in this direction. The language is both distorted and ossified, with platitudes and slogans used to express a simplistic and sterile Manicheism. However, the real problems 'are not black or white, for or against, but always in-between; this in-between is where a moral discussion takes place'.[49] And how could one fail to agree with Father Carrier (1986) when he declares: 'It is intolerable that "democracy", "fraternity", "love", "peace", "justice", "truth" have literally become double-edged weapons in the hands of enemy brothers!' Under these circumstances one can only deplore the fact that all too often in politics clichés are uttered (which of course delight the media but are otherwise of no interest) in the ping pong game of incantations and anathemas, systematically and *a priori* qualifying the reforms and laws envisioned by 'the other side' as 'wicked' or 'shameful'. Is this attitude really likely to sharpen people's critical spirit and refine their moral sense? Or to bring about, as many people more or less consciously wish, a 'new society' (as

---

[48] D. Ikeda and A. Peccei 1986.   [49] P. Ricoeur 1986.

announced by a 'rightist' Prime Minister) or a 'new citizenship' (as announced by a 'leftist' Prime Minister)?

If I have emphasized very general aspects (a common plan, future, culture, ethics, hope, will power) before contemplating more concrete measures which are necessarily more limited in scope, it is because the former, broadly speaking, condition the latter, making it possible or impossible for them to be used and more importantly, included in a coherent plan disposing of sufficient time and 'breath'. In the absence of such continuity and a plan to inspire and actuate it, we may still consider for a long time to come, like Fougeyrollas (1985), that 'economic, political and cultural life is strongly marked nowadays by resorting to palliatives'.

Concrete measures: education, prevention, repression

The concrete measures that could be envisaged are so many and so diverse that only a general outline will be given here. Nevertheless, three main areas or means of intervention may be distinguished somewhat arbitrarily. The first concerns education in its broadest sense, that is, all the influences exerted on a human being as he develops throughout his life, with the emphasis here on the social dimension of these influences and how they evolve. The second deals with preventing opportunities for violence, particularly those linked to urban lifestyles and the economic environment. The third area covers not so much methods of prevention as measures of repression and the specialized institutions that enforce them. In making this distinction which is only intended to more clearly outline a very involved subject, I may be giving the impression that there exists a natural or formal disparity between preventive action and measures of repression, whereas on the contrary, I am anxious to emphasize their close and necessary complementarity. The Study Committee on Violence, Criminality and Delinquency did so before me by declaring very clearly: 'Hence there is the conviction that crime prevention is just as important as punishing the criminal and that action against crime should include, in addition to repressive measures, and sometimes instead of them, measures of social prevention'.[50]

Educating the child: a fundamental step

Education starts within the family, continues at school, and is completed—if not replaced—by that provided, wittingly or not, by television. Ideally education should allow everyone to achieve and preserve full intellectual, affective, and moral maturity; to find by himself and for himself a meaning in life; and to create independently his own personal

---

[50] See Report of the working group 'Penal and penitentiary aspects' in A. Peyrefitte 1977.

ethic. It should therefore aim at preventing any fixation at an immature stage of development and in adults, any regression due to a process of 'infantilization'. Criminologists are well aware of the role played by immaturity, by 'that form of infantilism consisting of lack of foresight, impulsiveness and irresponsibility', and Cusson (1983a) quotes in this respect the words of Jean Genet: 'I do not know any hooligans who are not very childlike'. One could retort that there are all kinds of 'hooligans', some of whom are extremely intelligent. Yes, but they suffer from affective and moral infantilism. And it is precisely a well-balanced development, intellectual, affective, and moral alike, that education should endeavour to encourage.

I have stressed that right from the start a child has a vital need for affection and tenderness.[51] Affectionate, kindly, and reassuring parental attitudes will give him a warm and optimistic vision of the world and of others. He will not fail to learn later that this world also consists of spitefulness and ugliness; but it is infinitely more worthwhile to experience disappointment and lose a few illusions than not to have any from the start and to consider the world to be fundamentally hostile and evil. If they are given a positive outlook towards the world, with warm feelings, and guidelines and models provided by their educators (parents and teachers), children and adolescents learn to establish stable and satisfactory relationships, and to master their own freedom through respect for that of others. This naturally implies that parents and teachers should assume fully their role of educators and not 'withdraw'. Furthermore, it is highly desirable that trusting cooperation should grow up between parents and teachers in order to conceive and carry out an educational plan that is coherent and considerate of individual personalities. The beneficial effects of such cooperation are expressed by a clear decrease in violence in educational establishments. In the United States it is believed that 'the interaction (or lack thereof) between home and school may be an especially significant factor in school violence'.[52] In France a survey by *'Le Monde de l'Education'* (February 1986) states in its conclusions that in 'quiet' establishments the number of socio-educational activities is higher, the teachers more involved and 'there is a clear correlation between properly organized participation by parents and pupils, and the absence of violence'.

Of course the conditions of modern life do not make the task of educators any easier. But what is more worrying is that our 'modernity' or even 'post-modernity' seriously misjudges the basic needs of the child and the consequent obligations of those who felt it right to bring him into the world. When discussing problems of married couples, single-parent families, or even 'modern' procreation methods, it is always the 'rights' of each side which are debated and regrettably children

---

[51] See Chapter 7: The essential role of the family milieu, p. 162.
[52] B. Harootunian and S. J. Apter 1983.

and their future are the least important concern. And yet it is well known that deficient education within the family circle has serious consequences for the child: teachers know that the parents of 'difficult' children are often separated and those in charge of 'youth protection' acknowledge that the separation of the parents is a key factor in delinquency. We have no right to consider the child as an 'object' to be tossed about according to the whims we claim the right to satisfy. Aside from this refusal to take the child's needs into consideration and assume obligations towards him, it is unfortunately necessary to recall that cases of really ill-treated children are far more frequent than one would like to believe, and that ill-treatment has serious repercussions on the development of their socio-affective behaviour, in the form of marked aggressiveness or social 'withdrawal'.[53] The veil of secrecy surrounding child sexual abuse, which deeply traumatizes the child and leaves scars, must also be lifted.[54]

There is no need to 'moralize' or to declare (rather unrealistically) that everything must be 'sacrificed' for the child and his development, but one should be fully aware that the future of our human communities depends to a very large extent on the education they give their children. Short of everyone saying, 'after me the heavens can fall!' taking stock lucidly and frankly of all educational problems should be a top priority (which is far from being the case!). Since being a parent is the most important and most difficult of occupations, would it not be a good idea to help parents to perform it properly? Certain 'schools for parents' do a useful job in this respect, but their scope remains very limited. This is why the Study Committee on Violence, Criminality and Delinquency drew up the following recommendation: 'A certain amount of information could be given by qualified persons to parents or future parents, particularly through the intermediary of television, on certain educational factors, the child's development (the importance of early childhood and of the father–mother–children relationship), the ambivalence of adolescents (the desire for autonomy, independence, hence aggressiveness; but also the need to be protected and secure, hence the search for affection), and on mistakes that should not be made.'[55] Has this wise recommendation been heeded?

Reflections on television

Given the growing influence it exerts on the development of mentalities and behaviour, television occupies a privileged position that enables it to contribute in a large measure to the total development of a human being. The question is: does it really emphasize the bright aspects of life and the nobility of the world at least as much as the dark corners of the

[53] See H. R. Keller and D. Erne 1983.
[54] A. Miller 1986; M. Rouyer and M. Drouet 1986.
[55] See Report of the working group 'Protection of Youth', in A. Peyrefitte 1977.

human soul and the ugly aspects of the world; the positive values of love, justice, and hope at least as much as the 'values' linked to domination, contempt and despair? If this is not the case, it is because this 'tremendous potential tool of cultural democracy has become an uncontrolled machine of anti-cultural demagogy', 'insofar as culture is intelligence and dignity as opposed to the violence and magic of the spectacular'.[56] This severe pronouncement is not especially intended for television directors and producers, insofar as television is generally a reflection of a country's political and economic life. When the problems posed by audiovisual programmes are periodically and bitterly debated, it is claimed that political, economic, technological . . ., and cultural (quoted 'as a reminder'?) issues are at stake. It is certainly not broadcasts devoted to political life that teach us to respect the dignity of others and to take care to be tolerant. Turning political debate into a 'media show' only exacerbates party disputes and verbal civil war, with politicians seeking to achieve greater media efficacy than real efficacy; for they know, and probably deplore the fact that they are often judged more by the way they look and their 'murderous' platitudes than by their actions. Jean Delumeau (1985), who considers that a world without forgiveness is a world without hope, worries that 'on the pretext of ideology, each day political enemies are massacred to whom no word of hope or comfort has been spoken. Adversaries of the right or the left, they are regarded as irreconcilable, "misfits" and deserve no pity'. On the other hand, advertising is certainly not going to stimulate a critical sense or help to build a scale of values, since its aim is to charm, not to inform; it emphasizes whatever is futile and short-lived, with an irritating mix of false naivety and real turgidity. The 'television news' does not help either, often giving an anecdotal news item or the tendinitis of a sports star top priority at the expense of far more important events. What is more, it is rare for accidents, crimes, terrorist attacks, and wars not to form the main bulk of the subject matter dealt with.

One's vision of man and the world is inevitably influenced by the dominant tone cast on them day after day by television images. Neither is the language accompanying them harmless. In the area we are concerned with, the language used (especially in advertising) implies that aggressiveness, and even spitefulness, are signs of 'virility' and efficiency. As far as scenes of violence are concerned, which are very frequent in the films and telefilms screened, I have indicated[57] that many studies concur that the repeated displaying of these scenes had the effect of increasing the probability of the use of aggression as a means of expression and action. Singer (1984), who is critical of American television ('almost all of television in the U.S. is mindless, tasteless entertainment requiring nothing and adding nothing'), believes that it is not violence

[56] A. Girard 1986.
[57] In Chapter 7, under The essential role of the family milieu, p. 162.

as such that has harmful effects (that of 'Macbeth' or 'Little Red Riding Hood' does not worry him), but violence shown in a demeaning context. That is, violence associated with a degrading image of man and a total lack of respect for human life.

This question brings us back to a more general problem which I have already discussed earlier on when talking of psychotherapeutic or simply educational measures with the aim of structuring—or restructuring—the personality, that is, respect for freedom of conscience. It may certainly seem paradoxical that I advocate at the same time full respect for this freedom, a certain directivity in education, and in particular, the use of 'moral education' (which helps in acquiring, but does not impose, a set of moral principles). But a truly free conscience can only be an 'enlightened' conscience, and in children it cannot become enlightened by itself. Moreover, and in particular, it is constantly exposed to influences which may 'freely' exert a weakening or even degrading action. If I agree, in the name of 'freedom of expression' and the 'right to information' that everything should be said and shown, I cannot also agree, in the name of 'freedom of conscience', not to be allowed to say—to those for whom I feel responsible and to whoever cares to listen—what I believe conforms to the dignity of man and what in my opinion endangers it. Even if it is not always easy, children and adolescents can be helped to acquire, beyond the initial ethics of 'opportunism' and later on 'convention', the autonomous moral judgement that will enable them to truly assume their freedom. As far as television broadcasts are concerned, I think it is dangerous to allow the sole law of the 'market' to determine the quality of our television programmes. Didn't a former programming director of ABC (one of the three main American channels) declare: 'My mission consists of attracting a maximum number of devoted spectators while spending the minimum. Producing quality broadcasts is considered to be a chance accident—a happy one, of course!—but never a must!'[58]

Indispensable prevention

When contemplating the genesis and prevention of 'opportunities for violence', those factors which are linked to an urban lifestyle should not be separated from those pertaining to the economic environment. Large urban concentrations generate nuisances which often cause exasperation and hence, potentially, violence, while at the same time human relationships become impoverished and former close solidarities, which were a natural barrier to violence, fall apart. Moreover, the frustrations and temptations generously provided by the economic system thrive and are stimulated in big cities. It is therefore not surprising that in 1976 three-fifths of all major criminal acts took place in the seven French

[58] *Le Monde des Loisirs*, 7th September 1985.

departments where the most populated cities are located. Moreover, the height of buildings plays a major role, for a far higher crime rate has been noted in buildings of more than six storeys.[59] In big anonymous complexes there is both solitude and noise, the indifference and the tensions of collective life. It is in the big cities, too, that the rhythm of life is accelerated, stress increases, and the complexity and technicality of social functions gradually drain them of any human warmth. All this, together with an increasingly tight net of bureaucracy, is not likely to promote an individual's autonomy, facilitate the development of his sphere of freedom, and encourage him to become involved, one way or another, in community service. If efficient prevention of violence is to be achieved, threefold action[60] is required. As for urban space, it should be less dense and consist of districts with landmarks that render them familiar and facilitate individual appropriation. As to structures, it is important to promote cooperative links and more generally, any organized activity that encourages contacts, especially amongst young people. As for lifestyle, it can be improved by endeavouring to re-establish and enrich communication, reduce environmental drawbacks (in particular noise and traffic problems), and by encouraging the proliferation of individual houses.

A man needs calm and silence in familiar surroundings where he can feel at home. Only under such conditions can he stand aloof from outside influences and develop his own inner life through personal reflection. Otherwise, he will only live 'by proxy', without any real commitment, without any personal investment, through 'models' broadcast by the media. But is the search for an inner life, with personal convictions and aspirations, not a withdrawal into oneself and hence, an obstacle along the path towards a more fraternal and solidarity-minded human community? Quite the contrary, for to be truly solidarity-minded is to help others to become themselves, to achieve their full potential. How would I be able to do this without first achieving it for myself? It therefore only appears paradoxical that my inner life, which can only enrich others through its very substance, is the best bond for the community.

But it is not easy to preserve a degree of inner life when one is constantly tempted by all attractions used by a society that exalts the act of consumption, that further advocates it as a means of identifying oneself with 'models' who turn out to be first and foremost 'big consumers', and in so doing stimulates all kinds of desires. I would not feel that I had the right to criticize this society if looking around me, I saw that people were contented and happy. But I must say that far from creating joy and harmony, this dull economism creates on the contrary a general feeling of frustration and dissatisfaction which is expressed in

[59] See A. Peyrefitte 1977.
[60] See Report of the working group: 'Urbanisation, habitat and violence', in A. Peyrefitte 1977.

aggressiveness that is increasingly difficult to contain. People forget that 'material wealth is certainly something good provided one doesn't lose one's soul to it' and that 'the consumer spirit is not a bad thing as long as spiritual hunger is not stifled by it'.[61] Reducing a man to the single dimension of an 'economic factor' is to belittle his dignity.

Under the pretext of 'economic issues', which are real and the importance of which should not be underestimated, 'mercantile interests' increasingly invade and corrupt most human activities. Many people follow this trend, and 'clever' delinquency motivated by gain costs the community dear. The increase in gambling (horse racing, lotto, sports lotto . . .) is part of the same picture. The aim of gambling is immediate individual gain which can be obtained with much vaunted 'ease'. People justify its growing influence by declaring that it enables them to dream, without wondering whether we could not dream up, together, some other 'grand endeavour'. In short, encouraging gambling is to cultivate individualism, easiness, and mediocre ambition. The general fomenting of desires, the longing to have 'more and more' in the hope of being able to 'impress people', inevitably leads some to abuse their dominant positions, arousing defensive and often aggressive reactions. On the other hand, the impossibility of fully satisfying the need to possess, the reduced tolerance of perceived inequality and general complacency towards any kind of 'resourcefulness' (at least as long as one is not a victim oneself) also contribute to the development of violence. The intervening factors are so many and so interwoven, and violence of an economic source is so much of 'a phenomenon which to differing degrees, depending on the place or time, has marked world history', that it is easy to understand how an analysis, no matter how competent, of causes and remedies, is presented in a form which the authors themselves emphasize is one of 'modesty, limits and relativity'.[62] It would appear, in this sphere as in others, that the prevention of violence concerns the entire organization of society; which means that it concerns us all.

Needless to say, it is this very organization of society that is also in question, when one contemplates certain factors known to generate, facilitate, or aggravate violence: non-observance of the rules of the highway code, excessive alcohol consumption, abusive ownership of fire-arms. The car being both a means of expression and a vehicle, it is all too often used to flaunt an imaginary superiority, to show utter contempt, and to give vent to belligerence. So it is shocking to hear the director of a major automobile assocition speak of 'the State, the number one assassin on the roads', when pinpointing out the existence of certain deplorable 'black spots'. Leaving aside the quasi-criminal behaviour of some drivers, why should the State be blamed for the many cases where a car driver driving at excessive speed 'loses control of his car'? It is

---

[61] John Paul II, in Liechtenstein, September 1985.
[62] See Report of the working group 'Violence and economy', in A. Peyrefitte 1977.

probably in order not to 'upset' car drivers that care is taken to avoid placing signs along our motorways stating the authorized speed limit. As far as alcohol is concerned, encouraging its consumption by means of advertising cleverly circumvents the rules, and organizations responsible for fighting alcoholism are often obliged to admit that their efforts are useless. And yet, whether in road accidents or ill treatment inflicted within the family, alcohol is a great instigator of violence and misery. Moreover, under the influence of alcohol it becomes very easy to pull out one's .22 carbine. But if it is so easy to 'pull it out', this means it is within arm's reach, ready to kill, perhaps 'unintentionally'. Would it not be preferable, in many cases, that the weapon be absent, not only the intention to kill? But here again, any serious attempt to reduce the sale and stockpiling of fire-arms has proved completely fruitless. Because of the union of lobbies and demagogies, assuming responsibility on an individual and collective basis in these different areas is, unfortunately, not imminent.

About repression

Measures of dissuasion and repression have always aroused passionate debate as to whether they are necessary or legitimate and what practical methods should be used in applying them. As regards the need for them, it can be said in a down-to-earth way that if somebody wants something very badly and is not too full of scruples, only fear of the police can restrain him from taking it by means of aggression. A judge at the European Court for Human Rights will use more carefully chosen terms to allude to the consequences of repudiating the values resulting from religious doctrine, by stressing that 'because the law is often the means, in a pluralistic society, of imposing an obligation of social behaviour through rule of law and by means of sanctions, (it) has compensated for the weakening of ethical values.'[63] According to Cusson (1983a), measures of dissuasion and repression are required because therapeutic measures to re-educate and rehabilitate delinquents have proved incapable of reducing the level of repetition of the offence. I might add that in any case it is better to prevent through education what is apparently very difficult to correct subsequently through re-education.

Even so, one should not be blind to the ambiguous nature of penal sanctions. The Study Committee on Violence, Crime and Delinquency expresses this very clearly in its general report: 'Penal law is intended to express social disapproval of attacks on the ideals society holds dear. This function, which is fairly easy to maintain if the population adheres to universally acknowledged ethics, is infinitely more difficult in a divided society that is unsure of itself.' Social reprobation can only have any real dissuasive effect if it is authentic, in that it condemns any breach of the rules which the group respects, values, and effectively

[63] L. E. Pettiti 1986.

observes, and on condition that it is perceived as such by everyone, making them aware of the cost of cutting themselves off from this consensus. On the other hand, social reprobation loses its dissuasive power if it is hypocritical because members of the group cheat, defraud, and 'are smart', in bad faith, and if the punished offender comes to believe that his only mistake was to be caught and, that once caught, proper support would have enabled him to avoid punishment.

At a time when values are being questioned (and in the absence of an authentic sense of justice which can only result from a common ethic transcending the sole concern to satisfy our desire to possess, consume, and show off), attitudes and behaviour are very likely to be ruled by the needs of elementary opportunism: what serves my interests is 'good'; what might harm them is 'bad'. As regards the use of different forms of aggression and violence as means of expression and action, it is enough to look around us (and within ourselves!) to realize the influence of such intolerant opportunism. It is rather generally understood that the use of violent means of action is 'legitimate' so long as it aims to defend our personal interests or is the doing of people who think like us. Violence only becomes 'highly reprehensible' if it is likely to harm our interests or if it is used by people who do not think like us; we may then have a good conscience by condemning it with virtuous indignation. This attitude unconsciously supports the law of the strongest, against whose sway these measures of dissuasion and repression are precisely supposed to protect us. How can one be surprised, in such circumstances, at the ambiguities and weaknesses, the source of inefficiency, of all penal policies and of the institutions responsible for enforcing them in everyday life? It is absolutely unfair to attribute to their 'laxity' what is mainly due to our own lax attitudes and to hold them responsible for what has resulted to a large extent from our own irresponsibility.

Aggressiveness: fate or responsibility?

If I have so insistently brought up throughout this concluding chapter our individual and collective responsibilities in the genesis of opportunities for violence and how our attitudes are shaped towards them, it is in order to show that the answer to the question (fate or responsibility?) is perfectly clear. But it is obvious that even if we fully acknowledge the reality of these responsibilities (which is far from being the case!), the task to be accomplished, if peaceable and fraternal human communities are to be built, remains immense and enough to discourage the most determined of men. But the determination to work towards this can only stem from a solid faith in man and his dignity (as the inscription in the Antiquarium of the Residenz in Munich says so well, in its simplicity: *Voluntatem spes facit*). This is why it was essential to dispel the pessimistic vision which constantly holds up the spectre of a dark fate thought to be inherent in our biological heritage. In this respect I have tried to develop throughout this work arguments which are completely opposed to this way of looking at things and enable it to be replaced by a more optimistic vision that will enable us to take control of our own future. Each of us is free to prefer love to hatred and justice to iniquity, to prefer a society of mutual recognition and valorization to one of domination and scorn. Actions which express commitment based on personal reflexion are acts of freedom *par excellence*. I might add that in a recent study Paul Scott (1986) reached the conclusion that it was also man who invented the wars that have bloodied our planet and that on a purely biological level, there is nothing to stop him from re-inventing peace.

'Re-inventing' feelings

But it is not sufficient for us to know that we are unlikely to be building on sand and that the enterprise is not doomed from the start, for a fine building to rise spontaneously simply because we would like to live in it. The task will be arduous and long and it will allow no facility. But does man's greatness not lie in the free acceptance of the demands inherent in the meaning he wants to give his life? If this is so, we know exactly what materials we need. First of all, hope, whose flame may certainly waver but must never be allowed to go out. Our soulless technical civilization cannot maintain it. For technical progress may well make us richer and more powerful, freeing us from certain nuisances while in fact creating others. But it cannot make us 'better', that is warmer, more open, more tolerant, more just, more fraternal. While allowing our reason, that marvellous instrument of knowledge and lucidity, to fully invest all the space it can enlighten, it is important to rehabilitate feelings, for it is also (primarily?) on an affective plane that I grasp my own self and become attached to others. Respect for the dignity of others, an awareness of and consideration for their needs and aspirations, requires human warmth, without which these ideas have little chance of becoming a living reality. But affectivity is underestimated, repressed, even scorned; worse still, it is exploited in order to place constraint on others for purposes that have nothing to do with love of one's fellow man.

In praise of tolerance

Just as important is the spirit of tolerance implying the faculty to forgive, to make that difficult and worthy gesture of trust in human beings. It is a difficult gesture because it is generally regarded as a sign of weakness and therefore as a risk of 'losing face'. In reality sincere forgiveness shows strength of character, bringing freedom by eliminating resentment. But above all, forgiveness is both necessary and possible. Jean Delumeau (1985) was right in stating that 'there are only two possible solutions for the national and international antinomies that are tearing our world apart: either war or forgiveness'. And René Girard (1982) ends his analysis of the ravages caused by chasing after 'scapegoats' with these words: 'The time has come to forgive one another. If we wait any longer, we will have no more time.' And there is nothing impossible about such forgiveness. Before the last war, I was often together with other children of my age beside the Rhine, and we would throw stones at the youngsters we glimpsed on the other side, shouting at them 'dirty krauts!' (actually we used the Alsatian dialect which sounds confusingly similar to the one spoken on the other side). After the war and the years spent in an Alsace annexed by the Germans, my resentment was so great that I decided, together with my future wife, that our children would not learn the Germanic dialect our respective ancestors had spoken, from father to son, for centuries. But on a stained glass window in Strasbourg cathedral, a gift from the Council of Europe, an inscription states that the peoples of Europe have decided to put an end to their fighting. And in fact, the Bridge of Europe, which we now cross frequently and naturally, has ceased to be a frontier and become a place of coming together and exchanges. On a more general level, families and groups all over the world succeed in creating and preserving a serene atmosphere and satisfactory relationships. Why should those cases necessarily be considered as exceptions?

The new builders

In order to build, high ambition and great humility are required. A twofold ambition is indispensable, involving simultaneously the targeted objectives and the role we ourselves are to assume. For it is very ambitious (some may say, utopian) to hope to realize everybody's potential and to develop negotiated relationships at the expense of enforced ones, by recognizing the equal dignity of all human beings and endeavouring to give the fullest meaning, aside from the satisfactions brought about by consumption of material goods, to the deep joys so greatly intensified by sharing, those of affection, nature, and the various art forms. This must be done by all of us since this building effort, like any other, can only be undertaken from the bottom upwards. It would be quite useless to hope it could be made from the top since all power structures, whatever their underlying school of thought, are essentially conservative. For 'no school of thought can really think the thought that is capable of destroying it'.[64] Thus if we need the ambition to build, by ourselves and for ourselves, we also need to show patient humility. Firstly, humility in the face of the enormity of the task, in accepting that progress will be very slow, interspersed with many failures, and often rendered arduous by efforts that prove vain. But also a more personal and more difficult type of humility, which consists of recognizing one's

[64] R. Girard 1982.

own weaknesses and mistakes. In this enterprise, the essentials are not to keep a schedule, but to start the action in order to pass on to the younger generation some solid foundations and a reason to hope rather than a field of ruins.

In order to build and find a common direction, there is need for a goal that can be pursued with hope, ambition, tolerance, and humility, just as when sailing together. But these instruments are not sufficient to find one's bearings and stay the course if lies plunge us into darkness and if the fog of cheating and hypocrisy hide the light of the stars. There is no true freedom without truth, and this is why it is important ceaselessly to insist on the vital need to fight against the corruption of language which both reflects and accelerates the corruption of the mind and increasingly prevents us from debating the most vital issues with the necessary objectiveness, honesty, and serenity. No one has the right to talk of 'new space for freedom' or new 'laws of freedom' if these are notions that, like the 'freewheel', may well turn without the slightest effort, but do not lead us anywhere. The true freedom each of us most needs is not the kind that is begged for and granted, but the kind that is deserved and conquered as a result of a challenging personal quest. One must have deep respect for others and for their dignity to have the courage to tell them so and avoid deceiving them. Jean Cocteau warned us that 'to win the game' we would have to 'play hearts or cheat'. We have already cheated a great deal and the least that can be said is that we have won nothing. On the contrary, we cannot help, in our lucid moments, seeing the abyss open beneath our feet.

So only the first term of the alternative is left: to play hearts. Let us try to meet, now more than ever before, the greatest challenge of all: 'Love one another!'

GLOSSARY

Amygdala: subcortical complex of nuclei, located deep in the temporal lobe. This, the place where sensory information converges and is integrated, plays an important role in the association of affective attributes with the objective data of sensory information.

Broca's area of the cerebral cortex: located in the lower part of the lateral surface of the frontal lobe, in the dominant cerebral hemisphere (the left hemisphere in right-handed people). Area where the motor automatisms of spoken language are represented. Its destruction brings about a motor aphasia, or anarthria.

Catecholamines: a family of chemical substances characterized by the presence of a catechol nucleus on to which a lateral chain of amines is grafted. Neurotransmitters such as noradrenaline and dopamine belong to this family.

Cingulate cortex: cortical area located on the medial surface of the frontal lobe. Involved in the integration of 'motivational valence' to the internal representations of extra-personal space and the familiar social environment.

Corpus striatum: mass of grey matter located in the forebrain (telencephalon), containing the caudate nucleus and the lenticular nucleus. On to the corpus striatum project the fibres of the nigrostriatal dopaminergic pathway. The corpus striatum plays an important role in initiating movements and in adjusting postural tone.

Dopamine: neurotransmitter belonging to the catecholamine group. The cell bodies of the neurons which produce and release dopamine (the 'dopaminergic' neurons) are located mainly in the substantia nigra and in the adjacent area of the tegmentum of the mesencephalon.

Endorphins: polypeptide neurotransmitters (chains of amino acids) which function as the brain's internally generated pain-killers ('endogenous morphines').

Fornix: great efferent pathway of the hippocampus, relating the latter directly with the septum, the thalamus, the hypothalamus, and the anterior region of the mesencephalon.

GABA: gamma-aminobutyric acid. An amino acid acting as an inhibitory neurotransmitter.

GABAergic transmission: neurotransmission mechanism making use of gamma-aminobutyric acid (GABA) in the contact region between two nerve cells.

Hippocampus: primitive cerebral cortex, located on the medial surface of the temporal lobe. Being a site where sensory information converges and is integrated, it plays a role in memory processes, particularly in the memorization of the spatio-temporal attributes of sensory information.
Hypothalamus: group of nuclei (formations of grey matter) located in the middle part of the base of the brain. Through the control it exercises over the vegetative nervous system and the endocrine system, the hypothalamus plays an important role in regulating the vegetative functions and in integrating the different components of emotional reactions.

Limbic system: group of brain structures that developed over the most distant period of mammalian evolutionary history (the 'palaeomammalian brain' in MacLean's terminology). These are cortical areas (particularly the cortex of the internal, medial, surface of the cerebral hemispheres) and sub-cortical nuclear formations (of the amygdala and the septum). Interacting closely with the rest of the brain, the limbic system plays an essential role in memory processes and in the genesis of emotions.

Medial forebrain bundle: heterogeneous nerve pathway running through the lateral hypothalamus and providing reciprocal connections between structures in the mesencephalon and the telencephalon (forebrain).
Mesencephalic tegmentum: the ventral part of the mesencephalon, located below the level of the aqueduct of Sylvius. It contains in particular the cell bodies in which originate the dopaminergic fibres that project on to various structures of the telencephalon.
Mesencephalic tectum: or 'roof' of the mesencephalon. The dorsal part of the mesencephalon, located above the level of the aqueduct of Silvius. Includes the anterior and posterior corpora quadrigemina (superior and inferior colliculi), together with the dorsal portion of the grey matter surrounding the aqueduct.
Mesencephalon: midbrain, the highest part of the brain stem. Along it passes a channel, the aqueduct of Sylvius, which is surrounded by the periaqueductal grey matter.
Motor neurons: nerve cells providing motor innervation to the muscles. Their cell bodies are located in the ventral horns of the grey matter of the spinal cord and in the nuclei where the cranial motor nerves originate.

Neocortex: phylogenetically the most recent part of the cerebral cortex. Developed progressively in mammals, attaining its full development in the human species.
Neomammalian brain: areas of the brain which were developed mainly

during the most recent period of mammalian evolutionary history. The region chiefly involved is the cerebral cortex of the dorso-lateral surface of the brain hemispheres. In MacLean's 'triune brain' conception, the neomammalian brain is phylogenetically the most recent component, corresponding to a 'brain for cognitive activity'.

Neostriatum: phylogenetically the most recent part of the corpus striatum, it includes the caudate nucleus and the lateral part of the lenticular nucleus, the putamen. The palaeostriatum, phylogenetically older, takes in the medial part of the lenticular nucleus, the globus pallidus, and the substantia nigra.

Neurotransmitter: chemical substance permitting a neural signal (the basic unit in a neural message) to be sent via the contact region, or 'synapse', between two nerve cells, or between one nerve cell and an 'effector' cell, either muscular or glandular.

Nigro-striatal dopaminergic pathway: this is made up of nerve fibres which release dopamine within the corpus striatum and whose originating cell bodies are located in the substantia nigra. Cutting this pathway bilaterally in animals causes a loss of motor initiative. A deficiency in the release of dopamine within the corpus striatum gives rise in man to the clinical symptoms of Parkinson's disease.

Nucleus accumbens: a body of grey matter located in the most anterior part of the corpus striatum, near the septum. It receives (as does the neostriatum) afferent projections from the amygdala and hippocampus. Through its projections on to mesencephalic structures it plays a part in controlling motor activities, such as locomotion.

Ontogenesis: development of an individual, from the fertilization of the egg to the adult stage.

Palaeomammalian brain: areas of the brain which were developed during the earliest part of mammalian evolutionary history. It is a group of cortical and subcortical structures classed together as the 'limbic system'. For MacLean this, the oldest mammal brain (in phylogenetic terms), corresponds to a 'brain for the emotions' or 'emotional brain'.

Periaqueductal grey matter: surrounds the aqueduct of Silvius, the channel running along the mesencephalon. Plays a part in the genesis of aversive emotions and the behaviour—defence or flight—expressing them.

Phylogenesis: the way that animal species have been formed; their development through evolution.

Prefrontal cortex: cerebral cortex of the most anterior portion of the frontal lobe, particularly well developed in man. Because of its reciprocal 'modular' connections with the hippocampus, the corpus striatum and the posterior parietal association areas, it plays an essential role in the choice of an appropriate behavioural strategy, locating it correctly in space and time.

Progesterone: a hormone secreted by the corpus luteum (which is formed in the ovary during each ovarian cycle and during pregnancy) and by the placenta. This hormone, which prepares the mucous membrane of the uterus for the implantation of the fertilized egg, is also necessary to the continuation of pregnancy.

Pyramidal tract: motor pathway linking the cerebral cortex directly to the motor neurons of the spinal cord. It is also the means whereby the cortex exercises control over the upward transmission of sensory impulses.

Raphe nuclei: small masses of grey matter located close to the midline of the brain stem, running from the bulb to the mesencephalon. They consist mainly of cell bodies of the central neurons that synthesize and release serotonin.

Reptilian brain: phylogenetically the oldest areas of the brain, already present in reptiles. In the absence of palaeo- and neomammalian structures, this primitive brain organizes elementary and stereotyped behaviour, without the complex cognitive–affective elaborations that typify the global behaviour of the human brain.

Septum: medial partition and grey-matter nucleus, located in the anterior part of the cerebral hemispheres. Plays a part in processes that moderate emotional reactions and inhibit behaviour.

Serotonin: a neurotransmitter synthesized from an essential amino acid, tryptophan. The cell bodies of the neurons which produce and release serotonin (the 'serotonergic' neurons) are mainly located in the raphe nuclei.

Steroid hormones: a group of hormones whose basic structure consists of a polycyclic alcohol or sterol. Male and female sexual hormones belong to this group, as do the hormones secreted by the suprarenal cortex.

Substantia nigra: also called 'locus niger'. A mass of grey matter located in the anterior (or ventral) area of the mesencephalon. Contains the cell bodies in which originate the nerve fibres that make up the nigrostriatal dopaminergic pathway.

Superior colliculus: one of the two rounded protuberances, also known as the anterior corpora quadrigemina, occupying the anterior portion of the dorsal surface of the mesencephalon. They play a part in controlling eye movements.

Suprerenal cortex: peripheral part of the suprarenal gland. Produces steroid hormones, releasing them into the circulating blood; these hormones not only play a role in the regulation of the hydro-mineral and carbohydrate metabolisms but also act directly on the brain.

Suprarenal glands: endocrine glands located at the top of the kidneys and made up of two distinct parts: a central part (the suprarenal medulla) and a peripheral part (the suprarenal cortex).

Testosterone: male sexual hormone secreted by the testicles, stimulating the development of male genital organs and secondary sexual characteristics; also plays a role in male sexual behaviour by acting directly on the brain.

Thalamus: the most voluminous of the central grey-matter formations, located in the intermediate brain or diencephalon (between the mesencephalon and the telencephalon). Numerous thalamic nuclei act as relays along the sensory or motor pathways; others have an associative or a more general activating function.

Wernicke's area of the cerebral cortex: located in the upper part of the lateral surface of the temporal lobe, in the dominant cerebral hemisphere (the left hemisphere in right-handed people). Area where the auditory images of spoken words are represented. Its destruction brings about a sensory aphasia or loss of comprehension of the spoken language.

REFERENCES

Adamec, R. E. (1978). Normal and abnormal limbic system mechanisms of emotive biasing. In *Limbic mechanisms* (ed. K. E. Livingston and O. Horny-kiewicz), pp. 405–55. Plenum Publishing Corporation, New York.

Adamec, R. E. and Stark-Adamec, C. (1984). The contribution of limbic connectivity to stable behavioural characteristics of aggressive and defensive cats. In *Modulation of sensorimotor activity during alterations in behavioural states*, (ed. R. Bandler), pp. 325–39. Alan Liss, New York.

Adams, D. B. (1971). Defence and territorial behaviour dissociated by hypothalamic lesions in the rat. *Nature (Lond.)*, **232**, 573–4.

Adams, D. B. (1980). Motivational systems of agonistic behavior in muroid rodents: a comparative review and neural model. *Aggressive behavior*, **6**, 295-346.

Adams, D. B. (1983). Hormone-brain interactions and their influence on agonistic behavior. In *Hormones and aggressive behavior*, (ed. B. Svare), pp. 223–45. Plenum Press, New York.

Adams, D. (1984). There is no instinct for war. *Psychological journal* (Academy of Sciences of USSR, Moscow) **5**, 140–4, in Russian.

Adams, D. and Flynn, J. P. (1966). Transfer of an escape response from tail shock to brain-stimulated attack behavior. *J. exp. anal. behav.*, **9**, 401–10.

Ajuriaguerra, J. de (1977). Ontogenèse de la motricité. In *Du contrôle moteur à l'organisation du geste* (ed. H. Hécaen and M. Jeannerod), pp. 133–57. Masson, Paris.

Akil, H., Watson, S. J., Young, E., Lewis, M. E., Khachaturian, H., and Walker, J. M. (1984). Endogenous opioids: biology and function. *Ann. rev. neurosci.*, **7**, 223–55.

Albano, J. E., Mishkin, M., Westbrook, L. E., and Wurtz, R. H. (1982). Visuomotor deficits following ablation of monkey superior colliculus. *J. neurophysiol.*, **48**, 338–51.

Albert, D. J. and Brayley, K. N. (1979). Mouse killing and hyperreactivity following lesions of the medial hypothalamus, the lateral septum, the bed nucleus of the stria terminalis, or the region ventral to the anterior septum. *Physiol. behav.*, **23**, 439–43.

Albert, D. J. and Walsh, M. L. (1984). Neural systems and the inhibitory modulation of agonistic behavior: a comparison of mammalian species. *Neurosci. biobehav. rev.*, **8**, 5–24.

Albert, D. J., Walsh, M. L., and White, R. (1984). Rearing rats with mice prevents induction of mouse killing by lesions of the septum but not lesions of the medial hypothalamus or medial accumbens. *Physiol. behav.*, **32**, 143–5.

Albert, D. J., Walsh, M. L., and Longley, W. (1985). Group rearing abolishes hyperdefensiveness induced in weanling rats by lateral septal or medial accumbens lesions but not by medial hypothalamic lesions. *Behav. neural biol.*, **44**, 101–9.

Alès, C. (1984). Violence et ordre social dans une société amazonienne. Les Yanomami du Venezuela. *Études rurales*, **95–6**, 89–114.

Amir, S., Brown, Z. W., and Amit, Z. (1980). The role of endorphins in stress: evidence and speculations. *Neurosci. biobehav. rev.*, **4**, 77–86.

Ancel, M. (1985). *La Défense sociale*. Que sais-je? No. 2204. Presses universitaires de France, Paris.

Anderson, W. F. (1986). Le traitement des maladies génétiques. *La Recherche*, **17** (176), 458–68.

Ardrey, R. (1961). *African genesis*. Collins, London.

Argyle, M., Furnham, A., and Graham, J. A. (1981). *Social situations*. Cambridge University Press, Cambridge.

Armstrong, E. (1986). Enlarged limbic structures in the human brain: the anterior thalamus and medial mamillary body. *Brain research*, **362**, 394–7.

Aron, C. (1984). La neurobiologie du comportement sexuel des mammifères. In *Neurobiologie des comportements* (ed. J. Delacour), pp. 57–108. Hermann, Paris.

Aronson, L. R. and Cooper, M. L. (1979). Amygdaloid hypersexuality in male cats re-examined. *Physiol. behav.*, **22**, 257–65.

Averill, J. R. (1982). *Anger and aggression. An essay on emotion*. Springer Verlag, New York.

Baenninger, R. (1970). Suppression of interspecies aggression in the rat by several aversive training procedures. *J. comp. physiol. psychol.*, **70**, 382–8.

Baenninger, R. (1978). Some aspects of predatory behavior. *Aggressive behavior*, **4**, 287–311.

Ballé, C. (1976). *La Menace, un langage de violence*. Éditions du CNRS, Paris.

Balleyguier, G. (1981). Le caractère de l'enfant en fonction de son mode de garde pendant les premières années. *Monographies françaises de psychologie*, No. 55. Éditions du CNRS, Paris.

Bandler, R. (1982). Neural control of aggressive behaviour. *Trends in neurosciences*, **5**, 390–4.

Bandler, R. and Abeyewardene, S. (1981). Visual aspects of centrally elicited attack behavior in the cat: 'patterned reflexes' associated with selection of an approach to a rat. *Aggressive behavior*, **7**, 19–39.

Bandler, R. and Chi, C. C. (1972). Effects of olfactory bulb removal on aggression: a reevaluation, *Physiol. behav.*, **8**, 207–11.

Bandler, R., Depaulis, A., and Vergnes, M. (1985). Identification of midbrain neurones mediating defensive behaviour in the rat by microinjections of excitatory amino acids. *Behav. brain res.*, **15**, 107–19.

Baré, J.-F. (1984). Fantômes de la violence: énigmes tahitiennes. *Études rurales*, **95–6**, 23–46.

Barfield, R. J. (1984). Reproductive hormones and aggressive behavior. In *Biological perspectives on aggression*, (ed. K. J. Flannelly, R. J. Blanchard, and D. C. Blanchard), pp. 105–34. Alan Liss, New York.

Barnett, S. A. (1981). Models and morals: biological images of man. In *Multidisciplinary approaches to aggression research* (ed. P. F. Brain and D. Benton), pp. 515–29. Elsevier, Amsterdam.

Barnett, S. A. (1983). Humanity and natural selection. *Ethology and sociobiology*, **4**, 35–51.

Barr, G. A. (1981). Effects of different housing conditions on intraspecies fighting between male Long-Evans hooded rats. *Physiol. behav.*, **27**, 1041–4.

Basbaum, A. I. and Fields, H. L. (1984). Endogenous pain control systems: brainstem spinal pathways and endorphin circuitry. *Ann. rev. neurosci.*, **7**, 309–38.

Bear, D. M. (1983). Hemispheric specialization and the neurology of emotion. *Arch. neurol.*, **40**, 195–202.

Beatty, W. W., Dodge, A. M., Traylor, K. L., Donegan, J. C., and Godding, P. R. (1982). Septal lesions increase play fighting in juvenile rats. *Physiol. behav.*, **28**, 649–52.

Bekoff, M. (1981). Development of agonistic behaviour: ethological and ecological aspects. In *Multidisciplinary approaches to aggression research* (ed. P. F. Brain and D. Benton), pp. 161–78. Elsevier, Amsterdam.

Bell, P. A. and Baron, R. A. (1981). Ambient temperature and human violence. In *Multidisciplinary approaches to aggression research* (ed. P. F. Brain and D. Benton), pp. 421–30. Elsevier, Amsterdam.

Beninger, R. J. (1983). The role of dopamine in locomotor activity and learning. *Brain res. rev.*, **6**, 173–96.

Benoit, J.-C. and Berta, M. (1973). *L'Activation psychothérapique*. Charles Dessart, Brussels.

Benowitz, L. I., Bear, D. M., Rosenthal, R., Mesulam, M. M., Zaidel, E., and Sperry, R. W. (1983). Hemispheric specialization in nonverbal communication. *Cortex*, **19**, 5–11.

Benton, D. (1981). The extrapolation from animals to man: the example of testosterone and aggression. In *Multidisciplinary approaches to aggression research* (ed. P. F. Brain and D. Benton), pp. 401–18. Elsevier, Amsterdam.

Benton, D. (1982). Is the concept of dominance useful in understanding rodent behaviour? *Aggressive behavior*, **8**, 104–7.

Berg, D. and Baenninger, R. (1974). Predation: separation of aggressive and hunger motivation by conditioned aversion. *J. comp. physiol. psychol.*, **86**, 601–6.

Berkowitz, L. (1974). External determinants of impulsive aggression. In *Determinants and origins of aggressive behavior* (ed. J. de Wit and W. W. Hartup), pp. 147–65. Mouton, The Hague.

Berkowitz, L. (1981). On the difference between internal and external reactions to legitimate and illegitimate frustrations: a demonstration. *Aggressive behavior*, **7**, 83–96.

Berkowitz, L. (1984). Physical pain and the inclination to aggression. In *Biological perspectives on aggression* (ed. K. J. Flannelly, R. J. Blanchard, and D. C. Blanchard), pp. 27–47. Alan Liss, New York.

Bernicot, J. (1981). Le développement des systèmes sémantiques de verbes d'action. *Monographies françaises de psychologie*, No. 53. Éditions du CNRS, Paris.

Bernstein, I. S. (1981). Dominance: the baby and the bathwater. *The behavioral and brain sciences*, **4**, 419–57.

Billig, M. (1984). Racisme, préjugés et discrimination. In *Psychologie sociale* (ed. S. Moscovici), pp. 449–72. Presses universitaires de France, Paris.

Björklund, A. and Stenevi, U. (1984). Intracerebral neural implants: neuronal replacement and reconstruction of damaged circuitries. *Ann. rev. neurosci.*, **7**, 279–308.

Blanc, M. (1984). L'histoire génétique de l'espèce humaine. *La Recherche*, **15**,(155), 654–69.

Blanchard, D. C. and Blanchard, R. J. (1984). Affect and aggression: an animal model applied to human behavior. In *Advances in the study of aggression*, Vol. I (ed. R. J. Blanchard and D. C. Blanchard), pp. 1–62. Academic Press, Orlando.

Blanchard, D. C., Blanchard, R. J., Lee, E. M. C., and Nakamura, S. (1979). Defensive behaviors in rats following septal and septal-amygdala lesions. *J. comp. physiol. psychol.*, **93**, 378–90.

Blanchard, R. J. (1984). Pain and aggression reconsidered. In *Biological perspectives on aggression* (ed. K. J. Flannelly, R. J. Blanchard, and D. C. Blanchard), pp. 1–26. Alan Liss, New York.

Blanchard, R. J., Kleinschmidt, C. K., Flannelly, K. J., and Blanchard, D. C. (1984). Fear and aggression in the rat. *Aggressive behavior*, **10**, 309–15.

Boisvert, J.-M. and Beaudry, M. (1984). Les difficultés interpersonnelles et l'entraînement aux habiletés sociales. In *Cliniques de thérapie comportementale* (ed. O. Fontaine, J. Cottraux, and R. Ladouceur), pp. 75–89. P. Mardaga, Liège.

Bourdieu, P. (1985). *Entretiens avec 'le Monde'* — 6. *La société*, pp. 101–14. La Découverte/Le Monde, Paris.

Bourricaud, F. (1985). *Entretiens avec 'le Monde'* — 6. *La société*, pp. 115–25. La Découverte/Le Monde, Paris.

Bouthoul, G. and Carrère, R. (1976). *Le Défi de la guerre (1740–1974)*. Presses universitaires de France, Paris.

Bowers, D. C. (1979). Mouse killing, intermale fighting, and conditioned emotional response in rats. *Aggressive behavior*, **5**, 41–9.

Bowlby. J. (1984). *Attachment and loss*, Vol. I, *Attachment*, (2nd edn). Penguin Books, Harmondsworth.

Box, S. (1981). *Deviance, reality and society*, (2nd edn). Holt, Rinehart and Winston, London.

Brain, P. F. (1981). Hormones and aggression in infra-human vertebrates. In *The biology of aggression* (ed. P. F. Brain and D. Benton), pp. 181–213. Sijthoff and Noordhoff, Alphen aan den Rijn.

Brain, P. F. (1984). Biological explanations of human aggression and the resulting therapies offered by such approaches: a critical evaluation. In *Advances in the study of aggression* (ed. R. J. Blanchard and D. C. Blanchard), pp. 63–102. Academic Press, Orlando.

Brain, P. F. and Benton, D. (1979). The interpretation of physiological correlates of differential housing in laboratory rats. *Life sciences*, **24**, 99–116.

Broca, P. (1878). Anatomie comparée des circonvolutions cérébrales. Le grand lobe limbique et la scissure limbique dans la série des mammifères. *Rev. anthropol.*, **1**, 385–498.

Bronstein, P. M. and Hirsch, S. M. (1976). Ontogeny of defensive reactions in norway rats. *J. comp. physiol. psychol.*, **90**, 620–9.

Bruce, C. J. and Goldberg, M. E. (1984). Physiology of the frontal eye fields. *Trends in neurosciences*, **7**, 436–41.

de Bruin, J. P. C. (1981). Prefrontal cortex lesions and social behaviour. In *Functional recovery from brain damage* (ed. M. W. van Hof and G. Mohn), pp. 239–58. Elsevier, Amsterdam.

de Bruin, J. P. C., van Oyen, H. G. M, and van de Poll, N. (1983). Behavioural changes following lesions of the orbital prefrontal cortex in male rats. *Behav. brain res.*, **10**, 209–32.

Cabanac, M. and Russek, M. (1982). *Régulation et contrôle en biologie*. Les Presses de l'université Laval, Québec.

Cairns, R. B. (1979). *Social development. The origins and plasticity of interchanges*. Freeman and Co, San Francisco.

References

Caprara, G. V., Renzi, P., Alcini, P., D'Império, G., and Travaglia, G. (1983). Instigation to aggress and escalation of aggression examined from a person-ological perspective: the role of irritability and emotional susceptibility. *Aggressive behavior*, **9**, 345–51.

Carrier, H. (1986). La parabole de l'anti-culture. George Orwell, *1984*. In *Big Brother, un inconnu familier*, (ed. F. Rosenstiel and S. G. Shoham), pp. 219–26. Conseil de l'Europe, Strasbourg, et l'Age d'Homme, Paris.

Carroll, D. and O'Callaghan, M. A. J. (1981). Psychosurgery and the control of aggression. In *The biology of aggression* (ed. P. F. Brain and D. Benton), pp. 457–71. Sijthoff and Noordhoff, Alphen aan den Rijn.

Cattarelli, M. and Chanel, J. (1979). Influence of some biologically meaningful odorants on the vigilance states of the rat. *Physiol. behav.*, **23**, 831–8.

Cattarelli, M., Vernet-Maury, E., and Chanel, J. (1977). Modulation de l'activité du bulbe olfactif en fonction de la signification des odeurs chez le rat. *Physiol. behav.* **19**, 381–7.

Chabrol, H. (1984). *Les Comportements suicidaires de l'adolescent*. Presses universi-taires de France, Paris.

Changeux, J.-P. (1983). *L'Homme neuronal*. Fayard, Paris.

Charnay, J.-P. (1981). *Terrorisme et Culture*, Cahier No 20, collection 'Les sept épées'. Fondation pour les études de défense nationale, Paris.

Charrier, Y. and Ellul, J. (1985). *Jeunesse délinquante*. Éditions de l'AREFPPI, Nantes.

Chaurand, J.-P., Vergnes, M., and Karli, P. (1972). Substance grise centrale du mésencéphale et comportement d'agression interspécifique du rat. *Physiol. behav.*, **9**, 475–81.

Chaurand, J.-P., Schmitt, P., and Karli, P. (1973). Effets de lésions du tegmen-tum ventral du mésencéphale sur le comportement d'agression rat-souris. *Physiol. behav.*, **10**, 507–15.

Check, J. V. P. (1985). Pornography and erotica: influences on aggressive atti-tudes. Paper presented at the ISRA meeting, Parma.

Chomsky, N. (1980). Rules and representations. *The behavioral and brain sciences*, **3**, 1–61.

Chorover, S. L. (1980). Violence: a localizable problem? In *The psychosurgery debate* (ed. E. S. Valenstein), pp. 334–47. W. H. Freeman and Co, San Francisco.

Codol, J. P. (1982). Differentiating and non-differentiating behavior: a cognitive approach to the sense of identity. In *Cognitive analysis of social behavior* (ed. J. P. Codol and J. P. Leyens), pp. 267–93. Martinus Nijhoff, The Hague.

Coe, C. L., Wiener, S. G., Rosenberg, L. T., and Levine, S. (1985). Endocrine and immune responses to separation and maternal loss in nonhuman pri-mates. In *The psychobiology of attachment and separation* (ed. M. Reite and T. Field), pp. 163–99. Academic Press, Orlando.

Comstock, G. (1983). Media influences on aggression. In *Prevention and control of aggression* (ed. A. P. Goldstein), pp. 241–72. Pergamon Press, New York.

Cools, A. R. (1980). Role of the neostriatal dopaminergic activity in sequencing and selecting behavioural strategies: facilitation of processes involved in selecting the best strategy in a stressful situation. *Behav. brain res.*, **1**, 361–78.

Cools, A. (1981*a*). Psychopharmacology and aggression: an appraisal of the current situation. In *The biology of aggression* (ed. P. F. Brain and D. Benton), pp. 131–45. Sijthoff and Noordhoff, Alphen aan den Rijn.

Cools, A. (1981*b*). Aspects and prospects of the concept of neurochemical and cerebral organization of aggression: introduction of new research strategies in 'brain and behaviour' studies. In *The biology of aggression* (ed. P. F. Brain and D. Benton), pp. 405–25. Sijthoff and Noordhoff, Alphen aan den Rijn.

Coppens, Y. (1983). *Le Singe, l'Afrique et l'Homme*. Fayard, Paris.

Coppens, Y. (1984). *Leçon inaugurale de la chaire de paléoanthropologie et préhistoire du Collège de France*. Collège de France, Paris.

Corey, D. T. (1978). The determinants of exploration and neophobia. *Neurosci. biobehav. rev.*, **2**, 235–53.

Corkin, S. (1980). A prospective study of cingulotomy. In *The psychosurgery debate* (ed. E. S. Valenstein), pp. 164–204. W. H. Freeman and Co, San Francisco.

Cottraux, J. (1978). Les thérapies comportementales. Bases théoriques et tendances actuelles. In *Psychothérapies médicales*. 1. *Aspects théoriques, techniques et de formation* (ed. J. Guyotat), pp. 171–99. Masson, Paris.

Cusson, M. (1983*a*). *Le Contrôle social du crime*. Presses universitaires de France, Paris.

Cusson, M. (1983*b*). *Why delinquency?* University of Toronto Press, Toronto.

Damasio, A. R. and Geschwind, N. (1984). The neural basis of language. *Ann. rev. neurosci.*, **7**, 127–47.

Dantzer, R. (1981). Le stress des animaux d'élevage. *La Recherche*, **12**(120), 280–9.

Dantzer, R. (1984). Psychobiologie des émotions. In *Neurobiologie des comportements* (ed. J. Delacour), pp. 111–43. Hermann, Paris.

Daruna, J. H. and Kent, E. W. (1976). Comparison of regional serotonin levels and turnover in the brain of naturally high and low aggressive rats. *Brain res.*, **101**, 489–501.

Davis, M. (1980). Neurochemical modulation of sensory-motor reactivity: acoustic and tactile startle reflexes. *Neurosci. biobehav. rev.*, **4**, 241–63.

Debbasch, Ch. (1979). *La Théorie de la stigmatisation et la réalité criminologique*, (preface). Presses universitaires d'Aix-Marseille.

Deguy, M. and Dupuy, J.-P. (1982). *René Girard et le Problème du Mal*. Grasset, Paris.

Delacour, J. (1984). Neurobiologie de l'apprentissage. In *Neurobiologie des comportements* (ed. J. Delacour), pp. 217–57. Hermann, Paris.

Delgado, J. M. R. (1981). Brain stimulation and neurochemical studies on the control of aggression. In *The biology of aggression* (ed. P. F. Brain and D. Benton), pp. 427–55. Sijthoff and Noordhoff, Alphen aan den Rijn.

Delgado, J. M. R., Roberts, W. W., and Miller, N. E. (1954). Learning motivated by electrical stimulation of the brain. *Amer. J. physiol.*, **179**, 587–93.

Dell, P. (1976). Les systèmes non spécifiques à projections diffuses de l'encéphale. In *Physiologie* (ed. Ch. Kayser), t. II (3rd edn), pp. 317–76. Flammarion, Paris.

Delumeau, J. (1985). *Ce que je crois*. Grasset, Paris.

Denenberg, V. H., Hudgens, G. A., and Zarrow, M. X. (1964). Mice reared with rats: modification of behavior by early experience with another species. *Science*, **143**, 380–1.

Depaulis, A. and Vergnes, M. (1984). GABAergic modulation of mouse-killing in the rat. *Psychopharmacology*, **83**, 367–72.

Deschamps, J. C. (1982). Différenciations entre soi et autrui et entre groupes. In *Cognitive analysis of social behavior* (ed. J. P. Codol and J. P. Leyens), pp. 247–66. Martinus Nijhoff, The Hague.

De Sisto, M. J. and Zweig, M. (1974). Differentiation of hypothalamic feeding and killing. *Physiological psychology*, **2**, 67–70.

De Witte, P., Colpaert, F., and Schmitt, P. (1983). The self-regulation of hypothalamic rewarding brain stimulus after innoculation of Mycobacterium butyricum inducing chronic arthritis. *Physiological psychology*, **11**, 201–4.

Didiergeorges, F. and Karli, P. (1966). Stimulations 'sociales' et inhibition de l'aggressivité interspécifique chez le rat privé de ses afférences olfactives. *C.R. soc. biol.* (Paris), **160**, 2445–7.

Di Scala, G., Schmitt, P., and Karli, P. (1984). Flight induced by infusion of bicuculline methiodide into periventricular structures. *Brain res.*, **309**, 199–208.

Dixson, A. F. (1980). Androgens and aggressive behavior in primates: a review. *Aggressive behavior*, **6**, 37–67.

Doms, M. and Moscovici, S. (1984). Innovation et influence des minorités. In *Psychologie sociale* (ed. S. Moscovici), pp. 51–89. Presses universitaires de France, Paris.

Downer, J. L. de C. (1961). Changes in visual gnostic functions and emotional behaviour following unilateral temporal pole damage in the 'split-brain' monkey. *Nature*, **191**, 50–1.

Dupuy, J.-P. (1982). Mimésis et morphogenèse. In *René Girard et le Problème du Mal* (ed. M. Deguy and J.-P. Dupuy), pp. 225–78. Grasset, Paris.

Durant, J. R. (1981). The beast in man: an historical perspective on the biology of human aggression. In *The biology of aggression* (ed. P. F. Brain and D. Benton), pp. 17–46. Sijthoff and Noordhoff, Alphen aan den Rijn.

Durant, J. R. (1985). The science of sentiment: the problem of the cerebral localization of emotion. In *Perspectives in ethology, vol. 6: Mechanisms* (ed. P. P. G. Bateson and P. H. Klopfer), pp. 1–31. Plenum Press, New York.

Duyme, M. (1981). *Les Enfants abandonnés: rôle des familles adoptives et des assistances maternelles.* Monographies françaises de psychologie, No. 56. Éditions du CNRS, Paris.

Ebling, F. J. G. (1981). Ethical considerations in the control of human aggression. In *The biology of aggression* (ed. P. F. Brain and D. Benton), pp. 473–85. Sijthoff and Noordhoff, Alphen aan den Rijn.

Ebtinger, R. and Bolzinger, A. (1982). L'infantile en question. *L'Évolution psychiatrique*, **47**, 85–114.

Eclancher, F. (1983). Early VMH lesions in male and female rats: the role of hunger and olfactory bulbectomies in the initiation of the mouse-killing. *Aggressive behavior*, **9**, 111–12.

Eclancher, F. and Karli, P. (1971). Comportement d'agression interspécifique et comportement alimentaire du rat: effets de lésions des noyaux ventromédians de l'hypothalamus. *Brain res.*, **26**, 71–9.

Eclancher, F. and Karli, P. (1979a). Septal damage in infant and adult rats: effects on activity, emotionality, and muricide. *Aggressive behavior*, **5**, 389–415.

Eclancher, F. and Karli, P. (1979b). Effects of early amygdaloid lesions on the development of reactivity in the rat. *Physiol. behav.*, **22**, 1123–34.

Eclancher, F. and Karli, P. (1981). Environmental influences on behavioural effects of early and late limbic and diencephalic lesions in the rat. In *Functional recovery from brain damage* (ed. M. W. van Hof and G. Mohn), pp. 149–65. Elsevier, Amsterdam.

Eclancher, F., Schmitt, P., and Karli, P. (1975). Effets de lésions précoces de l'amygdale sur le développement de l'agressivité interspécifique du rat. *Physiol. behav.*, **14**, 277–83.

Eibl-Eibesfeldt, I. (1979). *The biology of peace and war*. Thames and Hudson, London.

Eiser, J. R. (1978). Cooperation and competition between individuals. In *Introducing social psychology* (ed. H. Tajfel and C. Fraser), pp. 151–75. Penguin Books, Harmondsworth.

Ekblad, S. (1984). Children's thoughts and attitudes in China and Sweden: impacts of a restrictive versus a permissive environment. *Acta psychiatrica scandinavica*, **70**, 578–90.

Emrich, H. M. (1981). *The role of endorphins in neuropsychiatry*. Karger, Basel.

Engel, J. and Caldecott-Hazard, S. (1984). Altered behavioral states in epilepsy: clinical observations and approaches to basic research. In *Modulation of sensorimotor activity during alterations in behavioral states* (ed. R. Bandler), pp. 471–86. Alan Liss, New York.

Eron, L. D. and Huesmann, L. R. (1984a). The control of aggressive behavior by changes in attitudes, values, and the conditions of learning. In *Advances in the study of aggression* (ed. R. J. Blanchard and D. C. Blanchard), Vol. 1, pp. 139–71. Academic Press, Orlando.

Eron, L. D. and Huesmann, L. R. (1984b). The relation of prosocial behavior to the development of aggression and psychopathology. *Aggressive behavior*, **10**, 201–11.

Essman, W. B. (1981). Drug effects upon aggressive behavior. In *Aggression and violence: a psychobiological and clinical approach* (ed. L. Valzelli and L. Morgese), pp. 150–75. Edizioni Saint-Vincent, Saint-Vincent.

Ettenberg, A. (1979). Conditioned taste preferences as a measure of brain-stimulation reward in rats. *Physiol. behav.*, **23**, 167–72.

Evarts, E. V., Kimura, M., Wurtz, R. H., and Hikosaka, O. (1984). Behavioral correlates of activity in basal ganglia neurons. *Trends in neurosciences*, **7**, 447–53.

Eysenck, H. J. (1980). The bio-social model of man and the unification of psychology. In *Models of man* (ed. A. J. Chapman and D. M. Jones), pp. 49–62. The British Psychological Society, Leicester.

Fernandez de Molina, A. and Hunsperger, R. W. (1962). Organization of the sub-cortical system governing defense and flight reactions in the cat. *J. Physiol., Lond.*, **160**, 200–13.

Feshbach, N. D. (1984). Empathy, empathy training and the regulation of aggression in elementary school children. In *Aggression in children and youth*, (ed. R. M. Kaplan, V. J. Konecni, and R. W. Novaco), pp. 192–208. Martinus Nijhoff Publishers, The Hague.

Feshbach, S. (1984). The catharsis hypothesis, aggressive drive, and the reduction of aggression. *Aggressive behavior*, **10**, 91–101.

Feyereisen, P. and Seron, X. (1984). Les troubles de la communication gestuelle. *La Recherche*, **15**(152), 156–64.

Field, T. (1985). Attachment as psychobiological attunement: being on the same wavelength. In *The psychobiology of attachment and separation* (ed. M. Reite and T. Field), pp. 415–54. Academic Press, Orlando.

Finger, S. and Almli, C. R. (1984). *Early brain damage*, Vol. 2: *Neurobiology and behavior*. Academic Press, Orlando.

Finger, S. and Stein, D. G. (1982). *Brain damage and recovery. Research and clinical perspectives*. Academic Press, New York.

Fitzgerald, R. E. and Burton, M. J. (1983). Neophobia and conditioned taste aversion deficits in the rat produced by undercutting temporal cortex. *Physiol. behav.*, **30**, 203–6.

Flannelly, K. J., Flannelly, L., and Blanchard, R. J. (1984). Adult experience and the expression of aggression: a comparative analysis. In *Biological perspectives on aggression* (ed. K. J. Flannelly, R. J. Blanchard, and D. C. Blanchard), pp. 207–59. Alan Liss, New York.

Flynn, J. P. (1976). Neural basis of threat and attack. In *Biological foundations of psychiatry* (ed. R. G. Grenell and S. Gabay), pp. 273–95. Raven Press, New York.

Fodor, J. A. (1980). Methodological solipsism considered as a research strategy in cognitive psychology. *The behavioral and brain sciences*, **3**, 63–109.

Fokkema, D. S. and Koolhaas, J. M. (1985). Acute and conditioned blood pressure changes in relation to social and psychosocial stimuli in rats. *Physiol. behav.*, **34**, 33–8.

Follick, M. J. and Knutson, J. F. (1978). Punishment of irritable aggression. *Aggressive behavior*, **4**, 1–17.

Fonberg, E. (1981). Specific versus unspecific functions of the amygdala. In *The amygdaloid complex* (ed. Y. Ben Ari), pp. 281–91. Elsevier, Amsterdam.

Fontaine, O. (1978). *Introduction aux thérapies comportementales*. Pierre Mardaga, Brussels.

Fougeyrollas, P. (1985). *Les Métamorphoses de la crise. Racismes et révolutions au xxᵉ siècle*. Hachette, Paris.

Franzen, E. A. and Myers, R. E. (1973). Neural control of social behavior: prefrontal and anterior temporal cortex. *Neuropsychologia*, **11**, 141–57.

Frégnac, Y. and Imbert, M. (1984). Development of neuronal selectivity in primary visual cortex of cat. *Physiol. rev.*, **64**, 325–434.

Freud, S. (1981). *Essais de psychanalyse*. Petite Bibliothèque Payot, No. 44, Paris.

Freund, J. (1983). *Sociologie du conflit*. Presses universitaires de France, Paris.

Furtos, J. (1978). Le mouvement de psychologie humaniste aux États-Unis. In *Psychothérapies médicales. I. Aspects théoriques, techniques et de formation* (ed. J. Guyotat), p. 201–28. Masson, Paris.

Fuster, J. M. (1984). Behavioral electrophysiology of the prefrontal cortex. *Trends in neurosciences*, **7**, 408–14.

Gainotti, G. (1983). Laterality of affect: the emotional behavior of right- and left-brain-damaged patients: In *Hemisyndromes* (ed. M. S. Myslobodsky), pp. 175–92. Academic Press, New York.

Galef, B. G. Jr. (1970). Aggression and timidity: responses to novelty in feral norway rats. *J. comp. physiol. psychol.*, **70**, 370–81.

Gandelman, R. (1981). Androgen and fighting behavior. In *The biology of aggression* (ed. P. F. Brain and D. Benton), pp. 215–30. Sijthoff and Noordhoff, Alphen aan den Rijn.

Gandelman, R. (1983). Gonadal hormones and sensory function. *Neuroscience and biobehavioral reviews*, **7**, 1–17.

Garbanati, J. A., Sherman, G. F., Rosen, G. D., Hofmann, M., Yutzey, D. A., and Denenberg, V. H. (1983). Handling in infancy, brain laterality and muricide in rats. *Behav. brain res.*, **7**, 351–9.

Gassin, R. (1979). Rapport introductif. In *La Théorie de la stigmatisation et la réalité criminologique*, pp. 23–30. Presses universitaires d'Aix-Marseille.

Gautier, J.-P. and Deputte, B. (1983). La communication vocale chez les singes, *La Recherche*, **14**(140), 53–63.

Gillan, D. J. (1982). Ascent of apes. In *Animal mind-human mind* (ed. D. R. Griffin), pp. 177–200. Springer Verlag, Berlin.

Gilly, M. (1984). Psychosociologie de l'éducation. In *Psychologie sociale* (ed. S. Moscovici), pp. 473–94. Presses universitaires de France, Paris.

Girard, A. (1986). L'enjeu culturel. In *Les Enjeux de la fin du siècle*, pp. 69–91. Desclée de Brouwer, Paris.

Girard, R. (1982). *Le Bouc émissaire*. Grasset et Fasquelle, Paris.

Giurgea, C. E. (1983). Les perspectives actuelles de la psychopharmacologie. *Bulletin de l'Académie royale de médecine de Belgique*, **138**, 431– 44.

Giurgea, C. E. (1984). Psychopharmacology tomorrow: '1984' or 'The little Prince'?, *Psychological medicine*, **14**, 491–6.

Glover, J. (1984). *What sort of people should there be? Genetic engineering, brain control and their impact on our future world*. Penguin Books, Harmondsworth.

Glowinski, J., Tassin, J. P., and Thierry, A. M. (1984). The mesocortico-prefrontal dopaminergic neurons. *Trends in neurosciences*, **7**, 415–18.

Goldblum, M.-C. and Tzavaras, A. (1984). La communication et ses troubles après lésion du système nerveux central. In *Neurobiologie des comportements* (ed. J. Delacour), pp. 177–202. Hermann, Paris.

Goldman-Rakic, P. S. (1984a). Modular organization of prefrontal cortex. *Trends in neurosciences*, **7**, 419–24.

Goldman-Rakic, P. S. (1984b). The frontal lobes: uncharted provinces of the brain. *Trends in neurosciences*, **7**, 425–9.

Goldstein, A. P. (1983a). *Prevention and control of aggression*. Pergamon Press, New York.

Goldstein, A. P. (1983b). Behavior modification approaches to aggression prevention and control. In *Prevention and control of aggression* (ed. A. P. Goldstein), pp. 156–209. Pergamon Press, New York.

Goldstein, A. P. and Keller, H. R. (1983). Aggression prevention and control: multitargeted, multichannel, multiprocess, multidisciplinary. In *Prevention and control of aggression* (ed. A. P. Goldstein), pp. 338–50. Pergamon Press, New York.

Goldstein, D. (1983). Spouse abuse. In *Prevention and control of aggression* (ed. A. P. Goldstein), pp. 37–65. Pergamon Press, New York.

Gorz, A. (1985). *Entretiens avec 'le Monde' — 6. La société*, pp. 202–13. La Découverte/Le Monde, Paris.

Gray, J. A. (1982). Précis of The neuropsychology of anxiety : an enquiry into the functions of the septo-hippocampal system. *The behavioral and brain sciences*, **5**, 469–534.

Green, S. and Marler, P. (1979). The analysis of animal communication. In *Handbook of behavioral neurobiology. — 3. Social behavior and communication* (ed. P. Marler and J. C. Vandenbergh) pp. 73–158. Plenum Press, New York.

Grossman, S. P. (1964). Behavioral effects of chemical stimulation of the ventral amygdala. *J. comp. physiol. psychol.*, **57**, 29–36.

Grossman, S. P. (1972). Aggression, avoidance, and reaction to novel environments in female rats with ventromedial hypothalamic lesions. *J. comp. physiol. psychol.*, **78**, 274–83.

Gully, K. J. and Dengerink, H. A. (1983). The dyadic interaction of persons with violent and nonviolent histories. *Aggressive behavior*, **9**, 13–20.

Gunnar, M. R. (1980). Contingent stimulation: a review of its role in early development. In *Coping and health* (ed. S. Levine and H. Ursin), pp. 101–19. Plenum Press, New York.

Guyotat, J. (1978). *Psychothérapies médicales — 1. Aspects théoriques, techniques et de formation*. Masson, Paris.

Halgren, E. (1981). The amygdala contribution to emotion and memory: current studies in humans. In *The amygdaloid complex* (ed. Y. Ben-Ari), pp. 395–408. Elsevier, Amsterdam.

Halsband, U. and Passingham, R. (1982). The role of premotor and parietal cortex in the direction of action. *Brain res.*, **240**, 368–72.

Hamburg, D. A. and van Lawick-Goodall, J. (1974). Factors facilitating development of aggressive behavior in chimpanzees and humans. In *Determinants and origins of aggressive behavior* (ed. J. de Wit and W. W. Hartup), pp. 59–85. Mouton, The Hague.

Hand, T. H. and Franklin, K. B. J. (1983). The influence of amphetamine on preference for lateral hypothalamic versus prefrontal cortex or ventral tegmental area self-stimulation. *Pharmacol. biochem. behav.*, **18**, 695–9.

Harootunian, B. and Apter, S. J. (1983). Violence in school. In *Prevention and control of aggression* (ed. A. P. Goldstein), pp. 66–83. Pegamon Press, New York.

Haug, M., Mandel, P. and Brain, P. F. (1981). Studies on the biological correlates of attack by group-housed mice on lactating intruders. In *The biology of aggression* (ed. P. F. Brain and D. Benton), pp. 509–17. Sijthoff and Noordhoff, Alphen aan den Rijn.

Heath, R. (1963). Electrical self-stimulation of the brain in man. *Amer. J. psychiat.*, **120**, 571–7.

Hein, A., Vital-Durand, F., Salinger, W., and Diamond, R. (1979). Eye movements initiate visual-motor development in the cat. *Science*, **204**, 1321–2.

Herrick, C. J. (1933). The functions of the olfactory parts of the cerebral cortex. *Proc. nat. Acad. sci., Washington*, **19**, 7–14.

Hinde, R. A. (1982). *Ethology. Its nature and relations with other sciences*. Oxford University Press, Oxford.

Hinde, R. A. (1983). *Primate social relationships*. Blackwell Scientific Publications, Oxford.

Hinton, J. W. (1981). Adrenal cortical and medullary hormones and their psychophysiological correlates in violent and in psychopathic offenders. In *The biology of aggression* (ed. P. F. Brain and D. Benton), pp. 291–300. Sijthoff and Noordhoff, Alphen aan den Rijn.

van Hof, M. W. and Mohn, G. (1981). *Functional recovery from brain damage*. Elsevier, Amsterdam.

Hood, K. E. (1984). Aggression among female rats during the estrus cycle. In *Biological perspectives on aggression* (ed. K. J. Flannelly, R. J. Blanchard, and D. C. Blanchard), pp. 181–8. Alan Liss, New York.

Hugelin, A. (1976). L'activation. In *Physiologie*, t. II (3rd edn). (ed. Ch. Kayser), pp. 1193–265. Flammarion, Paris.

Huston, J. P., Morgan, S., and Steiner, H. (1985). Behavioral correlates of plasticity in substantia nigra efferents. In *Brain plasticity, learning and memory* (ed. B. E. Will, P. Schmitt, and J. C. Dalrymple-Alford). Plenum Press, New York.

Hutchinson, R. R. (1983). The pain-aggression relationship and its expression in naturalistic settings. *Aggressive behavior*, **9**, 229–42.

Hyvärinen, J. (1982). Posterior parietal lobe of the primate brain. *Physiol. reviews*, **62**, 1060–129.

Ikeda, D. and Peccei, A. (1986). *Cri d'alarme pour le XXI⁰ siècle*. Presses universitaires de France, Paris.

Israël, L. (1984). *Initiation à la psychiatrie*. Masson, Paris.

Itil, T. M. (1981). Drug therapy in the management of aggression. In *Multidisciplinary approaches to aggression research* (ed. P. F. Brain and D. Benton), pp. 489–502. Elsevier, Amsterdam.

Jaccard, R. (1985). Les rhapsodies du moi. In *Entretiens avec 'le Monde'—5. L'individu*, pp. 5–13. La Découverte/Le Monde, Paris.

Jacob, F. (1981). *Le Jeu des possibles*. Fayard, Paris.

Jacobs, P. A., Brunton, M., Melville, M. M., Brittain, R. P., and McClemont, W. F. (1965). Aggressive behaviour, mental subnormality and the XYY male. *Nature*, **208**, 1351–2.

Jaspars, J. M. F. (1982). Social judgement and social behaviour. A dual representation model. In *Cognitive analysis of social behavior* (ed. J. P. Codol and J. P. Leyens), pp. 1–49. Martinus Nijhoff, The Hague.

Jaspers, R., Schwarz, M., Sontag, K. H., and Cools, A. R. (1984). Caudate nucleus and programming behaviour in cats: role of dopamine in switching motor patterns. *Behav. brain res.*, **14**, 17–28.

Jeannerod, M. (1983). *Le Cerveau-machine. Physiologie de la volonté*. Fayard, Paris.

Jeannerod, M. and Hécaen, H. (1979). *Adaptation et restauration des fonctions nerveuses*. Simep, Villeurbanne.

Jeavons, C. M. and Taylor, S. P. (1985). The control of alcohol-related aggression: redirecting the inebriate's attention to socially appropriate conduct'. *Aggressive behavior* **11**, 93–101.

Jenck, F., Schmitt, P., and Karli, P. (1983). Morphine applied to the mesencephalic central gray suppresses brain stimulation induced escape. *Pharmacol. biochem. behav.*, **19**, 301–8.

Jenck, F., Schmitt, P., and Karli, P. (1986). Morphine injected into the periaqueductal gray attenuates brain stimulation-induced aversive effects: an intensity discrimination study. *Brain res.*, **378**, 274–84.

Jodelet, D. (1984). Représentation sociale: phénomènes, concept et théorie. In *Psychologie sociale* (ed. S. Moscovici), pp. 357–78. Presses universitaires de France, Paris.

Jones, D. G. and Smith, B. J. (1980). The hippocampus and its response to differential environments. *Progr. neurobiol.*, **15**, 19–69.

Jouvet, M. (1984). Mécanismes des états de sommeil. In *Physiologie du sommeil*, (ed. O. Benoît), pp. 1–18. Masson, Paris.

Juraska, J. M., Greenough, W. T., and Conlee, J. W. (1983). Differential rearing affects responsiveness of rats to depressant and convulsant drugs. *Physiol. behav.*, **31**, 711–15.

Jürgens, U. (1982). Amygdalar vocalization pathways in the squirrel monkey, *Brain res.*, **241**, 189–96.

Jürgens, U. (1983). Control of vocal aggression in squirrel monkeys. In *Advances in vertebrate neuroethology* (ed. J. P. Ewert, R. R. Capranica, and D. J. Ingle), pp. 1087–102. Plenum Publishing Corporation, New York.

Kaas, J. H., Merzenich, M. M., and Killackey, H. P. (1983). The reorganization of somatosensory cortex following peripheral nerve damage in adult and developing mammals. *Ann. rev. neurosci.*, **6**, 325–56.

Kanki, J. P., Martin, T. L., and Sinnamon, H. M. (1983). Activity of neurons in the anteromedial cortex during rewarding brain stimulation, saccharin consumption and orienting behavior. *Behav. brain res.*, **8**, 69–84.

Kaplan, J.-C. (1983). Le génie génétique. In *Le Genre humain. —6. Les manipulations*, pp. 72–93. Fayard, Paris.

Karli, P. (1956). The Norway rat's killing-response to the white mouse. An experimental analysis. *Behaviour*, **10**, 81–103.

Karli, P. (1960). Effets de lésions expérimentales du septum sur l'agressivité interspécifique rat-souris. *C.R. soc. biol., Paris*, **154**, 1079–82.

Karli, P. (1971). Les conduites agressives. *La Recherche*, **2**(18), 1013–21.

Karli, P. (1976). Neurophysiologie du comportement. In *Physiologie* (ed. Ch. Kayser), t. II (3rd edn), pp. 1331–454. Flammarion, Paris.

Karli, P. (1981). Conceptual and methodological problems associated with the study of brain mechanisms underlying aggressive behaviour. In *The biology of aggression* (ed. P. F. Brain and D. Benton), pp. 323–61. Sijthoff and Noordhoff, Alphen aan den Rijn.

Karli, P. (1982). *Neurobiologie des comportements d'agression*. Presses universitaires de France, Paris.

Karli, P. (1984). Complex dynamic interrelations between sensorimotor activities and so-called behavioural states. In *Modulation of sensorimotor activity during alterations in behavioral states* (ed. R. Bandler), pp. 1–21. Alan Liss, New York.

Karli, P. (1986). Le cerveau de l'homme: source de contrainte ou instrument de liberté?. In *Big Brother, un inconnu familier* (ed. F. Rosenstiel and S. G. Shoham), pp. 251–67. Conseil de l'Europe, Strasbourg et l'Age d'Homme, Paris.

Karli, P. and Vergnes, M. (1964). Dissociation expérimentale du comportement d'agression interspécifique rat-souris et du comportement alimentaire. *C.R. soc. biol., Paris*, **158**, 650–3.

Karli, P., Vergnes, M., and Didiergeorges, F. (1969). Rat-mouse interspecific aggressive behaviour and its manipulation by brain ablation and by brain stimulation. In *Aggressive behaviour* (ed. S. Garattini and E. B. Sigg), pp. 47–55. Excerpta Medica Foundation, Amsterdam.

Karli, P., Vergnes, M., Eclancher, F., Schmitt, P., and Chaurand, J.-P. (1972). Role of the amygdala in the control of mouse-killing behavior in the rat. In *The neurobiology of the amygdala* (ed. B. E. Eleftheriou), pp. 553–80. Plenum Press, New York.

Karli, P., Eclancher, F., Vergnes, M., Chaurand, J.-P., and Schmitt, P. (1974). Emotional responsiveness and interspecific aggressiveness in the rat: interactions between genetic and experiential determinants. In *The genetics of behaviour* (ed. J. H. F. van Abeelen), pp. 291–319. North-Holland, Amsterdam.

Katz, R. J. (1981). Animal models and human depressive disorders. *Neurosci. biobehav. rev.*, **5**, 231–46.

Keller, H. R. and Erne, D. (1983). Child abuse: toward a comprehensive model. In *Prevention and control of aggression* (ed. A. P. Goldstein), pp. 1–36. Pergamon Press, New York.

Kelley, A. E. and Domesick, V. B. (1982). The distribution of the projection from the hippocampal formation to the nucleus accumbens in the rat: an anterograde–and retrograde-horseradish peroxydase study. *Neuroscience, 7,* 2321–35.

Kelley, A. E., Domesick, V. B., and Nauta, W. J. H. (1982). The amygdalostriatal projection in the rat: an anatomical study by anterograde and retrograde tracing methods. *Neuroscience, 7,* 615–30.

Kemble, E. D., Flannelly, K. J., Salley, H. and Blanchard, R. J. (1985). Mouse killing, insect predation, and conspecific attack by rats with differing prior aggressive experience. *Physiol. behav.*, **34**, 645–8.

Kesner, R. P. (1981). The role of the amygdala within an attribute analysis of memory. In *The amygdaloid complex* (ed. Y. Ben-Ari), pp. 331–42. Elsevier, Amsterdam.

Kesner, R. P. and Hardy, J. D. (1983). Long- term memory for contextual attributes: dissociation of amygdala and hippocampus. *Behav. brain res.*, **8**, 139–49.

Kling, A. (1972). Effects of amygdalectomy on social-affective behavior in nonhuman primates. In *The neurobiology of the amygdala* (ed. B. E. Eleftheriou), pp. 511–36. Plenum Press, New York.

Kling, A. (1981). Influence of temporal lobe lesions on radiotelemetered electrical activity of amygdala to social stimuli in monkey. In *The amygdaloid complex* (ed. Y. Ben-Ari), pp. 271–80. Elsevier, Amsterdam.

Kling, A. S., Perryman, K., and Parks, T. (1984). A telemetry system for the study of cortical-amygdala relationships in the squirrel monkey (S. sciureus). In *Modulation of sensorimotor activity during alterations in behavioral states* (ed. R. Bandler), pp. 351–65. Alan Liss, New York.

Klüver, H. and Bucy, P. C. (1937). 'Psychic blindness' and other symptoms following bilateral temporal lobectomy in rhesus monkeys. *Amer. J. physiol.*, **119**, 352–3.

Knutson, J. F. and Viken, R. J. (1984). Animal analogues of human aggression: studies of social experience and escalation. In *Biological perspectives on aggression* (ed. K. J. Flannelly, R. J. Blanchard, and D. C. Blanchard), pp. 75–94. Alan Liss, New York.

Kolb, B. (1984). Functions of the frontal cortex of the rat: a comparative review. *Brain research reviews*, **8**, 65– 98.

Koolhaas, J. M. (1978). Hypothalamically induced intraspecific aggressive behaviour in the rat. *Exp. brain res.*, **32**, 365–75.

Koolhaas, J. M. (1984). The corticomedial amygdala and the behavioral change due to defeat. In *Modulation of sensorimotor activity during alterations in behavioral states* (ed. R. Bandler), pp. 341–9. Alan Liss, New York.

Koolhaas, J. M., Schuurman, T., and Wiepkema, P. R. (1980). The organization of intraspecific agonistic behaviour in the rat. *Progr. neurobiol.*, **15**, 247–68.

Kornadt, H.-J. (1984). Development of aggressiveness: a motivation theory perspective. In *Aggression in children and youth* (ed. R. M. Kaplan, V. J. Konecni, and R. W. Novaco), pp. 73–87. Martinus Nijhoff Publishers, The Hague.

Kostowski, W., Plewako, M., and Bidzinski, A. (1984). Brain serotonergic neurons: their role in a form of dominance-subordination behavior in rats. *Physiol. behav.*, **33**, 365–71.

Kraemer, G. W. (1985). Effects of differences in early social experience on primate neurobiological-behavioral development. In *The psychobiology of attachment and separation* (ed. M. Reite and T. Field), pp. 135–61. Academic Press, Orlando.

Lagerspetz, K. M. J. and Lagerspetz, K. Y. H. (1974). Genetic determination of aggressive behaviour. In *The genetics of behaviour* (ed. J. H. F. van Abeelen), pp. 321–46. North-Holland, Amsterdam.

Lagerspetz, K. M. J. and Sandnabba, K. (1982). The decline of aggressiveness in male mice during group caging as determined by punishment delivered by the cage mates. *Aggressive behavior*, **8**, 319–34.

Lagerspetz, K. M. J. and Westman, M. (1980). Moral approval of aggressive acts: a preliminary investigation. *Aggressive behavior*, **6**, 119–30.

Lappuke, R., Schmitt, P., and Karli, P. (1982). Discriminative properties of aversive brain stimulation. *Behavioral and neural biology*, **34**, 159–79.

Lau, P. and Miczek, K. A. (1977). Differential effects of septal lesions on attack and defensive-submissive reactions during intraspecies aggression in rats. *Physiol. behav.*, **18**, 479–85.

Leauté, J. (1972). *Criminologie et Science pénitentiaire*. Presses universitaires de France, Paris.

Le Magnen, J. (1983). Body energy balance and food intake: a neuroendocrine regulatory mechanism. *Physiol. rev.*, **63**, 314–86.

Le Magnen, J. (1984). Bases neurobiologiques du comportement alimentaire. In *Neurobiologie des comportements* (ed. J. Delacour), pp. 3–54. Hermann, Paris.

Le Moal, M. (1984). Quelques aspects de la recherche biologique en psychiatrie. In *Neurobiologie des comportements* (ed. J. Delacour), pp. 147–74. Hermann, Paris.

Lenclud, G., Claverie, E., and Jamin, J. (1984). Une ethnographie de la violence est-elle possible? *Études rurales*, **95–6**, 9–21.

Leonard, C. M., Rolls, E. T., Wilson, F. A. W., and Baylis, G. C. (1985). Neurons in the amygdala of the monkey with responses selective for faces. *Behav. brain res.*, **15**, 159–76.

Leshner, A. I. (1981). The role of hormones in the control of submissiveness. In *Multidisciplinary approaches to aggression research* (ed. P. F. Brain and D. Benton), pp. 309–22. Elsevier, Amsterdam.

Lett, B. T. and Harley, C. W. (1974). Stimulation of lateral hypothalamus during sickness attenuates learned flavor aversions. *Physiol. behav.*, **12**, 79–83.

Levasseur, G. (1979). Rapport de synthèse. In *La Théorie de la stigmatisation et la réalité criminologique*, pp. 347–65. Presses universitaires d'Aix-Marseille.

Levinas, E. (1985). Religion et idée de l'infini. In *Douze Leçons de philosophie*, pp. 142–50. La Découverte/Le Monde, Paris.

Levinson, D. M., Reeves, D. L., and Buchanan, D. R. (1980). Reductions in aggression and dominance status in guinea pigs following bilateral lesions in the basolateral amygdala or lateral septum. *Physiol. behav.*, **25**, 963–71.

Lewis, M. E., Mishkin, M., Bragin, E., Brown, R. M., Pert, C. B. and Pert, A. (1981). Opiate receptor gradients in monkey cerebral cortex: correspondence with sensory processing hierarchies. *Science*, **211**, 1166–9.

Leyens, J.-P. (1979). *Psychologie sociale*. Pierre Mardaga, Brussels.

Leyens, J.-P. (1982). Implications des théories implicites de personnalité pour le diagnostic psychologique. In *Cognitive analysis of social behavior* (ed. J. P. Codol and J. P. Leyens), pp. 171–205. Martinus Nijhof, The Hague.

Lhermitte, F. (1983). 'Utilization behavior' and its relation to lesions of the frontal lobes. *Brain*, **106**, 237–55.

Livingston, R. B. (1978). A casual glimpse of evolution and development relating to the limbic system. In *Limbic mechanisms. The continuing evolution of the limbic system concept* (ed. K. E. Livingston and O. Hornykiewicz), pp. 17–21. Plenum Press, New York.

Livingstone, M. S. and Hubel, D. H. (1981). Effects of sleep and arousal on the processing of visual information in the cat. *Nature, Lond.*, **291**, 554–61.

Looney, T. A. and Cohen, P. S. (1982). Aggression induced by intermittent positive reinforcement. *Neurosci. biobehav. rev.*, **6**, 15–37.

Lore, R. and Takahashi, L. (1984). Postnatal influences on intermale aggression in rodents. In *Biological perspectives on aggression* (ed. K. J. Flannelly, R. J. Blanchard, and D. C. Blanchard), pp. 189–206. Alan Liss, New York.

Lorens, S. A. (1978). Some behavioral effects of serotonin depletion depend on method: a comparison of 5,7-dihydroxytryptamine, p-chlorophenylalanine, p-chlorophenylamphetamine and electrolytic raphe lesions. *Ann. N.Y. Acad. sci.*, **305**, 532–55.

Lorenz, K. (1967). *On aggression*. Methuen, London.

Luciano, D. and Lore, R. (1975). Aggression and social experience in domesticated rats. *J. comp. physiol. psychol.*, **88**, 917–23.

Mack, G. (1978). Contribution à l'étude des bases moléculaires du comportement muricide chez le rat. Ph.D. Thesis, Université Louis-Pasteur, Strasbourg.

Mac Lean, P. D. (1977). On the evolution of three mentalities. In *New dimensions in psychiatry: a world view*, Vol. 2. (ed. S. Arieti and G. Chrzanowski), pp. 306–28. J. Wiley and Sons, New York.

McCarthy, P. (1974). Youths who murder. In *Determinants and origins of aggressive behavior* (ed. J. de Wit and W. W. Hartup), pp. 589–94. Mouton, The Hague.

Mandel, P. (1978). Biochimie de l'agressivité, In *Exposés sur l'agressivité*, pp. 27–39. Publications de l'Académie des sciences, Institut 1978–3, Gauthier-Villars, Paris.

Mandel, P., Mack, G., and Kempf, E. (1979). Molecular basis of some models of aggressive behavior. In *Psychopharmacology of Aggression* (ed. M. Sandler), pp. 95–110. Raven Press, New York.

Mandell, A. J. (1983). Temporal and spatial dynamics underlying two neuropsychobiological hemisyndromes: hysterical and compulsive personality styles. In *Hemisyndromes. Psychobiology, neurology, psychiatry* (ed. M. S. Myslobodsky), pp. 327–46. Academic Press, New York.

Mark, V. H. and Carnahan W. A. (1980). Organic brain disease: a separate issue. In *The psychosurgery debate* (ed. E. S. Valenstein), pp. 129–38. W. H. Freeman and Company, San Francisco.

Markowitsch, H. J. (1982). Thalamic mediodorsal nucleus and memory: a critical evaluation of studies in animals and man. *Neurosci. biobehav. rev.*, **6**, 351–80.

Marler, P. (1976). On animal aggression: the roles of strangeness and familiarity. *Amer. psychol.*, **31**, 239–46.

Marler, P. and Vandenbergh, J. G. (1979). *Handbook of behavioral neurobiology,* Vol. 3: *Social behavior and communication.* Plenum Press, New York.

Mason, W. (1979). Ontogeny of social behavior. In *Social behavior and communication,* Vol. 3 of *Handbook of behavioral neurobiology* (ed. P. Marler and J. G. Vandenbergh), pp. 1–28. Plenum Press, New York.

Mason, W. A. (1982). Primate social intelligence: contributions from the laboratory. In *Animal mind-human mind* (ed. D. R. Griffin), pp. 131–44. Springer Verlag, Berlin.

Maxim, P. E. (1972). Behavioral effects of telestimulating hypothalamic reinforcement sites in freely moving rhesus monkeys. *Brain res.,* **42,** 243–62.

Maxson, S. C. (1981). The genetics of aggression in vertebrates. In *The biology of aggression* (ed. P. F. Brain and D. Benton), pp. 69–104. Sijthoff and Noordhoff, Alphen aan den Rijn.

Médioni, J. and Vaysse, G. (1984). La transmission des comportements. *La Recherche,* **15**(155), 698–712.

Meehan, W. P. and Henry, J. P. (1981). Social stress and the role of attachment behaviour in modifying aggression. In *Multidisciplinary approaches to aggression research* (ed. P. F. Brain and D. Benton), pp. 209–23. Elsevier, Amsterdam.

Memmi, A. (1985). *Ce que je crois.* Grasset, Paris.

Mesulam, M. M. (1981). A cortical network for directed attention and unilateral neglect. *Ann. neurol.,* **10,** 309–25.

Meyer-Bahlburg, H. F. L. (1981a). Sex chromosomes and aggression in humans. In *The biology of aggression* (ed. P. F. Brain and D. Benton), pp. 109–23. Sijthoff and Noordhoff, Alphen aan den Rijn.

Meyer-Bahlburg, H. F. L. (1981b). Androgens and human aggression. In *The biology of aggression* (ed. P. F. Brain and D. Benton), pp. 263–90. Sijthoff and Noordhoff, Alphen aan den Rijn.

Miczek, K. A. and Grossman, S. P. (1972). Effects of septal lesions on inter- and intraspecies aggression in rats. *J. comp. physiol. psychol.,* **79,** 37–45.

Miczek, K. A. and Krsiak, M. (1981). Pharmacological analysis of attack and flight. In *Multidisciplinary approaches to aggression research* (ed. P. F. Brain and D. Benton), pp. 341–54. Elsevier, Amsterdam.

Miczek, K. A. and Thompson, M. L. (1984). Analgesia resulting from defeat in a social confrontation: the role of endogenous opioids in brain. In *Modulation of sensorimotor activity during alterations in behavioral states* (ed. R. Bandler), pp. 431–56. Alan Liss, New York.

Miley, W. M. and Baenninger, R. (1972). Inhibition and facilitation of interspecies aggression in septal lesioned rats. *Physiol. behav.,* **9,** 379–84.

Miller, A. (1986). *L'Enfant sous terreur.* Aubier, Paris.

Milner, B. and Petrides, M. (1984). Behavioural effects of frontal-lobe lesions in man. *Trends in neurosciences,* **7,** 403–7.

Mishkin, M. (1978). Memory in monkeys severely impaired by combined but not by separate removal of amygdala and hippocampus. *Nature,* **273,** 297–8.

Mogenson, G. J. (1984). Limbic-motor integration — with emphasis on initiation of exploratory and goal-directed locomotion. In *Modulation of sensorimotor activity during alterations in behavioral states* (ed. R. Bandler), pp. 121–37. Alan Liss, New York.

Montagu, A. (1974). Aggression and the evolution of man. In *The neuropsychology of aggression* (ed. R. E. Whalen), pp. 1–32. Plenum Press, New York.

Montlibert, C. de (1984). Manifestations et violences à Longwy — 1979. In *Revue des sciences sociales de la France de l'Est*, No. 13 and 13 bis, pp. 73–96. Université des sciences humaines, Strasbourg.

Montmollin, G. de (1977). *L'Influence sociale. Phénomènes, facteurs et théories.* Presses universitaires de France, Paris.

Montmollin, G. de (1984). Le changement d'attitude. In *Psychologie sociale* (ed. S. Moscovici), pp. 91–138. Presses universitaires de France, Paris.

Moorcroft, W. H. (1971). Ontogeny of forebrain inhibition of behavioral arousal in the rat. *Brain res.*, **35**, 513–22.

Moorcroft, W. H., Lytle, L. D., and Campbell, B. A. (1971). Ontogeny of starvation–induced behavioral arousal in the rat. *J. comp. physiol. psychol.*, **75**, 59–67.

Morin, E. (1973). *Le Paradigme perdu: la nature humaine.* Éditions du Seuil, Paris.

Mos, J., Kruk, M. R., van der Poel, A. M. and Meelis, W. (1982). Aggressive behavior induced by electrical stimulation in the midbrain central gray of male rats. *Aggressive behavior*, **8**, 261–84.

Moscovici, S. (1982). The coming era of representations. In *Cognitive analysis of social behavior* (ed. J. P. Codol and J. P. Leyens), pp. 115–50. Martinus Nijhoff, The Hague.

Moscovici, S. (1985). *Entretiens avec 'le Monde'—6. La société*, pp. 67–77. La Découverte/Le Monde, Paris.

Mucchielli, A. (1981). *Les Motivations.* 'Que sais-je?', No. 1949. Presses universitaires de France, Paris.

Muraise, E. (1982). Le pouvoir des mythes. *Études polémologiques.* No. 25–26, pp. 75–88.

Myers, R. E., Swett, C., and Miller, M. (1973). Loss of social group affinity following prefrontal lesions in free-ranging macaques. *Brain res.*, **64**, 257–69.

Narabayashi, H. (1972). Stereotaxic amygdalectomy. In *The neurobiology of the amygdala* (ed. B. E. Eleftheriou), pp. 459–83. Plenum Press, New York.

Nauta, W. J. H. (1958). Hippocampal projections and related neural pathways to the midbrain in the cat. *Brain*, **81**, 319–40.

Nédoncelle, M. (1974). La manipulation des esprits. In *L'Homme manipulé*, pp. 53–67. Cerdic-Publications, Université des sciences humaines de Strasbourg.

Nuttin, J. M. (1972). Changement d'attitude et role playing. In *Introduction à la psychologie sociale*, Vol. 1 (ed. S. Moscovici), pp. 13–58. Larousse, Paris.

Nuttin, J. (1980). *Théorie de la motivation humaine.* Presses universitaires de France, Paris.

Olds, J. and Milner, P. (1954). Positive reinforcement produced by electrical stimulation of septal area and other regions of rat brain. *J. comp. physiol. psychol.*, **47**, 419–27.

Olgiati, V. R. and Pert, C. B. (1982). Visualization of maternal deprivation-induced alterations in opiate receptor brain distribution patterns in young rats. In *Neuroscience*, **7** (suppl.), S162.

Olweus, D. (1984). Development of stable aggressive reaction patterns in males. In *Advances in the study of aggression*, Vol. 1 (ed. R. J. Blanchard, and D. C. Blanchard), pp. 103–137. Academic Press, Orlando.

Olweus, D., Mattson, A., Schalling, D., and Lööw, H. (1980). Testosterone, aggression, physical and personality dimensions in normal adolescents. *Psychosomatic medicine*, **42**, 253–69.

Overmier, J. B., Patterson, J., and Wielkiewicz, R. M. (1980). Environmental contingencies as sources of stress in animals. In *Coping and health* (ed. S. Levine and H. Ursin), pp. 1–38, Plenum Press, New York.

Owens, D. J. and Straus, M. A. (1975). The social structure of violence in childhood and approval of violence as an adult. *Aggressive behavior*, **1**, 193–211.

Paddock, J. (1975). Studies on antiviolent and 'normal' communities. *Aggressive behavior*, **1**, 217–33.

Paddock, J. (1979). Symposium: faces of antiviolence. *Aggressive behavior*, **5**, 210–14.

Paddock, J. (1980). Learning non-aggression: the experience of non-literate societies, edited by Ashley Montagu. *Aggressive behavior*, **6**, 180–4.

Pager, J., Giachetti, I., Holley, A., and Le Magnen, J. (1972). A selective control of olfactory bulb electrical activity in relation to food deprivation and satiety in rats. *Physiol. behav.*, **9**, 573–9.

Paillard, J. (1978). The pyramidal tract: two million fibres in search of a function. *J. Physiol., Paris*, **74**, 155–62.

Paillard, J. (1982). Apraxia and the neurophysiology of motor control. *Phil. trans. R. soc. Lond.*, **B298**, 111–34.

Panksepp, J. (1979). A neurochemical theory of autism. *Trends neurosci.*, **2**, 174–7.

Panksepp, J. (1982). Toward a general psychobiological theory of emotions. *The Behavioural and Brain Sciences*, **5**, 407–67.

Panksepp, J., Herman, B. H., Vilberg, T., Bishop, P., and De Eskinazi, F. G. (1980). Endogenous opioids and social behavior. *Neurosci. biobehav. rev.*, **4**, 473–87.

Panksepp, J., Siviy, S., and Normansell, L. (1984). The psychobiology of play: theoretical and methodological perspectives. *Neurosci. biobehav. rev.*, **8**, 465–92.

Panksepp, J., Siviy, S. M., and Normansell, L. A. (1985). Brain opioids and social emotions. In *The psychobiology of attachment and separation* (ed. M. Reite and T. Field), pp. 3–49. Academic Press, Orlando.

Papez, J. W. (1937). A proposed mechanism of emotion. *Arch. neurol. psychiat.*, **38**, 725–43.

Passingham, R. E. (1982). *The human primate*. W. H. Freeman and Co, Oxford.

Patterson, G. R. (1984). Siblings: fellow travelers in coercive family processes. In *Advances in the study of aggression*, Vol. 1 (ed. R. J. Blanchard and D. C. Blanchard), pp. 173–215. Academic Press, Orlando.

Patterson, G. R., Dishion, T. J., and Bank, L. (1984). Family interaction: a process model of deviancy training. *Aggressive behavior*, **10**, 253–67.

Patterson, P. H. (1978). Environmental determination of autonomic neurotransmitter functions. *Ann. rev. neurosci.*, **1**, 1–17.

Paul, L., Miley, W. M., and Mazzagatti, N. (1973). Social facilitation and inhibition of hunger-induced killing by rats. *J. comp. physiol. psychol.*, **84**, 162–8.

Payne, A. P. (1974). The effects of urine on aggressive responses by male golden hamsters. *Aggressive behavior*, **1**, 71–9.

Penney, J. B. Jr. and Young, A. B. (1983). Speculations on the functional anatomy of basal ganglia disorders. *Ann. rev. neurosci.*, **6**, 73–94.

Penot, C. and Vergnes, M. (1976). Déclenchement de réactions d'agression interspécifique par lésion septale après lésion préalable de l'amygdale chez le rat, *Physiol. behav.*, **17**, 445–50.

Petter, J.-J. (1984). *Le Propre du singe*. Fayard, Paris.

Pettiti, L. E. (1986). La philosophie des droits de l'homme comme moyen de lutte contre les perversions étatiques. In *Big brother, un inconnu familier* (ed. F. Rosenstiel and S. G. Shoham), pp. 131–47. Conseil de l'Europe, Strasbourg et l'Age d'Homme, Paris.

Peyrefitte, A. (1977). *Réponses à la violence — 1: Rapport général — 2: Rapports des groupes de travail* (Rapport du comité présidé par A. Peyrefitte). Documentation française et Presses Pocket, Paris.

Phillips, A. G. (1984). Brain reward circuitry: a case for separate systems. *Brain res. bulletin*, **12**, 195–201.

Piaget, J. (1974). *Adaptation vitale et Psychologie de l'intelligence*. Hermann, Paris.

Picat, J. (1982). *Violences meurtrières et sexuelles*. Presses universitaires de France, Paris.

Pitkänen-Pulkkinen, L. (1981). Long-term studies on the characteristics of aggressive and non-aggressive juveniles. In *Multidisciplinary approaches to aggression research* (ed. P. F. Brain and D. Benton), pp. 225–43. Elsevier, Amsterdam.

Plon, M. (1972). Jeux et conflits. In *Introduction à la psychologie sociale*, Vol. 1 (ed. S. Moscovici), pp. 239–71. Larousse, Paris.

Ploog, D. (1981). Neurobiology of primate audio-vocal behavior. *Brain res. rev.*, **3**, 35–61.

Plotnik, R., Mir, D., and Delgado, J. M. R. (1971). Aggression, noxiousness and brain stimulation in unrestrained rhesus monkeys. In *Physiology of aggression and defeat* (ed. B. E. Eleftheriou and J. P. Scott), pp. 143–221. Plenum Press, New York.

Porrino, L. J., Crane, A. M., and Goldman-Rakic, P. S. (1981). Direct and indirect pathways from the amygdala to the frontal lobe in rhesus monkeys. *J. comp. neurol.*, **198**, 121–36.

Potegal, M. (1979). The reinforcing value of several types of aggressive behavior: a review. *Aggressive behavior*, **5**, 353–73.

Potegal, M., Yoburn, B., and Glusman, M. (1983). Disinhibition of muricide and irritability by intraseptal muscimol. *Pharmacol. biochem. behav.*, **19**, 663–9.

Price, J. L. (1981). The efferent projections of the amygdaloid complex in the rat, cat and monkey. In *The amygdaloid complex* (ed. Y. Ben-Ari), pp. 121–32. Elsevier, Amsterdam.

Pritzel, M. and Huston, J. P. (1983). Behavioral and neural plasticity following unilateral brain lesions. In *Hemisyndromes. Psychobiology, neurology, psychiatry* (ed. M. S. Myslobodsky), pp. 27–68. Academic Press, New York.

Pucilowski, O. and Kostowski, W. (1983). Aggressive behavior and the central serotonergic systems. *Behav. brain res.*, **9**, 33–48.

Pylyshyn, Z. W. (1980). Computation and cognition: issues in the foundations of cognitive science. *The behavioral and brain sciences*, **3**, 111–69.

Raab, A. and Oswald, R. (1980). Coping with social conflict: impact on the activity of tyrosine hydroxylase in the limbic system and in the adrenals. *Physiol. behav.*, **24**, 387–94.

Raufer, X. (1984). *Terrorisme, violence. Réponses aux questions que tout le monde se pose*. Carrère, Paris.

Rawlins, J. N. P., Feldon, J., and Butt, S. (1985). The effects of delaying reward on choice preference in rats with hippocampal or selective septal lesions. *Behav. brain res.*, **15**, 191–203.

Reid, L. D. and Siviy, S. M. (1983). Administration of opiate antagonists reveals endorphinergic involvement in reinforcement processes. In *The neurobiology of opiate reward processes* (ed. J. E. Smith and J. D. Lane), pp. 257–79. Elsevier, Amsterdam.

Renfrew, J. W. and Leroy, J. A. (1983). Suppression of shock elicited target biting by analgesic midbrain stimulation. *Physiol. behav.*, **30**, 169–72.

Restoin, A., Montagner, H., Rodriguez, D., Girardot, J. J., Casagrande, Ch., and Talpain, B. (1984). Ce que peut apporter l'éthologie à la connaissance du développement des comportements sociaux de l'enfant. *Bulletin de psychologie*, **37** (365), 603–19.

Reynolds, V. (1980). The rise and fall of human nature. In *Models of man* (ed. A. J. Chapman and D. M. Jones), pp. 35–47. The British Psychological Society, Leicester.

Ricœur, P. (1986). Entretien avec France de Lagarde. In *La Vie*, 31 juillet 1986.

Ristau, C. A. and Robbins, D. (1982). Cognitive aspects of ape language experiments. In *Animal mind-human mind* (ed. D. R. Griffin), pp. 299–331. Springer Verlag, Berlin.

Robert-Lamblin, J. (1984). L'expression de la violence dans la société ammassalimiut (côte orientale du Groenland). *Études rurales*, **95–96**, 115–29.

Robinson, T. E., Becker, J. B., and Camp, D. M. (1983). Sex differences in behavioral and brain asymmetries. In *Hemisyndromes* (ed. M. S. Myslobodsky), pp. 91–128. Academic Press, New York.

Rocher, G. (1968). *Introduction à la sociologie générale. — 3. Le changement social*, Collection Points. Éditions du Seuil, Paris.

Rodgers, R. J. (1981). Pain and aggression. In *The biology of aggression* (ed. P. F. Brain and D. Benton), pp. 519–27. Sijthoff and Noordhoff, Alphen aan den Rijn.

Rodgers, R. J. and Brown, K. (1976). Amygdaloid function in the central cholinergic mediation of shock-induced aggression in the rat. *Aggressive behavior*, **2**, 131–52.

Rogers, C. R. (1966). *Le Développement de la personne*. Dunod, Paris.

Rolls, E. T. (1976). The neurophysiological basis of brain-stimulation reward. In *Brain-stimulation reward* (ed. A. Wauquier and E. T. Rolls), pp. 65–87. North-Holland Publishing Company, Amsterdam.

Rolls, E. T., Burton, M. J., and Mora, F. (1980). Neurophysiological analysis of brain-stimulation reward in the monkey. *Brain res.*, **194**, 339–57.

Ropartz, P. (1978). Aspects éthologiques des conduites agressives. In *Exposés sur l'agressivité*, Publications de l'Académie des sciences, Institut, 1978–3, pp. 13–25. Gauthier-Villars, Paris.

Rose, J. D. (1986). Functional reconfiguration of midbrain neurons by ovarian steroids in behaving hamsters. *Physiol. behav.*, **37**, 633–47.

Rose, S., Kamin, L. J., and Lewontin, R. C. (1984). *Not in our genes*. Penguin Books, Harmondsworth.

Rosenzweig, S. (1981). The current status of the Rosenzweig picture-frustration study as a measure of aggression in personality. In *Multidisciplinary approaches to aggression research* (ed. P. F. Brain and D. Benton), pp. 113–25. Elsevier, Amsterdam.

Rouyer, M. and Drouet, M. (1986). *L'Enfant violenté. Des mauvais traitements à l'inceste*. Éditions du Centurion, Paris.

Sachser, N. and Pröve, E. (1984). Short-term effects of residence on the testosterone responses to fighting in alpha male guinea pigs. *Aggressive behavior*, **10**, 285–92.

Sandner, G., Schmitt, P., and Karli, P. (1982). Effect of medial hypothalamic stimulation inducing both escape and approach on unit activity in rat mesencephalon. *Physiol. behav.*, **29**, 269–74.

Sandner, G., Schmitt, P., and Karli, P. (1985). Effects of hypothalamic lesions on central gray stimulation induced escape behavior and on withdrawal reactions in the rat. *Physiol. behav.*, **34**, 291–7.

Sarason, I. G. and Sarason, B. R. (1984). Social and cognitive skills training: an antidote for adolescent acting out. In *Aggression in children and youth* (ed. R. M. Kaplan, V. J. Konecni, and R. W. Novaco), pp. 175–91. Martinus Nijhoff Publishers, The Hague.

Sbordone, R. J., Gorelick, D. A., and Elliott, M. L. (1981). An ethological analysis of drug-induced pathological aggression. In *Multidisciplinary approaches to aggression research* (ed. P. F. Brain and D. Benton), pp. 369–85. Elsevier, Amsterdam.

Schallert, T. and Whishaw, I. Q. (1978). Two types of aphagia and two types of sensorimotor impairment after lateral hypothalamic lesions: observations in normal weight, dieted, and fattened rats. *J. comp. physiol. psychol.*, **92**, 720–41.

Schmitt, P. and Karli, P. (1980). Escape induced by combined stimulation in medial hypothalamus and central gray. *Physiol. behav.*, **24**, 111–21.

Schmitt, P. and Karli, P. (1984). Interactions between aversive and rewarding effects of hypothalamic stimulations. *Physiol. behav.*, **32**, 617–27.

Schmitt, P., Abou-Hamed, H., and Karli, P. (1979a). Évolution de la fuite et de l'approche en fonction du point de stimulation dans l'hypothalamus médian. *Physiol. behav.*, **22**, 275–81.

Schmitt, P., Eclancher, F., and Karli, P. (1974). Étude des systèmes de renforcement négatif et de renforcement positif au niveau de la substance grise centrale chez le rat. *Physiol. behav.*, **12**, 271–9.

Schmitt, P., Paunovic, V. R., and Karli, P. (1979b). Effects of mesencephalic central gray and raphe nuclei lesions on hypothalamically induced escape. *Physiol. behav.*, **23**, 85–95.

Schmitt, P., Di Scala, G., Jenck, F., and Sandner, G. (1984). Periventricular structures, elaboration of aversive effects and processing of sensory information. In *Modulation of sensorimotor activity during alterations in behavioral states* (ed. R. Bandler), pp. 393–414. Alan Liss, New York.

Schuurman, T. (1980). Hormonal correlates of agonistic behavior in adult male rats. *Progr. brain res.*, **53**, 415–20.

Schwegler, H. and Lipp, H. P. (1983). Hereditary covariations of neuronal circuitry and behavior: correlations between the proportions of hippocampal synaptic fields in the regio inferior and two-way avoidance in mice and rats. *Behav. brain res.*, **7**, 1–38.

Scott, J. P. (1975). Violence and the disaggregated society. *Aggressive behavior*, **1**, 235–60.

Scott, J. P. (1981). The evolution of function in agonistic behavior. In *Multidisciplinary approaches to aggression research* (ed. P. F. Brain and D. Benton), pp. 129–57. Elsevier, Amsterdam.

Scott, J. P. (1984). Advances in aggression research: the future. In *Advances in the study of aggression*, Vol. 1 (ed. R. J. Blanchard and D. C. Blanchard), pp. 217–37. Academic Press, Orlando.

Scott, J. P. (1986). *The biological basis of warfare*. Report presented at a Symposium on the Psychobiology of Peace, Seville.

Searle, J. R. (1972). *Les Actes de langage. Essai de philosophie du langage*. Hermann, Paris.

Seifert, W. (1983). *Neurobiology of the hippocampus*. Academic Press, London.

Selmanoff, M. and Ginsburg, B. E. (1981). Genetic variability in aggression and endocrine function in inbred strains of mice. In *Multidisciplinary approaches to aggression research* (ed. P. F. Brain and D. Benton), pp. 247–68. Elsevier, Amsterdam.

Serafetinides, E. A. (1980). Epilepsy, cerebral dominance and behavior. In *Limbic epilepsy and the dyscontrol syndrome* (ed. M. Girgis, and L. G. Kiloh), pp. 29–39. Elsevier, Amsterdam.

Shibata, S., Yamamoto, T., and Ueki, S. (1982*a*). Differential effects of medial, central and basolateral amygdaloid lesions on four models of experimentally-induced aggression in rats. *Physiol. behav*, **28**, 289–94.

Shibata, S., Suwandi, D., Yamamoto, T., and Ueki, S. (1982*b*). Effects of medial amygdaloid lesions on the initiation and the maintenance of muricide in olfactory bulbectomized rats. *Physiol. behav.*, **29**, 939–41.

Shotter, J. (1980). Men the magicians: the duality of social being and the structure of moral worlds. In *Models of Man* (ed. A. J. Chapman and D. M. Jones), pp. 13–34. The British Psychological Society, Leicester.

Siegel, A. (1984). Anatomical and functional differentiation within the amygdala — Behavioral state modulation. In *Modulation of sensorimotor activity during alterations in behavioral states* (ed. R. Bandler), pp. 299–323. Alan Liss, New York.

Signoret, J.-L. (1984). Les troubles de mémoire chez l'homme. In *Neurobiologie des comportements* (ed. J. Delacour), pp. 205–14. Hermann, Paris.

Simon, H. (1981). Neurones dopaminergiques A10 et système frontal. *J. physiol., Paris*, **77**, 81–95.

Simon, N. G. (1979). The genetics of intermale aggressive behavior in mice: recent research and alternative strategies. *Neurosci. biobehav. rev.*, **3**, 97–106.

Singer, R. D. (1984). The function of television in childhood aggression. In *Aggression in children and youth* (ed. R. M. Kaplan, V. J. Konecni, and R. W. Novaco), pp. 263–80. Martinus Nijhoff Publishers, The Hague.

Smith, O. A. and De Vito, J. L. (1984). Central neural integration for the control of autonomic responses associated with emotion. *Ann. rev. neurosci*, **7**, 43–65.

Sodetz, F. J. and Bunnell, B. N. (1970). Septal ablation and the social behavior of the golden hamster. *Physiol. behav.*, **5**, 79–88.

Sorman, G. (1983). *La Révolution conservatrice américaine*. Fayard, Paris.

Soubrié, P. (1986). Reconciling the role of central serotonin neurons in human and animal behavior. *The behavioral and brain sciences*, **9**, 319–63.

Sperry, R. W. (1980). Mind-brain interaction: mentalism, yes; dualism, no. *Neuroscience*, **5**, 195–206.

Sperry, R. W. (1981). Changing priorities. *Ann. rev. neurosci.*, **4**, 1–15.

Spinelli, D. N. and Jensen, F. E. (1979). Plasticity: the mirror of experience. *Science*, **203**, 75–8.

Squire, L. R. (1982). The neuropsychology of human memory. *Ann. rev. neurosci.*, **5**, 241–73.

Stachnick, T. J., Ulrich, R., and Mabry, J. H. (1966). Reinforcement of intra and inter-species aggression with intracranial stimulation. *Amer. zoologist*, **6**, 663–8.

Stein, B. E. (1984). Development of the superior colliculus. *Ann. rev. neurosci.*, **7**, 95–125.

Steklis, H. D. and Kling, A. (1985). Neurobiology of affiliative behavior in nonhuman primates. In *The psychobiology of attachment and separation* (ed. M. Reite and T. Field), pp. 93–134. Academic Press, Orlando.

Stellar, J. R. and Stellar, E. (1985). *The neurobiology of motivation and reward.* Springer, New York.

Stoetzel, J. (1978). *La Psychologie sociale.* Flammarion, Paris.

Stokman, C. L. J. and Glusman, M. (1981). Directional interaction of midbrain and hypothalamus in the control of carbachol-induced aggression. *Aggressive behavior*, **7**, 131–44.

Stone, J., Dreher, B., and Rapaport, D. H. (1984). *Development of visual pathways in mammals.* Alan R. Liss, New York.

Storr, A. (1964). Possible substitutes for war. In *The natural history of aggression* (ed. J. D. Carthy and F. J. Ebling), pp. 137–44. Academic Press, London.

Swanson, L. W. (1983). The hippocampus and the concept of the limbic system. In *Neurobiology of the hippocampus* (ed. W. Seifert), pp. 3–19. Academic Press, London.

Tajfel, H. (1972). La catégorisation sociale. In *Introduction à la psychologie sociale*, Vol. 1 (ed. S. Moscovici), pp. 272–302. Larousse, Paris.

Tajfel, H. (1978). Intergroup behaviour. I. Individualistic perspectives. In *Introducing social psychology* (ed. H. Tajfel and C. Fraser), pp. 401–22. Penguin Books, Harmondsworth.

Tajfel, H. (1982). Experimental studies of intergroup behaviour. In *Cognitive analysis of social behaviour* (ed. J. P. Codol and J. P. Leyens), pp. 227–46. Martinus Nijhoff, The Hague.

Talairach, J. (1980). Aujourd'hui plus qu'hier, la neurochirurgie peut-elle réaliser une psychochirurgie rationnelle? *Cours publics organisés par le Mouvement universel de la responsabilité scientifique.* Sorbonne, Paris.

Teitelbaum, P., Schallert, T., and Whishaw, I. Q. (1983). Sources of spontaneity in motivated behavior. In *Handbook of behavioral neurobiology*, Vol. 6 (ed. E. Satinoff and P. Teitelbaum), pp. 23–65. Plenum Publishing Corporation, New York.

Theilgaard, A. (1981). Aggression in men with XYY and XXY karyotypes. In *The biology of aggression* (ed. P. F. Brain and D. Benton), pp. 125–30. Sijthoff and Noordhoff, Alphen aan den Rijn.

Thiébot, M. H., Hamon, M., and Soubrié, P. (1984). Serotonergic neurones and anxiety-related behaviour in rats. In *Psychopharmacology of the limbic system* (ed. M. R. Trimble and E. Zarifian), pp. 164–73. Oxford University Press.

Thierry, B. (1984). Étude comparée des interactions sociales chez trois espèces de macaque (M. mulatta, M. fascicularis, M. tonkeana). *Thèse de 3ᵉ cycle de neurosciences.* Université Louis-Pasteur, Strasbourg.

Thomas, E. and Evans, G. J. (1983). Septal inhibition of aversive emotional states. *Physiol. behav.*, **31**, 673–8.

Thompson, C. I. (1981). Long-term behavioral development of rhesus monkeys after amygdalectomy in infancy. In *The amygdaloid complex* (ed. Y. Ben-Ari), pp. 259–70. Elsevier, Amsterdam.

Thompson, M. E. and Thorne, B. M. (1975). The effects of colony differences and olfactory bulb lesions on muricide in rats. *Physiological psychol.*, **3**, 285–9.

Thor, D. H. and Holloway, W. R. (1984). Social play in juvenile rats: a decade of methodological and experimental research. *Neurosci. biobehav. rev.*, **8**, 455–64.

Thorpe, S. J., Rolls, E. T., and Maddison, S. (1983). The orbitofrontal cortex: neuronal activity in the behaving monkey. *Exp. brain res.*, **49**, 93–115.

Toates, F. (1980). *Animal behaviour. A systems approach.* John Wiley and Sons, Chichester.

Tort, P. (1985). L'histoire naturelle du crime. In *Le Genre humain. — 12. Les usages de la nature*, pp. 217–232. Éditions Complexe, Bruxelles.

Touzard, H. (1979). Les bases psychosociales de la théorie de la stigmatisation. In *La Théorie de la stigmatisation et la réalité criminologique*, pp. 31–45. Presses universitaires d'Aix-Marseille.

Tsuda, A., Tanaka, M., Nishikawa, T., and Hirai, H. (1983). Effects of coping behavior on gastric lesions in rats as a function of the complexity of coping tasks. *Physiol. behav.*, **30**, 805–8.

Turner, B. H., Mishkin, M., and Knapp, M. (1980). Organization of the amygdalopetal projections from modality-specific cortical association areas in the monkey. *J. comp. neurol.*, **191**, 515–43.

Ulrich, R. and Symannek, B. (1969). Pain as a stimulus for aggression. In *Aggressive behaviour* (ed. S. Garattini and E. B. Sigg), pp. 59–69. Excerpta Medica Foundation, Amsterdam.

Ursin, H. (1965). The effect of amygdaloid lesions on flight and defense behavior in cats. *Exper. neurol.*, **11**, 61–79.

Ursin, H. (1980). Personality, activation and somatic health. A new psychosomatic theory. In *Coping and health* (ed. S. Levine and H. Ursin), pp. 259–79. Plenum Press, New York.

Ursin, H. (1985). The instrumental effects of emotional behavior. In *Perspectives in ethology*, Vol. 6: *Mechanisms*, (ed. P. P. G. Bateson and P. H. Klopfer), pp. 45–62. Plenum Press, New York.

Ursin, H., Jellestad, F., and Cabrera, I. G. (1981). The amygdala, exploration and fear. In *The amygdaloid complex* (ed. Y. Ben-Ari), pp. 317– 29. Elsevier, Amsterdam.

Valenstein, E. S. (1980). Overview. 1. Historical perspective. 2. Rational and surgical procedures. In *The psychosurgery debate* (ed. E. S. Valenstein), pp. 11–75. W. H. Freeman and Company, San Francisco.

Van den Bercken, J. H. L. and Cools, A. R. (1982). Evidence for a role of the caudate nucleus in the sequential organization of behaviour. *Behav. brain res.*, **4**, 319–37.

Vaysse, G. and Médioni, J. (1982). *L'Emprise des gènes, et les modulations expérientielles du comportement.* Privat, Toulouse.

Vergnes, M. (1976). Contrôle amygdalien de comportements d'agression chez le rat. *Physiol. behav.*, **17**, 439–44.

Vergnes, M. (1981). Effect of prior familiarization with mice on elicitation of mouse-killing in rats: role of the amygdala. In *The amygdaloid complex* (ed. Y. Ben-Ari), pp. 293–304. Elsevier, Amsterdam.

Vergnes, M. and Karli, P. (1963). Déclenchement du comportement d'agression interspécifique rat-souris par ablation bilatérale des bulbes olfactifs. Action de l'hydroxyzine sur cette agressivité provoquée. *C. R. soc. biol., Paris*, **157**, 1061–3.

Vergnes, M. and Karli, P. (1969). Effets de l'ablation des bulbes olfactifs et de l'isolement sur le développement de l'agressivité interspécifique du rat. *C.R. soc. biol., Paris.*, **163**, 2704–6.

Vergnes, M. and Kempf, E. (1981). Tryptophan deprivation: effects on mouse-killing and reactivity in the rat. *Pharmacol. biochem. behav.*, **14** (suppl. 1), 19–23.

Vergnes, M. and Penot, C. (1976a). Agression intraspécifique induite par chocs électriques et réactivité après lésion du raphé chez le rat. Effets de la physostigmine'. *Brain res.*, **104**, 107–19.

Vergnes, M. and Penot, C. (1976b). Effets comportementaux des lésions du raphé chez des rats privés du septum. *Brain res.*, **115**, 154–9.

Vergnes, M., Boehrer, A., and Karli, P. (1974). Interspecific aggressiveness and reactivity in mouse-killing and nonkilling rats: compared effects of olfactory bulb removal and raphé lesions. *Aggressive behavior*, **1**, 1–16.

Vergnes, M., Depaulis, A., and Boehrer, A. (1986). Parachlorophenylalanine-induced serotonin depletion increases offensive but not defensive aggression in male rats. *Physiol. behav.*, **36**, 653–8.

Vernikos-Danellis, J. (1980). Adrenocortical responses of humans to group hierarchy, confinement and social interaction. In *Coping and health* (ed. S. Levine and H. Ursin), pp. 225–32. Plenum Press, New York.

Viken, R. J. and Knutson, J. F. (1982). The effects of negative reinforcement for irritable aggression on resident-intruder behavior. *Aggressive behavior*, **8**, 371–83.

Viken, R. J. and Knutson, J. F. (1983). Effects of reactivity to dorsal stimulation and social role on aggressive behavior in laboratory rats. *Aggressive behavior*, **9**, 287–301.

Vloebergh, A. (1984). Le langage des chimpanzés: la controverse rebondit. *La Recherche*, **15**(157), 1014–15.

Waal, F. de (1982). *Chimpanzee politics. Power and sex among apes*. Unwin Paperbacks, London.

Waldbillig, R. J. (1979). The role of electrically excitable mesencephalic behavioral mechanisms in naturally occurring attack and ingestive behavior. *Physiol. behav.*, **22**, 473–7.

Walletschek, H. and Raab, A. (1982). Spontaneous activity of dorsal raphe neurons during defensive and offensive encounters in the tree-shrew. *Physiol. behav.*, **28**, 697–705.

Wasman, M. and Flynn, J. P. (1962). Directed attack elicited from hypothalamus. *Arch. neurol.*, **6**, 220–7.

Waterhouse, B. D. and Woodward, D. J. (1980). Interaction of norepinephrine with cerebrocortical activity evoked by stimulation of somatosensory afferent pathways in the rat. *Exp. neurol.*, **67**, 11–34.

Watzlawick, P., Weakland, J., and Fisch, R. (1975). *Changements. Paradoxes et psychothérapie*, collection Points. Éditions du Seuil, Paris.

Weiss, J. M., Goodman, P. A., Losito, B. G., Corrigan, S., Charry, J. M., and Bailey, W. H. (1981). Behavioral depression produced by an uncontrollable stressor: relationship to norepinephrine, dopamine, and serotonin levels in various regions of rat brain. *Brain res. rev.*, **3**, 167–205.

Wheatley, M. D. (1944). The hypothalamus and affective behavior in cats. A study of the effects of experimental lesions, with anatomic correlations. *Arch. neurol. psychiat., Chicago*, **52**, 296–316.

Whitehead, T. (1981). Sex hormone treatment of prisoners. In *Multidisciplinary approaches to aggression research* (ed. P. F. Brain and D. Benton), pp. 503–11. Elsevier, Amsterdam.

Widlöcher, D. (1983*a*). *Les Logiques de la dépression*. Fayard, Paris.

Widlöcher, D. (1983*b*). *Le Ralentissement dépressif*, collection Nodules. Presses universitaires de France, Paris.

Widlöcher, D. (1984). Quel usage faisons-nous du concept de pulsion?. In *La Pulsion, pour quoi faire?* (ed. D. Anzieu, R. Dorey, J. Laplanche, and D. Widlöcher), pp. 29–42. Association psychanalytique de France, Paris.

Will, B. and Eclancher, F. (1984). Early brain damage and early environment. In *Early brain damage*, Vol. 2 *Neurobiology and behavior* (ed. S. Finger and C. R. Almli), pp. 349–67. Academic Press, Orlando.

Will, B. E., Schmitt, P., and Dalrymple-Alford, J. C. (1985). *Brain plasticity, learning, and memory*. Plenum Press, New York.

Willner, P. (1983). Dopamine and depression: a review of recent evidence. II. Theoretical approaches. *Brain res. rev.*, **6**, 225–36.

Wise, R. A. and Bozarth, M. A. (1984). Brain reward circuitry: four circuit elements 'wired' in apparent series. *Brain res. bulletin*, **12**, 203–8.

Witkin, H. A., Mednick, S. A., Schulsinger, F., Bakkeström, E., Christiansen, K. O., Goodenough, D. R., Hirschhorn, K., Lundsteen, C., Owen, D. R., Philip, J., Rubin, D. B., and Stocking, M. (1976). Criminality in XYY and XXY men. *Science*, **193**, 547–55.

Wood, D. J. (1980). Models of childhood. In *Models of man* (ed. A. J. Chapman and D. M. Jones), pp. 227–42. The British Psychological Society, Leicester.

Yadin, E., Guarini, V., and Gallistel, C. R. (1983). Unilaterally activated systems in rats self-stimulating at sites in the median forebrain bundle, medial prefrontal cortex, or locus coeruleus. *Brain res.*, **266**, 39–50.

Yamamoto, B. K., Lane, R. F., and Freed, C. R. (1982). Normal rats trained to circle show asymmetric caudate dopamine release. *Life sciences*, **30**, 2155–62.

Yoshimura, H. and Miczek, K. A. (1983). Separate neural sites for d-amphetamine suppression of mouse-killing and feeding behavior in rats. *Aggressive behavior*, **9**, 353–63.

Zagrodzka, J., Brudnias-Stepowska, Z., and Fonberg, E. (1983). Impairment of social behavior in amygdalar cats. *Acta neurobiol. exp.*, **43**, 63–77.

Zimbardo, P. G. (1972). La psychologie sociale: une situation, une intrigue et un scénario en quête de la réalité. In *Introduction à la psychologie sociale*, Vol. 1 (ed. S. Moscovici), pp. 82–102. Larousse, Paris.

Zimmerman, D. (1983). Moral education. In *Prevention and control of aggression* (ed. A. P. Goldstein), pp. 210–40. Pergamon Press, New York.

INDEX

acephalous societies 164
acting out 205, 206
activation of organism 78–80
active minorities 125
Adams, David, on war 4
adaptive behaviour 60–1
adolescents
  delinquency 174–5
  murderers 162
adoption of children 45
advertising 137–8, 244
adynamia 79
affection, parental 242
affective behaviour 46–7
affective connotations 83–4
affective factors 29
affiliative behaviour 217–18
African genesis (Ardrey) 8
Age of Enlightenment 26
akinesia 115
alcohol 172, 248
Ammassalimuit society 164
amygdala 253
  and cerebral cortex 212–13
  cholinergic activity 211
  connections 96
  lesions and sexual behaviour 98
  lesions and socio-affective behaviour 99
  and emotions 103–6
  and experience 206–20
  integration of affective significance 29,
    30, 85
  kindling and personality 211–12
  lesions 182, 204
  moderating arousal 81
  projections 97
  and raphe nuclei 211
  role in memory 102
  stimulation 186
  surgery 132–3, 223–5
analgesia and stress 82–3
androgens 113, 146, 226
anger 153
  vs fear 149, 150
  'legitimacy' 154
Angoulême, Comic Strip Salon 178
animals, extrapolations to man 118–21
anosmia 200
anterior corpora quadrigemina *see*
    superior colliculus
anti-androgens 226

anti-violence *see* Tahiti, communities 178
apes 31–2
  intelligence 41
  social behaviour 37–41
aphagia 90
appetence 29
  reward system 89–90
appetitive *vs* aversive reactions 199–200
approach and escape 85
appropriating property 171
aqueduct of Sylvius 254
archicortex 96
Ardrey, Robert, *African genesis* 8
area 7a of posterior parietal cortex 108
Arendt, Hannah 236, 239
arousal, behavioural 78
assassinations, Frappat on 5
assertiveness training 128
asymmetry, functional 36–7
attachment 179, 217
attitude and behaviour 125
Australopithecus 32
authoritarian character 73
automobile associations 247
autonomic nervous system 54–5
aversion 29
  system 86–7, 92–4, 198–9
aversive emotions 148–51, 195–206
aversive *vs* appetitive reactions 199–200

baccalauréat, question on morality 236
behaviour
  attitude and 125
  interactions with brain 21–51
  and meaning 52
  modification 122–38
    by social change 124
  priorities 63
  therapies 127, 232–5
behaviourism 23
Bell and Baron, examiner–candidate
    experiments 154
Berkowitz, Leonard, examiner–candidate
    experiments 153, 154
bicuculline 85–6, 197
biological *vs* cultural evolution 34
Blanchard, D. C., on aggression provoked
    by pain 153
blindness, psychic 94, 206
blood pressure *vs* passivity 214

body image and body scheme 71
Bogardus on social distance 169
bonding 14
  mother–child 231
born criminal, Lombroso on 155, 222
Bourdieu, rational utopism 220
Bourricaud, François, on cultural
    evolution 122
Bouthoul and Carrère on fronts of
    collective aggressiveness 19
brain 26–31
  chimpanzee 36
  as computer 48
  control mechanisms 56–7
  functional asymmetry 36–7
  interactions with behaviour 21–51
  lesions 56
  ontogenesis 42–51
  phylogenesis 27–30
  plasticity 42–5
  redundancy 56
  sexual differences (human) 113
  surgery 1, 2, 9, 58–9, 131–3, 223–4
  triune (Maclean) 30
  volume, weight 34
Brill, A. A., on catharsis 237
Broca's area 32, 36, 253
Brücke, E. (teacher of Freud) 15
buildings, height 246
Burckhardt, Gottlieb (psychosurgery) 131
Burgess, Anthony, *Clockwork orange* 1

Caprara, G. V., et al., examiner–candidate
    experiments 157
carbachol 197
Carrier, Father, quoted 240
Carroll and O'Callaghan, amygdalectomy
    224
cars 172, 247–8
castration 191, 226
catecholamines 114–16, 253
catharsis 237
caudate nucleus 194
  connections with prefrontal cortex 111
  stimulation 187
cercopithecinae 38
cerebral cortex
  amygdala 212–13
  and sensorimotor function 29
  see also specific areas
cerebral hemisphere, right 110–11
Charnay, Jean-Paul, on terrorism 18–19
Charrier and Ellul on preventive
    action 233–4
Chatila and Sabra massacres, Lebanon
    8–9

Check, J. V. P., on pornography 178
children
  adoption 45
  aggressive phases 17
  education 241–3
  hyperkinetic, surgery to amygdala 133
Chimpanzee politics (de Waal) 41
chimpanzees 31–2, 167
  brains 36
  symbolic communication 38–41
Chinese *vs* Swedish children 70, 162
cholinergic activity in amygdala 211
Chomsky, Noam, on development of
    language 47
chromosomes
  Y 143, 144
  YY configuration 2, 222
cingulate cortex 29, 30, 85, 253
  surgery 132
circling 43
circularity of depressive illness 134–5
cities 245–8
classification of persons 25–6, 73
Claverie, Elisabeth, *Ethnography of
    violence* 165
Clockwork orange (Kubrick) 1
cognitive psychotherapies 129
cognitive realities 72
cognitive system, consonant or dissonant
    elements 71
collective aggressiveness, fronts of 19
collective consciousness 229
collective identity 228–9
colliculus, superior 29, 256
Comic Strip Salon (Angoulême) 178
communication, symbolic, by
    chimpanzees 38–41
comparative neurobiology 75–6
competition 68–9, 202
  and children 17
computer, brain as 48
congruence (Rogers) 126
connotations, affective 83–4
conscience, freedom of 245
consciousness, collective 229
consonant or dissonant elements in
    cognitive system 71
consumption 171, 246–7
  vs culture 237
  see also covetousness
continuous lighting experiment 157–9
control
  emotions and impulses 216–17
  environment 76–8
control mechanisms in brain 56–7
coordination 54
Coppens, Y., on hominization 32, 34
Corkins, S., on cingulotomy 132

corpora quadrigemina, anterior *see*
    superior colliculus
corpus striatum 253
cortex *see* cerebral cortex
corticosteroids 113–14
corticosterone 147
cost
    assessment 176
    foreseeable, of aggression 215–16
Council of Europe 251
covetousness 171
    see also consumption
crime 18
    mental deficiency 144–5
    restraint 177
'crime chromosome' 144–5, 222
Crime in the United States (FBI) 172
criminal, born (Lombroso) 155, 222
criminology 24–5, 238
Cry of alarm for the twenty-first century
    (Peccei) 240
cultural development of children 45
cultural evolution 122
    vs biological evolution 34
    Reynolds on 34
culture 163–72
    vs consumption 237
    monkeys 40
Cusson, M., on delinquency in
    adolescents 174–5
cutaneous nerve, section 44

Darwinism, social 221
decision *vs* motivation 111–12
defeat, experience of 208
defences, social 238; *see also* police
defensive aggression 170–1, 196–7
    vs offensive aggression 149–50, 185, 194
dehumanization 178
delayed responses, role of dopaminergic
    pathways 116
Delgado, J. M. R. 225
delinquency in adolescents 174–5
Delumeau, Jean, on politics 244
depression, behavioural 77
depressive illness 134–6
desensitization 128
determinism 21
development
    behavioural arousability (rats) 80, 210
    human potential 129
    language 47
    self-awareness 45–6, 50
    social 157–60
deviancy 230
deviant behaviour, therapy 24
directive education *vs* freedom 234, 245

discharge patterns in limbic system 101
discomfort 154
disintegration, social 164
dispositional *vs* situational factors 22–6
dissonant or consonant elements in
    cognitive system 71
dominance 68–9
    social systems based on 167
dopamine 115–16, 253
    as neuromodulator 86, 91
dopaminergic pathways 43–4
    nigrostriatal 255
    role in delayed responses 116
double Y chromosome 2, 222
dyscontrol syndrome 132

Ebling, F. J. G., *Ethical considerations in
    control of human aggression* 228
economism 246–7
education 126–7, 241–5
    of moral sense 235–7
ego defences 17
Eibl-Eibesfeldt, I. 4, 11
Ekblad, S., on cultural development of
    children 45
electrodermal activity 157
Emmanuel, Pierre, quoted 240
emotional mind *vs* rational mind 30
emotional reactivity 156–7
emotions 46–7, 65–7
    and aggression 139–40
    amygdala 103–6
    aversive 148–51, 195–206
    control 216–17
    false 179
    septum 106–7
emotive language 66
encephalization in evolution 29
end-directed behaviour 62–4
endocrine state and intraspecific
    aggression 191
endomorphines 253
endorphins 91, 114, 117–18, 218
Enlightenment, Age of 26
environment
    causes of aggression 16–17
    effect on socio-affective behaviour
        69–70, 81–2
    vs heredity 22–6
    mastery 76–8
epilepsy, temporal lobe, and
    personality 212
epileptogenic stimulation of limbic
    system 101
equilibria, restoration 60, 64
ergotropic action 54–5
escape and approach 85

Ethical considerations in control of human aggression (Ebling) 228
ethnicity 34
ethnocentrism 169
Ethnography of violence (Claverie) 165
Europe, Council of 251
evolution, encephalization 29
evolution (cultural) 122
  vs biological evolution 34
examiner–candidate experiments 153, 154, 157, 235
experience 81–2
  and amygdala 206–20
  of defeat 208
  human 217–18
  integration by limbic system 100
experimental brain surgery 58–9
extrapolations from animals to man 118–21
Eysenck, H. J. 22

facilitating and inhibiting structures 182
factors affecting aggression 139–80
false emotions 179
familiarity 151, 188, 207
fear
  vs anger 149, 150
  vs need 237
Federal Bureau of Investigation, *Crime in the United States* 172
feeding behaviour 186
  effect of limbic lesions 98
  motivation 63–4
fetal substantia nigra, implantation 43
firearms 19, 171–2, 248
Fodor, J. A., on brain as computer 48
food odours, response 82
forensic psychiatry 24–5
foreseeable cost of aggression 215–16
Forman, Milos, *One flew over the cuckoo's nest* 1
fornix 31, 253
Frappat, Bruno, on assassinations 5
freedom 49
  vs directive education 234, 245
  Skinner on 234–5
French Psychoanalytical Association, *La pulsion, pour quoi faire* 15
Freud, Sigmund 14–15, 128
frontal cortex
  and amygdala 212–13
  bilateral ablation 182
  development 35
  lesions 83, 110
fronts of collective aggressiveness 19
frustration 153–4, 155

function
  vs mechanism 53
  structuring effect of 42–5
fundamentalism 169–70
futures 240

GABA 253
  agonists, effect at periaqueductal grey matter 84, 149
GABAergic transmission 253–4
  blockade 85–6
  stimulation 188–9
Galef, B. C., Jr, on neophobia 151
gambling 247
games 179–80
gastric ulcers 214
generators of spontaneous activity 84–6
genetic engineering 2, 130–1, 222–3
genetic factors 143–5
Genet, Jean, on hooligans 242
Girard, Augustin, quoted 240
Girard, René
  on mimetic desire 171
  on scapegoating 232
glucoreceptors 80
gnoses 35
Goldstein, Arnold, on reward and punishment in behaviour therapies 233
grammar, universal 47
guns 19, 171–2, 248

height of buildings 246
helplessness, learned 63, 77, 135–6
heredity *vs* environment 22–6
Herrick, C. J., and limbic system 94
hierarchical social systems 167
hippocampal formation 96
hippocampus 96–7, 254
  effect of lesions on sexual behaviour 98
  interactions with prefrontal cortex 110
  role in memory 102, 103
history, individual 122
homeostasis 60, 64, 67
hominization 31–4
Homo erectus 33
Homo habilis 32
homology in comparative neurobiology 75
hooligans, Genet on 242
hormones 112–18, 145–8
  early influence in humans 148
humanism 220, 249–52
humanist psychology 129
human potential, development 129
humoral responses 62–3
hunger, and perception of odours 192

hyperkinetic children, surgery to
  amygdala 133
hypersensitivity, and aggression 193–4
hypothalamus 65, 85, 89–90, 92, 254
  lateral, stimulation 215
  lesions 186
  and reactivity levels 199
  stimulation 193–5
  surgery in sexual deviance 133
  ventromedial, lesions 203

'I' and social Ego 49–50
ideas, killing for 19
identity 23
  collective 228–9
  individual 72–4
ideology, and firearms 19
Ikor, Roger, on Sabra and Chatila
  massacres in Lebanon 8–9
illness, mental 226–7
image of politicians 244
immersion 128
Immorality Act (South Africa) 170
implantation of fetal substantia nigra 43
impulses and emotions, control 216–17
inborn errors of metabolism 222
individual
  history 122
  identity 72–4
  responsibility, *vs* collectivism 230
infantilism 242
inhibiting and facilitating structures 182
inhibition, behavioural 80–1
  and serotoninergic pathways 117
injury to brain 56
inner life 246
instinct, aggression as 8–9, 10–16
instrumental value of aggression 172–4,
  213–15
intelligence
  apes 41
  human 61
  vs instinct 13–14
interactionism 229
interactionist hypothesis 58
interactions
  brain with behaviour 21–51
  with mother 159, 160–3
  prefrontal cortex with hippocampus 110
  reward system with aversion
    system 93–4
interdisciplinary approach 4
interiorization 48
internal representations 107–11, 122
intraspecific aggression
  endocrine state 191
  vs muricidal behaviour 189

intruders *vs* residents 149
irritability 157
isolation, social 157, 159–60
Israël, L., on psychotherapies 126

Jacob, François, on cultural evolution 123
Jacobsen, Carlyle (psychosurgery) 131
Jacobs, P. A., on XYY configuration 144
Java macaques 167

Karli, Pierre
  experiments on neurobiology of
    aggression 181–218
  on rat–mouse aggression 181–90
Kesey, Ken, *One flew over the cuckoo's
  nest* 1
killer rats 146, 151, 176, 182, 187
killing
  for ideas 19
  in war 5
kindling in amygdala 106
  and personality 211–12
Klüver and Bucy, and limbic system 94,
  206
Kohlberg and Turiel, examiner–candidate
  experiments 235
Kubrick, Stanley, *Clockwork orange* 1

labelling 73, 230–1
lactation, and intraspecific aggression 191
language 33–4, 35–6, 50–1, 52
  in aggression 168–9
  development 47
  emotive 66
  ossified 240–1
  and violence 9–10
La pulsion, pour quoi faire (French
  Psychoanalytical Association) 15
La révolution conservatrice américaine
  (Sorman) 19
lateralization of stimuli 197
law, penal 248; *see also* social defences;
  police
lawlessness 247–8
learned helplessness 63, 77, 135–6
learning, limbic lesions 101–2
Lebanon, Sabra and Chatila massacres,
  Ikor on 8–9
'legitimacy' of anger 154
Le Monde de l'éducation 242
Le petit prince (Saint-Exupèry) 137
lesions of brain 56
  see also specific structures
levels of integration 28–30
lexigrams 39

Leyens, J.-P., on contexts of
    psychologists 25
Lhermitte, François, utilization
    behaviour 35, 83
lighting, continuous, experiment 157–9
limbic system 30–1, 86, 94–107, 254
    epileptogenic stimulation 101
    integration of experience 100
    Klüver and Bucy on 206
    mediation of reinforcement 100
    memory 102–4
    see also specific components
lobotomy, prefrontal 132
locus coeruleus, activation 114, 115
locus niger 256
Lombroso on born criminal 155, 222
Lorenz, Konrad
    on aggression as instinct 11–15
    Montague on 4
    Natural history of evil 2–3
    On aggression 8
lotteries, state 18
love ceremonies 14
loyalty 236

macaques, Thierry on 167
Maclean, P. D., triune brain 30
macrosmatic species 191–3
manipulation of minds 137–8
Mark and Ervin
    and amygdalectomy 225
    psychosurgery 132
Maslow, A. H. 129
massacres in Lebanon 8–9
mastery 76–8, 213–14
materialism 179–80
    see also consumption; covetousness
meaning 238–9
    and behaviour 52
mechanism vs function 53
medial fore brain bundle 254
mediator, brain as 26–7
medicalisation 227
medio-dorsal nucleus of thalamus 102
medulla, suprarenal 80
medulla oblongata 34
Memmi, Albert, on fear vs need 237
memory 46
    and limbic system 102–4
mental deficiency and crime 144–5
mental illness 226–7; see also depressive
    illness
mesencephalon 254
    dorsolateral, stimulation 186
mesocortical system 115
mesolimbic system 115
metabolism, inborn errors 222

milk see lactation
mimetic desire, Girard on 171
minorities, active 125
mnesic activities and limbic system 102–4
modelling 174–5
modification of behaviour 122–38
modulation of responses and sensory
    stimuli 82–3
Moniz, Egas (psychosurgery) 131
monkeys
    culture 40
    vs man, limbic systems compared 97
    social learning 40
Montague, Ashley, on Lorenz 4
Montaigne, M. Eyquem, seigneur de, on
    wars of religion 19
Montlibert, Christian de, on violent social
    movements 18
morality, baccalauréat question 236
moral madness 222
moral sense, education of 235–7
Morin, Edgar, on reductionism 17
Morris, Desmond, on chimpanzees 167
Moscow, Psychological Journal 4
mother, interactions with 159, 160–3
mother–child bonding 231
mother–child relation 67–8
motivation 52–74
    vs decision 111–12
    for feeding behaviour 63–4
motivational systems 182–4
motor cars 172, 247–8
motor function 29
motor neurons 254
movements, social, violent 18
μ-like opiate receptors 118
murder 163
murderers, adolescent 162
muricidal behaviour 181–90
muscimol 203
muscles
    postural activity 28
    transposition 54
mydriasis 93
myotatic reflex 63
myths and violence 9

Narabayashi, H., and amygdalectomy 224
Natural history of evil (Lorenz) 2–3
natural selection 221
Nauta, W. J. H., neural circuit 95
Neanderthal man 33
need vs fear, Memmi on 237
negative reinforcement 87
neglect syndromes 46, 108
negotiation vs power 237
neocortex 254

neomammalian brain 255
neophobia 151–2
neostriatum 255
neurobiology
    of aggression 181–218
    comparative 75–6
    social role 1–4, 9
neuromodulators 86, 91, 112–18
neuronal aversion system 92–4, 198–9
neuron discharge patterns in limbic
    system 101
neurotransmitters 255
news, television 244
nigrostriatal pathway 43–4, 115, 194, 255
1984 (Orwell) 137
nociceptive stimulation 63
noise 246
noradrenaline, cerebral 114
nuclear arms, proliferation 19–20
nucleus accumbens 255
Nuttin, J.
    behavioural resources 17
    Theory of human motivation 58

Oaxaca Valley, communities 178
obesity 63
objectives and strategies 57–8, 62
oculomotor area of cerebral cortex 84–5
oestrogens 226
oestrus, attenuation of aggression in
    rats 147
offensive *vs* defensive aggression 149–50,
    185, 194
olfactory bulbs
    ablation 200–1, 208
    lesions 183, 188
    in macrosmatic species 192–3
    and reactivity levels 199
Olweus, D., on aggressive
    personality 160
On aggression (Lorenz) 8
One flew over the cuckoo's nest (Forman) 1
ontogenesis 42–51, 255
opiate receptors 213
opiate systems 218
orbito-frontal cortex 213
orthosympathetic nervous system 54–5
Orwell, George, *1984* 137
otherness 169
over-protectiveness 68

Paddock, John, on anti-violence in
    Tahiti 178
pain
    aggression provoked by 152–3
    attenuation by self-stimulation 89

palaeocortex 96
palaeomammalian brain 255
Papez, J. W., neural circuit 94–5, 103
parachlorophenylalanine 211
parents 242–3
    interactions with 160–3
Parkinson's disease 115
passivity *vs* blood pressure 214
pathological activation 79
peace pipe 13
Peccei, Aurelio, *Cry of alarm for the
    twenty-first century* 240
Penal Code, Article 64, responsibility 228
penal law 248
perception and aggression 191–5
performance, specifications 57
periaqueductal grey matter 29, 81, 92,
    196–9, 255
    effect of GABA agonists 84, 149
permissiveness *vs* restriction 70
persecution 178
personality 154–63
    aggressive 160
    kindling in amygdala 211–12
    restructuring, *vs* social change 228–30
PG region of posterior parietal cortex 108
pheromones 113
phobias, behaviour therapies 127–8
phonation 85
phylogenesis 27–30, 255
Piaget, Jean, on development of
    language 47
pigs, effect of prior social interactions 141
pipe of peace 13
plasticity 42–5
play
    fighting 202
    social 214, 217
police 248
politics on television 244
pornography, Check on 178
positive reinforcement 87, 173–4, 187–8,
    215, 234–5
posterior parietal cortex, PG region 108
postural activity of muscles 28
power *vs* negotiation 237
Pravda 170
praxes 35
predatory behaviour 186
prediction *vs* 'prophetism' 219–20
prefrontal cortex 109–10, 255
    connections with caudate nucleus 111
    effect of lesions on socio-affective
        behaviour 99
    self-stimulation 91
prefrontal lobotomy 132
pressure groups 125
presubicular cortex 96

pre-tuning of sensory pathways 62
prevention 245–8
  vs repression 241
preventive action (Charrier and Ellul) 233–4
priorities in behaviour 63
progesterone 147, 148, 256
property, appropriating 171
'prophetism' *vs* prediction 219–20
prosocial behaviour 161–2
protectiveness 68
psychiatric abuse 137
psychic blindness 94, 206
psychoanalysis 23
Psychobiological Laboratory (Curt
  Richter) 181
psychodynamics 128
Psychological Journal (Moscow) 4
psychologists, contexts of 25
psychometry 22
psychopharmacology 2, 133–6, 225–6
psychosurgery 1, 2, 9, 58–9, 131–3, 223–4
psychotherapies 2, 126, 232
  cognitive 129
punishment 153
  in behaviour therapies 233
Pylyshyn, Z. W., on brain as
  computer 48–9
pyramidal tract 30–1, 35, 256

question on morality (*baccalauréat*) 236

race 34
racism 169
rage, social role 66
rape 178
raphe nuclei 256
  and amygdala 211
  lesions 202–3, 204–6
  and reactivity levels 199
rapid eye movements 84–5
rational mind *vs* emotional mind 30
rational utopism (Bourdieu) 220
rat–mouse aggression, author on 181–90
rats
  attenuation of aggression at oestrus 147
  castrated male 191
  development of behavioural arousability
    80, 210
  killer 146, 151, 176, 182, 187
Raufer, X., on terrorism 18
reactivity 77–8, 80
  emotional 156–7
  levels 199–206
realities, cognitive 72
receptivity, enhanced, and aggression
  193–4

reciprocal structuring 26
recorder, brain as 59
reductionism 3, 17
redundancy in brain 56
'reflex' aggression 152
reflexes, spinal 63
rehabilitation after injury 56
reification of aggression 10
reinforcement 216
  mediation by limbic system 100
  negative 87
  positive 87, 173–4, 187–8, 215, 234–5
religion, wars, Montaigne on 19
representations, internal 107–11, 122
repression *vs* prevention 241
reptilian brain 256
residents *vs* intruders 149
resistivity of skin 157
resources, behavioural (Nuttin) 17
respect 179
responses, modulation 83
responsibility 50, 219–20, 229
  Article 64 of Penal Code 228
restoration to equilibria 60, 64
restraint
  on aggression 175
  on crime 177
restriction *vs* permissiveness 70
reticular activating system 55, 78–9
reversibility of interventions 226
reward and punishment in behaviour
  therapies, Goldstein on 233
reward system 86–90, 91–2
  interactions with aversion system 93–4
Reynolds, V., on cultural evolution 34
rhesus monkeys 167
rhinencephalon 96
Richter, Curt, Psychobiological Laboratory
  181
Ricoeur, Paul, on meaning 239
Rift Valley 32
right cerebral hemisphere 110–11
risk assessment 176
ritualization of aggression 12–14
Rocher, Guy, definition of social change
  124–5
Rogers, C. R. 129–30
  congruence 126
Rosenzweig frustration test 155
Rose, S., et al., on neurobiology 181

Sabra and Chatila massacres in Lebanon
  8–9
saccades 84–5
Saint-Exupèry, Antoine de, *Le petit prince*
  137
scapegoating 232, 251

Scott, Paul, on aggression 142
Searle, J. R., on development of language 47
self-affirmation training 128
self-awareness, development 45–6, 50
self-stimulation 87–94
sensorimotor function 29
sensory neglect 90
sensory pathways, pre-tuning 62
sensory stimuli, modulation 82–3
septum 81, 97, 256
    and emotions 106–7
    lesions 183, 201–3
    and reactivity levels 199
    and social interactions 107
serotonergic activity, reduction 205–6
serotonin 116–17, 256
    and sleep 79
    turnover 200
sex hormones 112–13, 146, 226
sexuality
    behaviour, effect of limbic lesions 98–9
    deviance, surgery to hypothalamus 133
    differences in human brain 113
    and intraspecific aggression 191
    response in hamster 83
siblings, role 161
sign language and chimpanzees 38
sincerity 236
Singer, R. D., on television in United States 244–5
situational determinants of action 71–2
situational *vs* dispositional factors 22–6
situations
    vs personality 155–6
    provoking aggression 165–6, 237
skin, resistivity 157
Skinner, B. F., on freedom 234–5
sleep 79
social behaviour
    apes 37–41
    plasticity 44–5
social change
    behaviour modification by 124
    vs restructuring personality 228–30
social Darwinism 221
social defences 17, 238
    see also police
social development 157–60
social disintegration 164
social distance, Bogardus on 169
social effects of lesions of frontal cortex 213
social Ego and 'I' 49–50
social factors on aggression 23, 163–72
social interactions
    prior, effect in pigs 141
    septum 107

social learning in monkeys 40
social movements, violent 18
social play 214, 217
social skills 232–3
social stress 158
social systems 166–8
socio-affective behaviour 66–70, 202
    effect of environment 69–70, 81–2
    effect of limbic lesions 99
sociobiology 222
Sorman, G., *La révolution conservatrice américaine* 19
South Africa, Immorality Act 170
specifications of performance 57
spinal reflexes 63
spinal shock 29–30
spontaneous activity 78, 80
    generators 84–6
state lotteries 18
steroids 113–14, 256
stigmatization 73, 230–1
Storr, Anthony 8
strangeness 151, 169, 187
strategies and objectives 57–8, 62
stress
    analgesia 82–3
    social 158
structuring, reciprocal 26
structuring effect of function 42–5
Study Committee on Violence, Criminality and Delinquency 231, 241, 243
    on penal law 248
    Violence, the answer to frustration 170
subicular area 110
substantia nigra 256
    fetal, implantation 43
    stimulation 187
suicide 17–18
superior colliculus 29, 84–5, 256
suprarenal cortex 80, 147, 256
    hormones 113–14
surgery to brain 1, 2, 9, 58–9, 131–3, 223–4
Swedish *vs* Chinese children 70, 162
Sylvius, aqueduct of 254
symbolic communication by chimpanzees 38–41
sympathetic nervous system 54–5

Tahiti, communities 165, 177–8
teachers and parents 242
tectum, mesencephalic 254
tegmentum, mesencephalic 254
telencephalon 95–6
television 179, 243–5
    violence 162–3, 175
temporal cortex and amygdala 212–13

temporal lobes
  bilateral removal 94, 206
  effect of lesions on sexual behaviour 98
  epilepsy, and personality 212
  pathways from visual cortex 100
temporal (time) information, integration
  109
territorial social systems 167
terrorism 18–19
testosterone 113, 146, 226, 257
thalamus 257
  medio-dorsal nucleus 102
Theory of human motivation (Nuttin) 58
therapy for deviant behaviour 24; *see also*
    psychoanalysis; psychopharmacology;
    psychosurgery; psychotherapies
Thierry, B., on macaques 167
time *see* temporal information, integration
'toilet-flush' theory 11
tolerance 250–1
Tonkean macaques 167
tranquillizers 225
transitional cortex 96
transposition of muscles 54
trauma to brain 56
triune brain (Maclean) 30
2-deoxyglucose marker 91

ulcers, gastric 214
unilateral neglect 46
United States, television 244–5
universal grammar 47
urban environment 245–8
utilization behaviour (Lhermitte) 35, 83
utopism, rational (Bourdieu) 220

value judgements 5
value of aggression 172–4, 213–15

vegetative nervous system 54–5
victories 173
videos, violence 175
violence on television 162–3, 175, 244–5
Violence, the answer to frustration (Study
    Committee on Violence, Criminality
    and Delinquency) 170
violent social movements 18
visual cortex, pathways to temporal lobes
  100
visual signals of apes 38
visual system 44
vocal behaviour of apes 37–8
vocalization 85
volume of brain 34

Waal, F. de, on chimpanzees 167
  Chimpanzee politics 41
waking 54
war 19, 177
  Adams on 4
  killing in 5
weapons
  firearms, ownership 171–2
  nuclear, proliferation 19–20
weight of brain 34
Wernicke's area 36, 257
Widlöcher, Daniel, on instinct 15–16
women, dehumanization 178

xenophobia 169–70
XYY configuration 2, 144, 222

Yanomami society 164, 165, 178
Y chromosome 143, 144